Computer Methods for Ordinary Differential Equations and Differential–Algebraic Equations

Computer Methods for Ordinary Differential Equations and Differential–Algebraic Equations

Uri M. Ascher
University of British Columbia
Vancouver, British Columbia, Canada

Linda R. Petzold
University of California, Santa Barbara
Santa Barbara, California

siam
Society for Industrial and Applied Mathematics
Philadelphia

10 9 8 7 6

Library of Congress Cataloging-in-Publication Data

Ascher, U. M. (Uri M.), 1946-
Computer methods for ordinary differential equations and differential-algebraic equations / Uri M. Ascher, Linda R. Petzold.
p. cm.
Includes bibliographical references and index.
ISBN 978-0-898714-12-8 (pbk.)
1. Differential equations–Data processing. 2. Differential -algebraic equations–Data processing. I. Petzold, Linda Ruth. II. Title.
QA372.A78 1998
515'.352'0285–dc21 98-21535

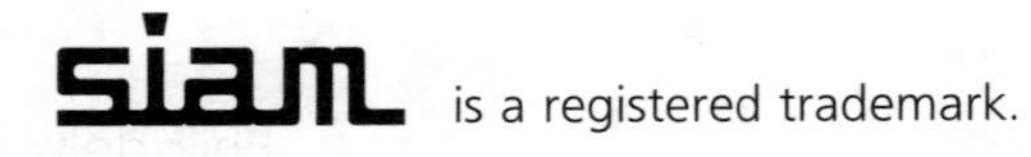

Contents

List of Figures

List of Tables

Preface

This book was developed from course notes that we wrote, having repeatedly taught courses on the numerical solution of ordinary differential equations (ODEs) and related problems. We have taught such courses at a senior undergraduate level as well as at the level of a first graduate course on numerical methods for differential equations. The audience typically consists of students from mathematics, computer science, and a variety of disciplines in engineering and the sciences such as mechanical, electrical, and chemical engineering, physics, and earth sciences.

The material that this book covers can be viewed as a first course on the numerical solution of differential equations. It is designed for people who want to gain a practical knowledge of the techniques used today. The course aims to achieve a thorough understanding of the issues and methods involved and of the reasons for the successes and failures of existing software. On one hand, we avoid an extensive, thorough, theorem-proof–type exposition: we try to get to current methods, issues, and software as quickly as possible. On the other hand, this is not a quick recipe book, as we feel that a deeper understanding than can usually be gained by a recipe course is required to enable students or researchers to use their knowledge to design their own solution approaches for any nonstandard problems they may encounter in future work. The book covers initial value and boundary value problems, as well as differential-algebraic equations (DAEs). In a one-semester course, we have typically been covering over 75% of the material it contains.

We wrote this book partially as a result of frustration at not being able to assign a textbook adequate for the material that we have found ourselves covering. There is certainly excellent, in-depth literature around. In fact, we are making repeated references to exhaustive texts which, combined, cover almost all the material in this book. Those books contain the proofs and references which we omit. They span thousands of pages, though, and the time commitment required to study them in adequate depth may be more than many students and researchers can afford to invest. We have tried to stay within a 350-page limit and to address all three ODE-related areas mentioned above. A significant amount of additional material is covered in the exercises that conclude all but the first chapter. Other additional important topics are referred to in brief sections titled "Notes and References." Software is an important and well-developed part of this subject. We have attempted to cover the most fundamental software issues in the text. Much of the excellent and publicly available software is described in the "Software" sections at the end of the relevant chapters, and available codes are cross-referenced in the index. Review material is highlighted and presented in the text when needed, and it is also cross-referenced in the index.

Traditionally, numerical ODE texts have spent a great deal of time developing families of higher-order methods, e.g., Runge–Kutta and linear multistep methods, applied first to nonstiff problems and then to stiff problems. Initial value problems and boundary value problems have been treated in separate texts, although they have much in common. There have been fundamental differences in approach, notation, and even basic definitions between ODE initial value problems, ODE boundary value problems, and partial differential equations (PDEs).

We have chosen instead to focus on the classes of problems to be solved, mentioning wherever possible applications which can lend insight into the requirements and the potential sources of difficulty for numerical solution. We begin by outlining the relevant mathematical properties of each problem class, then carefully develop the lower-order numerical methods and fundamental concepts for the numerical analysis. Next we introduce the appropriate families of higher-order methods, and finally we describe in some detail how these methods are implemented in modern adaptive software. An important feature of this book is that it gives an integrated treatment of ODE initial value problems, ODE boundary value problems, and DAEs, emphasizing not only the differences between these types of problems but also the fundamental concepts, numerical methods, and analysis which they have in common. This approach is also closer to the typical presentation for PDEs, leading, we hope, to a more natural introduction to that important subject.

Knowledge of significant portions of the material in this book is essential for the rapidly emerging field of numerical dynamical systems. These are numerical methods employed in the study of the long-term, qualitative behavior of various nonlinear ODE systems. We have emphasized and developed in this work relevant problems, approaches, and solutions. But we avoided developing further methods which require deeper, or more specific, knowledge of dynamical systems, which we did not want to assume as a prerequisite.

The plan of the book is as follows. Chapter 1 is an introduction to the different types of mathematical models which are addressed in the book. We use simple examples to introduce and illustrate initial and boundary value problems for ODEs and DAEs. We then introduce some important applications where such problems arise in practice.

Each of the three parts of the book which follow starts with a chapter which summarizes essential theoretical or analytical issues (i.e., before applying any numerical method). This is followed by chapters which develop and analyze numerical techniques. For initial value ODEs, which comprise roughly half of this book, Chapter 2 summarizes the theory most relevant for computer methods, Chapter 3 introduces all the basic concepts and simple methods (relevant also for boundary value problems and for DAEs), Chapter 4 is devoted to one-step (Runge–Kutta) methods, and Chapter 5 discusses multistep methods.

Chapters 6–8 are devoted to boundary value problems for ODEs. Chapter 6 discusses the theory which is essential to understanding and making effective use of the numerical methods for these problems. Chapter 7 briefly considers shooting-type methods, and Chapter 8 is devoted to finite difference approximations and related techniques.

The remaining two chapters consider DAEs. This subject has been researched and solidified only very recently (in the past 15 years). Chapter 9 is concerned with background material and theory. It is much longer than Chapters 2 and 6 because understanding the relationship between ODEs and DAEs, and the questions regarding reformulation of DAEs, is essential and already suggests a lot regarding computer approaches. Chapter 10 discusses numerical methods for DAEs.

Various courses can be taught using this book. A 10-week course can be based on the first 5 chapters, with an addition from either one of the remaining two parts. In a 13-week course (or shorter in a more advanced graduate class) it is possible to comfortably cover Chapters 1–5 and either Chapters 6–8 or Chapters 9–10, with a more superficial coverage of the remaining material.

The exercises vary in scope and level of difficulty. We have provided some hints, or at least warnings, for those exercises that we (or our students) have found more demanding.

Many people helped us with the tasks of shaping up, correcting, filtering, and refining the material in this book. First and foremost are our students in the various classes we taught on this subject. They made us acutely aware of the difference between writing with the desire to explain and writing with the desire to impress. We note, in particular, G. Lakatos, D. Aruliah, P. Ziegler, H. Chin, R. Spiteri, P. Lin, P. Castillo, E. Johnson, D. Clancey, and D. Rasmussen. We have benefitted particularly from our earlier collaborations on other, related books with K. Brenan, S. Campbell, R. Mattheij, and R. Russell. Colleagues who have offered much insight, advice, and criticism include E. Biscaia, G. Bock, C. W. Gear, W. Hayes, C. Lubich, V. Murata, N. Nedialkov, D. Negrut, D. Pai, J. B. Rosen, L. Shampine, and A. Stuart. Larry Shampine, in particular, did an incredibly extensive refereeing job and offered many comments which have helped us to significantly improve this text. We have also benefitted from the comments of numerous anonymous referees.

U. M. Ascher
L. R. Petzold

Part I: Introduction

Chapter 1

Ordinary Differential Equations

Ordinary differential equations (ODEs) arise in many instances when using mathematical modeling techniques for describing phenomena in science, engineering, economics, etc. In most cases the model is too complex to allow one to find an exact solution or even an approximate solution by hand: an efficient, reliable computer simulation is required.

Mathematically, and computationally, a first cut at classifying ODE problems is with respect to the additional or side conditions associated with them. To see why, let us look at a simple example. Consider

$$u''(t) + u(t) = 0, \qquad 0 \le t \le b,$$

where t is the independent variable (it is often, but not always, convenient to think of t as "time"), and $u = u(t)$ is the unknown, dependent variable. Throughout this book we use the notation

$$u' = \frac{du}{dt}, \qquad u'' = \frac{d^2u}{dt^2},$$

etc. We shall often omit explicitly writing the dependence of u on t.

The general solution of the ODE for u depends on two parameters α and β,

$$u(t) = \alpha \sin(t + \beta).$$

We can therefore impose two side conditions.

- ***Initial value problem*** (IVP): Given values $u(0) = c_1$ and $u'(0) = c_2$, the pair of equations

$$\alpha \sin\beta = u(0) = c_1,$$
$$\alpha \cos\beta = u'(0) = c_2$$

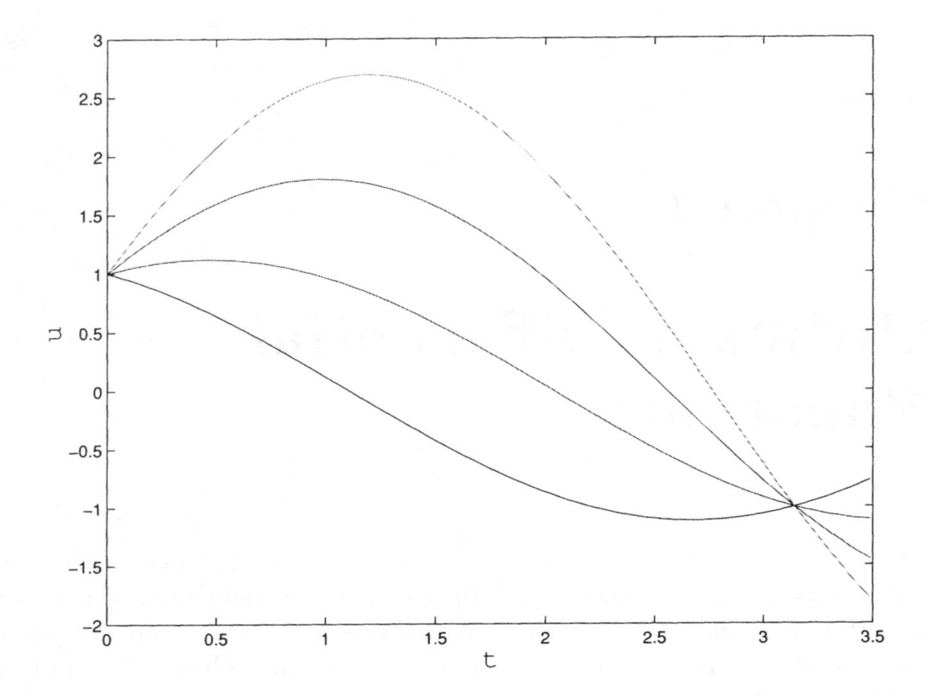

Figure 1.1: *u vs. t for $u(0) = 1$ and various values of $u'(0)$.*

can always be solved uniquely for $\beta = \tan^{-1}\frac{c_1}{c_2}$ and $\alpha = \frac{c_1}{\sin\beta}$ (or $\alpha = \frac{c_2}{\cos\beta}$; at least one of these is well defined). The IVP has a unique solution for any initial data $\mathbf{c} = (c_1, c_2)^T$. Such solution curves are plotted for $c_1 = 1$ and different values of c_2 in Figure 1.1.

- ***Boundary value problem*** (BVP): Given values $u(0) = c_1$ and $u(b) = c_2$, it appears from Figure 1.1 that for $b = 2$, say, if c_1 and c_2 are chosen carefully then there is a unique solution curve that passes through them, just like in the initial value case. However, consider the case where $b = \pi$: now different values of $u'(0)$ yield the same value $u(\pi) = -u(0)$ (see again Figure 1.1). So, if the given value of $u(b) = c_2 = -c_1$ then we have infinitely many solutions, whereas if $c_2 \neq -c_1$ then no solution exists.

This simple illustration already indicates some important general issues. For IVPs, one starts at the initial point with all the solution information and marches with it (in "time")—the process is *local*. For BVPs the entire solution information (for a second-order problem this consists of u and u') is not locally known anywhere, and the process of constructing a solution is *global* in t. Thus we may expect many more (and different) difficulties with the latter, and this is reflected in the numerical procedures discussed in this book.

1.1 IVPs

The general form of an IVP that we shall discuss is

$$\begin{aligned} \mathbf{y}' &= \mathbf{f}(t, \mathbf{y}), \qquad 0 \le t \le b, \\ \mathbf{y}(0) &= \mathbf{c} \ \ \text{(given)}. \end{aligned} \tag{1.1}$$

Here $\mathbf{y}$ and $\mathbf{f}$ are vectors with m components, $\mathbf{y} = \mathbf{y}(t)$, and $\mathbf{f}$ is in general a nonlinear function of t and $\mathbf{y}$. When $\mathbf{f}$ does not depend explicitly on t, we speak of the *autonomous* case. When describing general numerical methods we shall often assume the autonomous case simply in order to carry less notation around. The simple example from the beginning of this chapter is in the form (1.1) with $m = 2$, $\mathbf{y} = (u, u')^T$, $\mathbf{f} = (u', -u)^T$.

In (1.1) we assume, for simplicity of notation, that the starting point for t is 0. An extension of everything which follows to an arbitrary interval of integration $[a, b]$ is obtained without difficulty.

Before proceeding further, we give three examples which are famous for being very simple on one hand and for representing important classes of applications on the other hand.

Example 1.1 (simple pendulum)
Consider a tiny ball of mass 1 attached to the end of a rigid, massless rod of length 1. At its other end the rod's position is fixed at the origin of a planar coordinate system (see Figure 1.2).

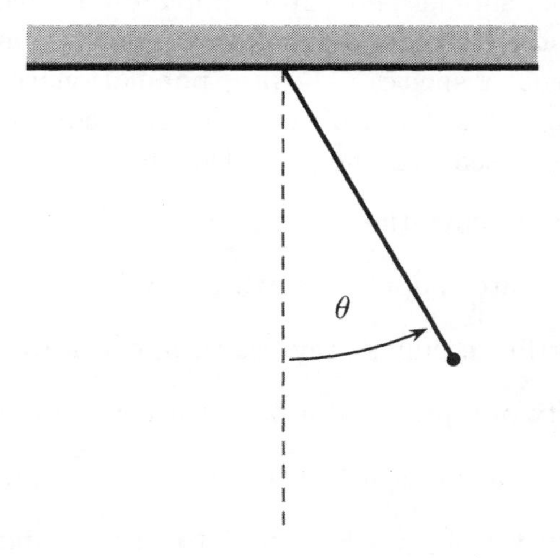

Figure 1.2: *Simple pendulum.*

Denoting by θ the angle between the pendulum and the y-axis, the friction-free motion is governed by the ODE (cf. Example 1.5 below)

$$\theta'' = -g \sin\theta, \tag{1.2}$$

where g is the (scaled) constant of gravity. This is a simple, nonlinear ODE for θ. The initial position and velocity configuration translates into values for $\theta(0)$ and $\theta'(0)$. The linear, trivial example from the beginning of this chapter can be obtained from an approximation of (a rescaled) (1.2) for small displacements θ. ♦

The pendulum problem is posed as a second-order scalar ODE. Much of the software for IVPs is written for first-order systems in the form (1.1). A scalar ODE of order m,

$$u^{(m)} = g(t, u, u', \ldots, u^{(m-1)}),$$

can be rewritten as a first-order system by introducing a new variable for each derivative, with $y_1 = u$:

$$\begin{aligned} y_1' &= y_2, \\ y_2' &= y_3, \\ &\vdots \\ y_{m-1}' &= y_m, \\ y_m' &= g(t, y_1, y_2, \ldots, y_m). \end{aligned}$$

Example 1.2 (predator-prey model)
Following is a basic, simple model from population biology which involves differential equations. Consider an ecological system consisting of one prey species and one predator species. The prey population would grow unboundedly if the predator were not present, and the predator population would perish without the presence of the prey. Denote

- $y_1(t)$—the prey population at time t;
- $y_2(t)$—the predator population at time t;
- α—prey's birthrate minus prey's natural death rate ($\alpha > 0$);
- β—probability of a prey and a predator coming together;
- γ—predator's natural growth rate (without prey; $\gamma < 0$);
- δ—increase factor of growth of predator if prey and predator meet.

Typical values for these constants are $\alpha = .25$, $\beta = .01$, $\gamma = -1.00$, $\delta = .01$.

Writing

$$\mathbf{y} = \begin{pmatrix} y_1 \\ y_2 \end{pmatrix}, \qquad \mathbf{f} = \begin{pmatrix} \alpha y_1 - \beta y_1 y_2 \\ \gamma y_2 + \delta y_1 y_2 \end{pmatrix}, \tag{1.3}$$

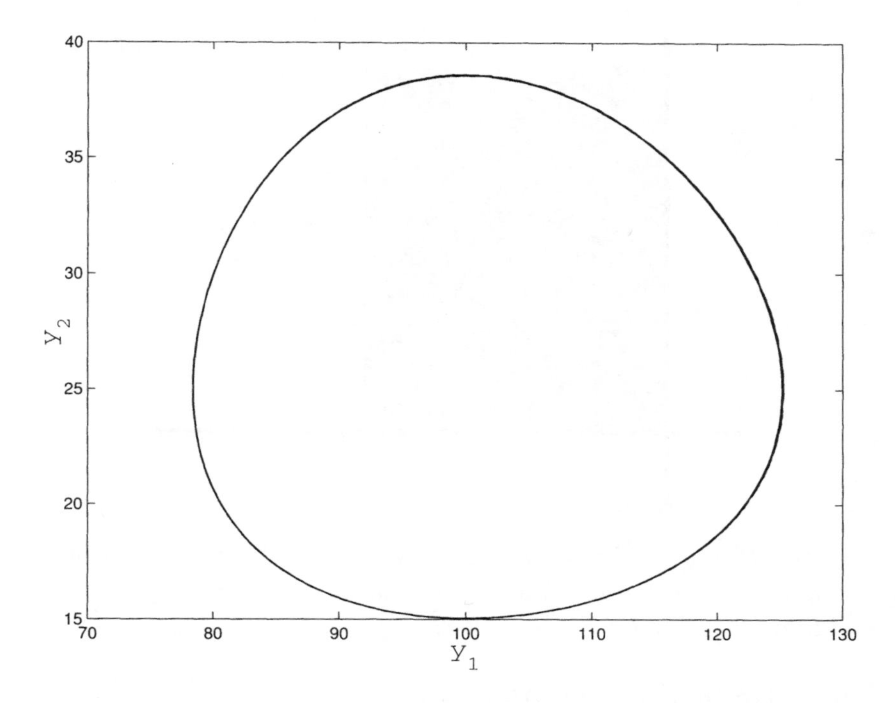

Figure 1.3: *Periodic solution forming a cycle in the $y_1 \times y_2$ plane.*

we obtain an ODE in the form (1.1) with $m = 2$ components, describing the time-evolution of these populations.

The qualitative question here is, starting from some initial values $\mathbf{y}(0)$ out of a set of reasonable possibilities, will these two populations survive or perish in the long run? As it turns out, this model possesses periodic solutions: starting, say, from $\mathbf{y}(0) = (80, 30)^T$, the solution reaches the same pair of values again after some time period T, i.e., $\mathbf{y}(T) = \mathbf{y}(0)$. Continuing to integrate past T yields a repetition of the same values, $\mathbf{y}(T + t) = \mathbf{y}(t)$. Thus, the solution forms a cycle in the phase plane (y_1, y_2) (see Figure 1.3). Starting from any point on this cycle, the solution stays on the cycle for all time. Other initial values not on this cycle yield other periodic solutions with a generally different period. So, under these circumstances the populations of the predator and prey neither explode nor vanish for all future times, although their number never becomes constant.[1] ♦

[1]In other examples, such as the Van der Pol equation (7.13), the solution forms an attracting *limit cycle*: starting from any point on the cycle the solution stays on it for all time, and starting from points near the solution, it tends in time towards the limit cycle.

The neutral stability of the cycle in our current example, in contrast, is one reason why this predator-prey model is discounted among mathematical biologists as being too simple.

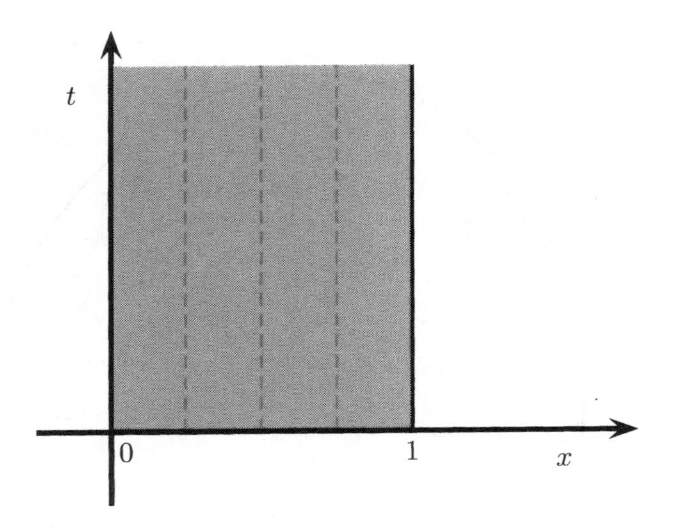

Figure 1.4: *Method of lines. The shaded strip is the domain on which the diffusion PDE is defined. The approximations $y_i(t)$ are defined along the dashed lines.*

Example 1.3 (a diffusion problem)
A typical diffusion problem in one space variable x and time t leads to the partial differential equation (PDE)

$$\frac{\partial u}{\partial t} = \frac{\partial}{\partial x}\left(p\frac{\partial u}{\partial x}\right) + g(x,u),$$

for an unknown function $u(t,x)$ of two independent variables defined on a strip $0 \le x \le 1$, $t \ge 0$. For simplicity, assume that $p = 1$ and g is a known function. Typical side conditions which make this problem well posed are

$$\begin{aligned} u(0,x) = q(x),\ 0 \le x \le 1 &\qquad \text{(initial conditions)}, \\ u(t,0) = \alpha(t),\ u(t,1) = \beta(t),\ t \ge 0 &\qquad \text{(boundary conditions)}. \end{aligned}$$

To solve this problem numerically, consider discretizing in the space variable first. For simplicity assume a uniform mesh with spacing $\Delta x = 1/(m+1)$, and let $y_i(t)$ approximate $u(x_i, t)$, where $x_i = i\Delta x$, $i = 0, 1, \ldots, m+1$. Then replacing $\frac{\partial^2 u}{\partial x^2}$ by a second-order central difference, we obtain

$$\frac{dy_i}{dt} = \frac{y_{i+1} - 2y_i + y_{i-1}}{\Delta x^2} + g(x_i, y_i), \qquad i = 1, \ldots, m,$$

with $y_0(t) = \alpha(t)$ and $y_{m+1}(t) = \beta(t)$ given. We have obtained an initial value ODE problem of the form (1.1) with the initial data $c_i = q(x_i)$.

This technique of replacing spatial derivatives by finite difference approximations and solving an ODE problem in time is referred to as the *method of lines*. Figure 1.4 illustrates the origin of the name. Its more general form is discussed further in Example 1.7 below. ♦

We now return to the general IVP for (1.1). Our intention in this book is to keep the number of theorems down to a minimum: the references which we quote have them all in great detail. But we will nonetheless record those which are of fundamental importance, and the one just below captures the essence of the (relative) simplicity and locality of initial value ODEs. For the notation that is used in this theorem and throughout the book, we refer to Section 1.6.

Theorem 1.1 *Let* $\mathbf{f}(t, \mathbf{y})$ *be continuous for all* $(t, \mathbf{y})$ *in a region* $\mathcal{D} = \{0 \le t \le b, |\mathbf{y}| < \infty\}$. *Moreover, assume Lipschitz continuity in* $\mathbf{y}$: *there exists a constant* L *such that for all* $(t, \mathbf{y})$ *and* $(t, \hat{\mathbf{y}})$ *in* $\mathcal{D}$,

$$|\mathbf{f}(t, \mathbf{y}) - \mathbf{f}(t, \hat{\mathbf{y}})| \le L|\mathbf{y} - \hat{\mathbf{y}}|. \tag{1.4}$$

Then

1. *For any* $\mathbf{c} \in \Re^m$ *there exists a unique solution* $\mathbf{y}(t)$ *throughout the interval* $[0, b]$ *for the IVP* (1.1). *This solution is differentiable.*

2. *The solution* $\mathbf{y}$ *depends continuously on the initial data: if* $\hat{\mathbf{y}}$ *also satisfies the ODE (but not the same initial values) then*

$$|\mathbf{y}(t) - \hat{\mathbf{y}}(t)| \le e^{Lt}|\mathbf{y}(0) - \hat{\mathbf{y}}(0)|. \tag{1.5}$$

3. *If* $\hat{\mathbf{y}}$ *satisfies, more generally, a perturbed ODE*

$$\hat{\mathbf{y}}' = \mathbf{f}(t, \hat{\mathbf{y}}) + \mathbf{r}(t, \hat{\mathbf{y}}),$$

where $\mathbf{r}$ *is bounded on* $\mathcal{D}$, $\|\mathbf{r}\| \le M$, *then*

$$|\mathbf{y}(t) - \hat{\mathbf{y}}(t)| \le e^{Lt}|\mathbf{y}(0) - \hat{\mathbf{y}}(0)| + \frac{M}{L}(e^{Lt} - 1). \tag{1.6}$$

Thus we have solution existence, uniqueness, and continuous dependence on the data—in other words, a *well-posed problem*—provided that the conditions of the theorem hold. Let us check these conditions: if $\mathbf{f}$ is differentiable in $\mathbf{y}$ (we shall automatically assume this throughout), then the constant L can be taken as a bound on the first derivatives of $\mathbf{f}$ with respect to $\mathbf{y}$. Denote by $\mathbf{f}_{\mathbf{y}}$ the *Jacobian matrix*,

$$(\mathbf{f}_{\mathbf{y}})_{ij} = \frac{\partial f_i}{\partial y_j}, \qquad 1 \le i, j \le m.$$

We can write

$$\mathbf{f}(t,\mathbf{y}) - \mathbf{f}(t,\hat{\mathbf{y}}) = \int_0^1 \frac{d}{ds}\mathbf{f}(t,\hat{\mathbf{y}} + s(\mathbf{y}-\hat{\mathbf{y}}))\, ds$$
$$= \int_0^1 \mathbf{f}_{\mathbf{y}}(t,\hat{\mathbf{y}} + s(\mathbf{y}-\hat{\mathbf{y}}))\,(\mathbf{y}-\hat{\mathbf{y}})\, ds.$$

Therefore, we can choose $L = \sup_{(t,\mathbf{y})\in\mathcal{D}} \|\mathbf{f}_{\mathbf{y}}(t,\mathbf{y})\|$.

In many cases we must restrict $\mathcal{D}$ in order to be assured of the existence of such a (finite) bound L. For instance, if we restrict $\mathcal{D}$ to include bounded $\mathbf{y}$ such that $|\mathbf{y}-\mathbf{c}| \leq \gamma$, and on this $\mathcal{D}$ both the Lipschitz bound (1.4) holds and $|\mathbf{f}(t,\mathbf{y})| \leq M$, then a unique existence of the solution is guaranteed for $0 \leq t \leq \min(b, \gamma/M)$, giving the basic existence result a more local flavor.

For further theory and proofs, see, for instance, Mattheij and Molnaar [67].

Note: Before continuing our introduction, let us remark that a reader who is interested in getting to the numerics of IVPs as soon as possible may skip the rest of this chapter and the next, at least on first reading.

1.2 BVPs

The general form of a BVP which we consider is a nonlinear first-order system of m ODEs subject to m independent (generally nonlinear) boundary conditions,

$$\mathbf{y}' = \mathbf{f}(t,\mathbf{y}), \tag{1.7a}$$
$$\mathbf{g}(\mathbf{y}(0),\mathbf{y}(b)) = \mathbf{0}. \tag{1.7b}$$

We have already seen in the beginning of the chapter that in those cases where solution information is given at both ends of the integration interval (or, more generally, at more than one point in time), nothing general like Theorem 1.1 can be expected to hold. Methods for finding a solution, both analytically and numerically, must be global, and the task promises to be generally harder than for IVPs. This basic difference is manifested in the current status of software for BVPs, which is much less advanced or robust than what is available for IVPs.

Of course, well-posed BVPs do arise on many occasions.

Example 1.4 (vibrating spring)
The small displacement u of a vibrating spring obeys a linear differential equation

$$-(p(t)u')' + q(t)u = r(t),$$

where $p(t) > 0$ and $q(t) \geq 0$ for all $0 \leq t \leq b$. (Such an equation also describes many other physical phenomena in one space variable t.) If the spring is fixed at one end and is left to oscillate freely at the other end, then we get the boundary conditions

$$u(0) = 0, \qquad u'(b) = 0.$$

We can write this problem in the form (1.7) for $\mathbf{y} = (u, u')^T$. Better still, we can use

$$\mathbf{y} = \begin{pmatrix} u \\ pu' \end{pmatrix},$$

obtaining

$$\mathbf{f} = \begin{pmatrix} p^{-1}y_2 \\ qy_1 - r \end{pmatrix}, \qquad \mathbf{g} = \begin{pmatrix} y_1(0) \\ y_2(b) \end{pmatrix}.$$

This BVP has a unique solution (which gives the minimum for the energy in the spring), as shown and discussed in many books on finite element methods, e.g., Strang and Fix [90]. ♦

Another example of a BVP is provided by the predator-prey system of Example 1.2, if we wish to find the periodic solution (whose existence is evident from Figure 1.3). We can specify $\mathbf{y}(0) = \mathbf{y}(b)$. However, note that b is unknown, so the situation is more complex. Further treatment is deferred to Chapter 6 and Exercise 7.5. A complete treatment of the topic of finding periodic solutions for ODE systems falls outside the scope of this book.

What can generally be said about existence and uniqueness of solutions to a general BVP (1.7)? We may consider the associated IVP (1.1) with the initial values $\mathbf{c}$ as a parameter vector to be found. Denoting the solution for such an IVP by $\mathbf{y}(t; \mathbf{c})$, we wish to find the solution(s) for the nonlinear algebraic system of m equations

$$\mathbf{g}(\mathbf{c}, \mathbf{y}(b; \mathbf{c})) = \mathbf{0}. \tag{1.8}$$

However, in general, there may be one, many, or no solutions for a system like (1.8). We delay further discussion to Chapter 6.

1.3 Differential-Algebraic Equations

Both the prototype IVP (1.1) and the prototype BVP (1.7) refer to an *explicit ODE* system

$$\mathbf{y}' = \mathbf{f}(t, \mathbf{y}). \tag{1.9}$$

A more general form is an *implicit ODE*

$$\mathbf{F}(t, \mathbf{y}, \mathbf{y}') = \mathbf{0}, \tag{1.10}$$

where the Jacobian matrix $\frac{\partial \mathbf{F}(t,\mathbf{u},\mathbf{v})}{\partial \mathbf{v}}$ is assumed to be nonsingular for all argument values in an appropriate domain. In principle, it is then often possible to solve for $\mathbf{y}'$ in terms of t and $\mathbf{y}$, obtaining the explicit ODE form (1.9). However, this transformation may not always be numerically easy or cheap to realize (see Example 1.6 below). Also, in general there may be additional questions of existence and uniqueness; we postpone further treatment until Chapter 9.

Consider next another extension of the explicit ODE, that of an *ODE with constraints*:

$$\mathbf{x}' = \mathbf{f}(t, \mathbf{x}, \mathbf{z}), \tag{1.11a}$$

$$\mathbf{0} = \mathbf{g}(t, \mathbf{x}, \mathbf{z}). \tag{1.11b}$$

Here the ODE (1.11a) for $\mathbf{x}(t)$ depends on additional algebraic variables $\mathbf{z}(t)$, and the solution is forced in addition to satisfy the algebraic constraints (1.11b). The system (1.11) is a *semi-explicit* system of *differential-algebraic equations* (DAEs). Obviously, we can cast (1.11) in the form of an implicit ODE (1.10) for the unknown vector $\mathbf{y} = \binom{\mathbf{x}}{\mathbf{z}}$; however, the obtained Jacobian matrix

$$\frac{\partial \mathbf{F}(t, \mathbf{u}, \mathbf{v})}{\partial \mathbf{v}} = \begin{pmatrix} I & 0 \\ 0 & 0 \end{pmatrix}$$

is no longer nonsingular.

Example 1.5 (simple pendulum revisited)
The motion of the simple pendulum of Figure 1.2 can be expressed in terms of the Cartesian coordinates (x_1, x_2) of the tiny ball at the end of the rod. With $z(t)$ a Lagrange multiplier, Newton's equations of motion give

$$x_1'' = -zx_1,$$
$$x_2'' = -zx_2 - g,$$

and the fact that the rod has a fixed length 1 gives the additional constraint

$$x_1^2 + x_2^2 = 1.$$

After rewriting the two second-order ODEs as four first-order ODEs, we obtain a DAE system of the form (1.11) with four equations in (1.11a) and one in (1.11b).

In this very simple case of a multibody system, the change of variables $x_1 = \sin\theta$, $x_2 = -\cos\theta$ allows elimination of z by simply multiplying the ODE for x_1 by x_2 and the ODE for x_2 by x_1 and subtracting. This yields the simple ODE (1.2) of Example 1.1. Such a simple elimination procedure is usually impossible in more general situations, though. ♦

The difference between an implicit ODE (with a nonsingular Jacobian matrix) and a DAE is fundamental. Consider the simple example

$$\begin{aligned} x' &= z, \\ 0 &= x - t. \end{aligned}$$

Clearly, the solution is $x = t$, $z = 1$, and no initial or boundary conditions are needed. In fact, if an arbitrary initial condition $x(0) = c$ is imposed, it may well be inconsistent with the DAE (unless $c = 0$, in which case this initial condition is just superfluous). We refer to Chapter 9 for more on this. Another point to note is that even if consistent initial values are given, we cannot expect a simple, general existence and uniqueness theorem like Theorem 1.1 to hold for (1.11). The nonlinear equations (1.11b) alone may have any number of solutions. Again we refer the reader to Chapter 9 for more details.

1.4 Families of Application Problems

Initial and boundary value problems for ODE and DAE systems arise in a wide variety of applications. Often an application generates a family of problems which share a particular system structure and/or solution requirements. Here we briefly mention three families of problems from important applications. The notation we use is typical for these applications and is not necessarily consistent with (1.1) or (1.11).

Note: You *don't need* to understand the details given in this section in order to follow the rest of the text; this material is supplemental.

Example 1.6 (mechanical systems)
A fast, reliable simulation of the dynamics of multibody systems is needed in order to simulate the motion of a vehicle for design or to simulate safety

tests in physically based modeling in computer graphics, and in a variety of instances in robotics. The system considered is an assembly of rigid bodies (e.g., comprising a car suspension system). The kinematics define how these bodies are allowed to move with respect to one another. Using generalized position coordinates $\mathbf{q} = (q_1, \dots, q_n)^T$ for the bodies, with m (so-called holonomic) constraints $g_j(t, \mathbf{q}(t)) = 0$, $j = 1, \dots, m$, the equations of motion can be written as

$$\frac{d}{dt}\left(\frac{\partial L}{\partial q_i'}\right) - \frac{\partial L}{\partial q_i} = 0, \quad i = 1, \dots, n,$$

where $L = T - U - \sum \lambda_j g_j$ is the Lagrangian, T is the kinetic energy, and U is the potential energy. See almost any book on classical mechanics, for example, Arnold [1] or the lighter Marion and Thornton [65]. The resulting equations of motion can be written as

$$\mathbf{q}' = \mathbf{v}, \tag{1.12a}$$

$$M(t, \mathbf{q})\mathbf{v}' = \mathbf{f}(t, \mathbf{q}, \mathbf{v}) - G^T(t, \mathbf{q})\boldsymbol{\lambda}, \tag{1.12b}$$

$$\mathbf{0} = \mathbf{g}(t, \mathbf{q}), \tag{1.12c}$$

where $G = \frac{\partial \mathbf{g}}{\partial \mathbf{q}}$, M is a positive definite generalized mass matrix, $\mathbf{f}$ are the applied forces (other than the constraint forces), and $\mathbf{v}$ are the generalized velocities. The system sizes n and m depend on the chosen coordinates $\mathbf{q}$. Typically, using relative coordinates (describing each body in terms of its near neighbor) results in a smaller but more complicated system. If the topology of the multibody system (i.e., the connectivity graph obtained by assigning a node to each body and an edge for each connection between bodies) does not have closed loops, then with a minimal set of coordinates one can eliminate all the constraints (i.e., $m = 0$) and obtain an implicit ODE in (1.12). For instance, Example 1.1 uses a minimal set of coordinates for a particular multibody system without loops, while Example 1.5 does not. If the multibody system contains loops (e.g., a robot arm, consisting of two links, with the path of the "hand" prescribed, as in Example 10.9), then the constraints cannot be totally eliminated in general, and a DAE must be considered in (1.12) even if a minimal set of coordinates is employed. ♦

Example 1.7 (method of lines)
The diffusion equation of Example 1.3 is an instance of a time-dependent PDE in one space dimension,

$$\frac{\partial u}{\partial t} = f\left(t, u, \frac{\partial u}{\partial x}, \frac{\partial^2 u}{\partial x^2}\right). \tag{1.13}$$

Time-dependent PDEs naturally arise in more than one space dimension as well, with higher-order spatial derivatives and as systems of PDEs. The process described in Example 1.3 is general: such a PDE can be transformed into a large system of ODEs by replacing the spatial derivatives in one or more dimensions by a discrete approximation (via finite difference, finite volume, or finite element methods; see texts on numerical methods for PDEs, e.g., Strikwerda [91], [90]). Typically, we obtain an IVP. This technique of semidiscretizing in space first and solving an initial value ODE problem in time is referred to as the *method of lines*. It makes sense when two conditions are satisfied. i) The "time" variable t is sufficiently different from the "space" variables to warrant a special treatment. ii) There is no sharp front in the solution that moves rapidly as a function of both space and time; i.e., the rapid moving fronts (if there are any) can be reasonably well decoupled in time and space. Typically, the method of lines is more suitable for parabolic PDEs than for hyperbolic ones.

Remaining with the prototype diffusion problem considered in Example 1.3, in some situations the "special" independent variable is not time but one of the spatial variables. This is the case in some interface problems. Another way to convert a PDE to an ODE system is then to replace the time derivative by a difference approximation. Replacing the time derivative by a simple backward difference approximation using time step Δt in the diffusion equation yields

$$\frac{u^n - u^{n-1}}{\Delta t} = \frac{\partial^2 u^n}{\partial x^2} + g(x, u^n),$$

and using $u^0 = q(x)$ and the given boundary conditions yields a BVP in x for each n. This technique of replacing the time derivative by a difference approximation and solving the BVP in space is called the *transverse method of lines.* ♦

Example 1.8 (optimal control)

A rather large number of applications give rise to *optimal control* problems. For instance, the problem may be to plan a route for a vehicle traveling between two points (and satisfying equations of motion) such that fuel consumption is optimized or the travel time is minimized. Another instance is to optimize the performance of a chemical processing plant. Typically, the state variables of the system, $\mathbf{y}(t)$, satisfy an ODE system which involves a control function $\mathbf{u}(t)$,[2]

$$\mathbf{y}' = \mathbf{f}(t, \mathbf{y}, \mathbf{u}), \qquad 0 \le t \le b. \tag{1.14a}$$

This system may be subject to some side conditions, e.g.,

$$\mathbf{y}(0) = \mathbf{c}, \tag{1.14b}$$

[2]The dimension of $\mathbf{u}(t)$ is generally different from that of $\mathbf{y}(t)$.

but it is possible that $\mathbf{y}(b)$ is prescribed as well, or that there are no side conditions at all. The control $\mathbf{u}(t)$ must be chosen so as to optimize some criterion (or cost) function, say,

$$\text{minimize } J = \phi(\mathbf{y}(b), b) + \int_0^b L(t, \mathbf{y}(t), \mathbf{u}(t))dt \tag{1.15}$$

subject to (1.14).

The *necessary conditions* for an optimum in this problem are found by considering the Hamiltonian function

$$H(t, \mathbf{y}, \mathbf{u}, \boldsymbol{\lambda}) = \sum_{i=1}^{m} \lambda_i f_i(t, \mathbf{y}, \mathbf{u}) + L(t, \mathbf{y}, \mathbf{u}),$$

where $\lambda_i(t)$ are *adjoint variables*, $i = 1, \ldots, m$. The conditions

$$y_i' = \frac{\partial H}{\partial \lambda_i}, \qquad i = 1, \ldots, m,$$

yield the state equations (1.14a), and in addition we have ODEs for the adjoint variables,

$$\lambda_i' = -\frac{\partial H}{\partial y_i} = -\sum_{j=1}^{m} \lambda_j \frac{\partial f_j}{\partial y_i} - \frac{\partial L}{\partial y_i}, \qquad i = 1, \ldots, m, \tag{1.16}$$

and

$$0 = \frac{\partial H}{\partial u_i}, \qquad i = 1, \ldots, m_u. \tag{1.17}$$

This gives a DAE in general; however, $\mathbf{u}(t)$ can often be eliminated from (1.17) in terms of $\mathbf{y}$ and $\boldsymbol{\lambda}$, yielding an ODE system. Additional side conditions are required as well:

$$\lambda_i(b) = \frac{\partial \phi}{\partial y_i}(b), \qquad i = 1, \ldots, m. \tag{1.18}$$

The system (1.14), (1.16), (1.17), (1.18) comprises a boundary value ODE (or DAE).

An *indirect approach* for solving this optimal control problem involves the numerical solution of the BVP just prescribed. The techniques described in Chapters 7 and 8 are directly relevant. In contrast, a *direct approach* involves the discretization of (1.14), (1.15) and the subsequent numerical solution of the resulting large, sparse (but finite-dimensional) constrained optimization problem. The techniques described in this book are relevant for this approach too, although less directly. Each of these two approaches has its advantages (and fans). Note that, even though (1.14) is an IVP, the direct approach does not yield a local process, which would have allowed a simple marching algorithm, because a change in the problem anywhere has a global effect, necessitating a global solution process (as needed for a BVP).

A closely related family of applications involves *parameter estimation* in an ODE system. Given a set of solution data in time (usually obtained by experiment), the problem is to choose the parameters to minimize a measure of the distance between the data and the solution of the ODE (or DAE) depending on the parameters.

We note, furthermore, that optimal control applications often require, in addition to the above model, inequality (algebraic) constraints on the controls $\mathbf{u}(t)$ and on the state variables $\mathbf{y}(t)$ (e.g., a maximum speed or acceleration which must not, or cannot, be exceeded in the vehicle route planning application). Such inequality constraints complicate the analysis yielding necessary conditions, but we do not pursue this further. There are many books on optimal control and parameter estimation, e.g., Bryson and Ho [22]. ♦

1.5 Dynamical Systems

Recent years have seen an explosion of interest and efforts in the study of the long-term, *qualitative* behavior of various *nonlinear* ODE systems. Typically, one is interested in the behavior of the flow of a system $\mathbf{y}' = \mathbf{f}(t, \mathbf{y})$, not only in one trajectory for a given initial value $\mathbf{c}$. Attention is often focussed then on *limit sets* (a limit set is a special case of an *invariant set*, i.e., a set of initial data that is mapped into itself by the flow).

Although most of this book is concerned with the accurate and reliable simulation of solution trajectories, and the reader is not assumed to possess a background in dynamical systems, the techniques we explore are essential for numerical dynamical systems. Moreover, various additional challenges arise when considering the simulation of such qualitative properties. In some cases these additional challenges can be addressed using simple tricks (e.g., for finding a periodic solution or for projecting onto a given invariant defined by algebraic equations), while on other occasions the challenge is rather more substantial (e.g., finding an invariant set in general or numerically integrating an ODE over a very long time period).

Throughout this book we will pay attention to such additional considerations, especially when they extend our investigation in a natural way. We will certainly not attempt to do a complete job; rather, we will point out problems, some solutions, and some directions. For much more, we refer the reader to Stuart and Humphries [93].

1.6 Notation

Throughout the book, we use the following conventions for notation.

- Scalar variables and constants are denoted by Roman and Greek letters, e.g., t, u, y, K, L, N, α, β, etc.

- Vectors are denoted by boldface letters, e.g., $\mathbf{f}$, $\mathbf{y}$, $\mathbf{c}$, etc. The ith component of the vector $\mathbf{y}$ is denoted y_i. (Distinguish this from the notation $\mathbf{y}_n$, which will be used later on to denote a vector approximating $\mathbf{y}$ at position t_n.) Two special vectors are $\mathbf{0} = (0, \dots, 0)^T$ and $\mathbf{1} = (1, \dots, 1)^T$.

- The maximum norm of a vector is denoted just like the absolute value of a scalar: $|\mathbf{y}| = \max_{1 \le i \le m} |y_i|$. Occasionally, the Euclidean vector norm $|\mathbf{y}|_2 = \sqrt{\mathbf{y}^T\mathbf{y}}$ proves more convenient than the maximum norm; we may drop the subscript when the precise vector norm used does not matter or is obvious.

- Capital Roman letters are used for matrices. The induced norms of matrices are denoted by double bars:

$$\|A\| = \sup_{|\mathbf{x}|=1} \frac{|A\mathbf{x}|}{|\mathbf{x}|}.$$

  Occasionally, a boldface capital Roman letter, e.g., $\mathbf{A}$, is used for large matrices consisting of blocks which are themselves matrices. The letter I is reserved for an identity matrix.

- The (sup) norms of functions are denoted as follows:

$$\|\mathbf{y}\| = \sup_{0 \le t \le b} |\mathbf{y}(t)|.$$

- Letters from other alphabets, e.g., $\mathcal{D}$, $\mathcal{L}$, $\mathcal{N}_h$, are used to denote domains and operators. Also, $\mathcal{R}e$ and $\mathcal{I}m$ denote the real and imaginary parts of a complex scalar, and $\Re$ is the set of real numbers.

- For a vector function $\mathbf{g}(\mathbf{x})$, where $\mathbf{g}$ has n components and $\mathbf{x}$ has k components ($\mathbf{g}$ may depend on other variables too, e.g., $\mathbf{g} = \mathbf{g}(t, \mathbf{x}, \mathbf{y})$), we denote the *Jacobian matrix*, i.e., the $n \times k$ matrix of first partial derivatives of $\mathbf{g}$ with respect to $\mathbf{x}$, by $\mathbf{g}_\mathbf{x}$ or by $\frac{\partial \mathbf{g}}{\partial \mathbf{x}}$:

$$\left(\frac{\partial \mathbf{g}}{\partial \mathbf{x}}\right)_{i,j} \equiv (\mathbf{g}_\mathbf{x})_{i,j} = \frac{\partial g_i}{\partial x_j}, \qquad 1 \le i \le n,\ 1 \le j \le k.$$

  We use the Jacobian matrix notation frequently in this book, and occasionally find one of these common notational forms to be clearer than the other in a particular context. Hence we keep them both.

- The *gradient* of a scalar function of k variables $g(\mathbf{x})$, denoted $\nabla g(\mathbf{x})$, is its one-row Jacobian matrix transposed into a vector function:

$$\nabla g(\mathbf{x}) = g_\mathbf{x}^T.$$

The *divergence* of a vector function $\mathbf{g}(\mathbf{x})$, where $\mathbf{g}$ and $\mathbf{x}$ both have k components, is the scalar function denoted by $\nabla\mathbf{g}(\mathbf{x})$ and given by

$$\nabla\mathbf{g}(\mathbf{x}) = \sum_{i=1}^{k} \frac{\partial g_i}{\partial x_i}.$$

Part II: Initial Value Problems

Chapter 2

On Problem Stability

The term *stability* has been used in the literature for a large variety of different concepts. The basic, qualitative idea is that a model that produces a solution (output) for given data (input) should possess the property that if the input is perturbed by a small amount then the output should also be perturbed by only a small amount. But the precise application of this idea to initial value ordinary differential equations (ODEs), to boundary value ODEs, and to numerical methods has given rise to a multitude of definitions. The reader should therefore be careful, when speaking of stability, to distinguish between stability of problems and of numerical methods, and between stability of initial value problems (IVPs) and boundary value problems (BVPs).

In this chapter we briefly discuss the stability of IVPs. No numerical solutions or methods are discussed yet; that will start only in the next chapter. Matrix eigenvalues play a central role here, so we also include a quick review below.

2.1 Test Equation and General Definitions

Consider at first the simple scalar ODE, often referred to later as the *test equation*

$$y' = \lambda y, \tag{2.1}$$

where λ is a constant. We allow λ to be complex, because it represents an eigenvalue of a system's matrix. The solution for $t \geq 0$ is

$$y(t) = e^{\lambda t} y(0).$$

If $y(t)$ and $\hat{y}(t)$ are two solutions of the test equation, then their difference for any t depends on their difference at the initial time:

$$|y(t) - \hat{y}(t)| = |(y(0) - \hat{y}(0))e^{\lambda t}| = |y(0) - \hat{y}(0)|e^{\mathcal{R}e(\lambda)t}.$$

Review: Matrix eigenvalues. Given an $m \times m$ real matrix A, an *eigenvalue* λ is a scalar which satisfies

$$A\mathbf{x} = \lambda\mathbf{x}$$

for some vector $\mathbf{x} \neq \mathbf{0}$. In general, λ is complex, but it is guaranteed to be real if A is symmetric. The vector $\mathbf{x}$, which is clearly determined only up to a scaling factor, is called an *eigenvector*. Counting multiplicities, A has m eigenvalues.

A *similarity transformation* is defined, for any nonsingular matrix T, by

$$B = T^{-1}AT\,.$$

The matrix B has the same eigenvalues as A and the two matrices are said to be similar. If B is diagonal, $B = \operatorname{diag}\{\lambda_1, \ldots, \lambda_m\}$, then the displayed λ_i are the eigenvalues of A, the corresponding eigenvectors are the columns of T, and A is said to be *diagonalizable*. Any symmetric matrix is diagonalizable, in fact, by an *orthogonal matrix* (i.e., T can be chosen to satisfy $T^T = T^{-1}$). For a general matrix, however, an orthogonal similarity transformation can only bring A to a matrix B in upper triangular form (which, however, still features the eigenvalues on the main diagonal of B).

For a general A there is always a similarity transformation into a *Jordan canonical form*,

$$B = \begin{pmatrix} \Lambda_1 & & & & 0 \\ & \Lambda_2 & & & \\ & & \cdot & & \\ & & & \cdot & \\ 0 & & & & \Lambda_s \end{pmatrix}; \quad \Lambda_i = \begin{pmatrix} \lambda_i & 1 & & 0 \\ & \lambda_i & 1 & \\ & & \ddots & 1 \\ 0 & & & \lambda_i \end{pmatrix}, \quad i = 1, \ldots, s\,.$$

We may consider $y(t)$ as the "exact" solution sought, and $\hat{y}(t)$ as the solution where the initial data has been perturbed. Clearly, then, if $\mathcal{R}\mathrm{e}(\lambda) \le 0$ this perturbation difference remains bounded at all later times, if $\mathcal{R}\mathrm{e}(\lambda) < 0$ it decays in time, and if $\mathcal{R}\mathrm{e}(\lambda) > 0$ the difference between the two solutions grows unboundedly with t. These possibilities correspond to a *stable*, an *asymptotically stable*, and an *unstable* solution, respectively.

The precise definition for a general ODE system

$$\mathbf{y}' = \mathbf{f}(t, \mathbf{y}) \tag{2.2}$$

is more technical, but the spirit is the same as for the test equation. We consider (2.2) for *all* $t \ge 0$ and define a solution (or trajectory) $\mathbf{y}(t)$ to be

- *stable* if given any $\epsilon > 0$ there is a $\delta > 0$ such that any other solution $\hat{\mathbf{y}}(t)$ satisfying the ODE (2.2) and

$$|\mathbf{y}(0) - \hat{\mathbf{y}}(0)| \le \delta$$

  also satisfies

$$|\mathbf{y}(t) - \hat{\mathbf{y}}(t)| \le \epsilon \qquad \text{for all } t \ge 0;$$

- *asymptotically stable* if, in addition to being stable,

$$|\mathbf{y}(t) - \hat{\mathbf{y}}(t)| \to 0 \quad \text{as } t \to \infty.$$

It would be worthwhile for the reader to compare these definitions to the bound (1.5) of the fundamental Existence and Uniqueness Theorem 1.1. Note that the existence theorem speaks of a finite, given integration interval.

These definitions are given with respect to perturbations in the initial data. What we really need to consider are perturbations at any later time and in the right-hand side of (2.2) as well. These correspond to the bound (1.6) in Theorem 1.1 and lead to slightly stronger requirements. But the spirit is already captured in the simple definitions above, and the more complete definitions are left to ODE texts.

Example 2.1
Suppose we integrate a given IVP exactly for $t \ge 0$; then we perturb this trajectory at a point $t_0 = h$ by an amount $\delta = \delta(h)$ and integrate the IVP exactly again for $t \ge t_0$, starting from the perturbed value. This process, which resembles the effect of a numerical discretization step, is now repeated a few times. The question is, then, how do the perturbation errors propagate? In particular, how far does the value of the last trajectory computed at $t = b$ ($\gg h$) get from the value of the original trajectory at $t = b$?

For the test equation (2.1), we can calculate everything precisely. If $y(t_0) = c$ then $y(t) = ce^{\lambda(t-t_0)}$. So, starting from $y(0) = 1$, we calculate the trajectories

$$y(t) = y_I(t) = e^{\lambda t},$$
$$y_{II}(t) = (e^{\lambda h} - \delta)e^{\lambda(t-h)} = e^{\lambda t} - \delta e^{\lambda(t-h)},$$
$$y_{III}(t) = e^{\lambda t} - \delta e^{\lambda(t-h)} - \delta e^{\lambda(t-2h)},$$
$$\vdots$$

For each such step we can define the error due to the jth perturbation,

$$e_j(t) = \delta e^{\lambda(t-jh)}.$$

So, after n steps the difference between the original trajectory and the last one computed at $t \geq nh$ is

$$e(t) = \sum_{j=1}^{n} e_j(t).$$

Apparently from the form of $e_j(t)$, the errors due to perturbations tend to decrease in time for asymptotically stable problems and to increase in time for unstable problems. This effect is clearly demonstrated in Figure 2.1, where we took $h = 0.1$, $\delta = 0.05$ and plotted curves for the values $\lambda = -1, 1, 0$. Note that the instability of $y' = y$ can really generate a huge deviation for large t (e.g., $t = 30$). ♦

2.2 Linear, Constant-Coefficient Systems

Here we consider the extension of the test equation analysis to a simple ODE system,

$$\mathbf{y}' = A\mathbf{y}, \tag{2.3}$$

where A is a constant $m \times m$ matrix. The solution for $t \geq 0$ is

$$\mathbf{y}(t) = e^{At}\mathbf{y}(0). \tag{2.4}$$

Denote the eigenvalues of A by $\lambda_1, \lambda_2, \ldots, \lambda_m$, and let

$$\Lambda = \text{diag}\{\lambda_1, \lambda_2, \ldots, \lambda_m\}$$

be the diagonal $m \times m$ matrix having these eigenvalues as its diagonal elements. If A is diagonalizable then there exists a similarity transformation that carries it into Λ, viz.

$$T^{-1}AT = \Lambda.$$

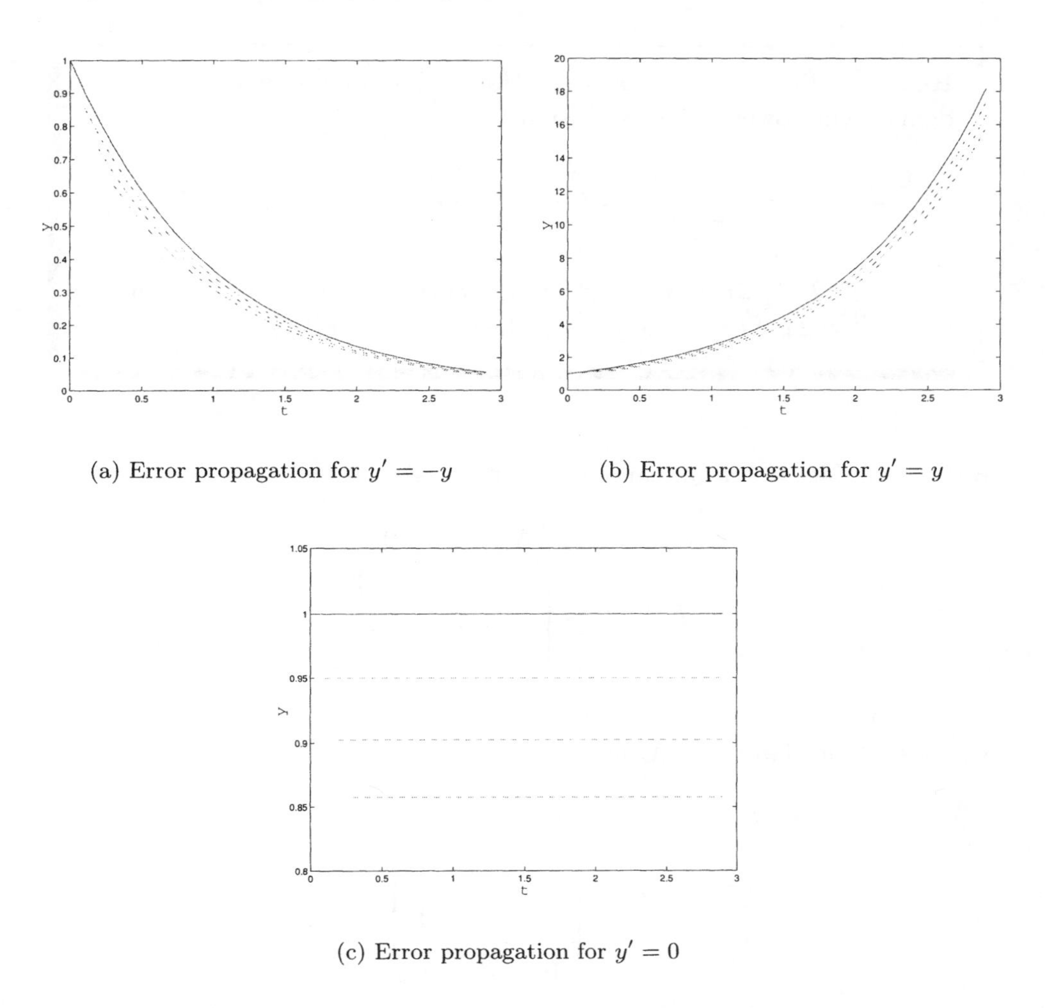

(a) Error propagation for $y' = -y$

(b) Error propagation for $y' = y$

(c) Error propagation for $y' = 0$

Figure 2.1: *Errors due to perturbations for stable and unstable test equations. The original, unperturbed trajectories are in solid curves, the perturbed in dashed. Note that the y-scales in* (a) *and* (b) *are* not *the same.*

Then the change of variables

$$\mathbf{w} = T^{-1}\mathbf{y}$$

yields the ODE for $\mathbf{w}$:

$$\mathbf{w}' = \Lambda\mathbf{w}.$$

The system for $\mathbf{w}$ is decoupled: for each component w_i of $\mathbf{w}$ we have a test equation $w_i' = \lambda_i w_i$. Therefore, the stability for $\mathbf{w}$, hence also for $\mathbf{y}$, is determined by the eigenvalues: stability is obtained if $\mathcal{R}e(\lambda_i) \leq 0$, for all $i = 1, \ldots, m$, and asymptotic stability holds if the inequalities are all strict.

In the more general case, A may not be similar to any diagonal matrix.

Review: The matrix exponential. The matrix exponential is defined via power series expansion by

$$e^{At} = \sum_{n=0}^{\infty} \frac{t^n A^n}{n!} = I + tA + \frac{t^2 A^2}{2} + \frac{t^3 A^3}{6} + \cdots .$$

If $A = T\Lambda T^{-1}$, where Λ is a diagonal matrix, then it is easy to show that $e^{At} = Te^{\Lambda t}T^{-1}$, where $e^{\Lambda t} = \text{diag}\{e^{\lambda_1 t}, \ldots, e^{\lambda_m t}\}$.

Rather, we face a Jordan canonical form:

$$T^{-1}AT = \begin{pmatrix} \Lambda_1 & & 0 \\ & \ddots & \\ 0 & & \Lambda_l \end{pmatrix},$$

where each Jordan block Λ_i has the form

$$\Lambda_i = \begin{pmatrix} \lambda_i & 1 & & 0 \\ & \ddots & \ddots & \\ & & \ddots & 1 \\ 0 & & & \lambda_i \end{pmatrix}.$$

A little more is required then. A short analysis, which we omit, establishes that in general, the solution of the ODE (2.3) is

- *stable* iff all eigenvalues λ of A satisfy either $\mathcal{R}\text{e}(\lambda) < 0$ or $\mathcal{R}\text{e}(\lambda) = 0$ and λ is simple (i.e., it belongs to a 1×1 Jordan block),
- *asymptotically stable* iff all eigenvalues λ of A satisfy $\mathcal{R}\text{e}(\lambda) < 0$.

Example 2.2
Consider the second-order ODE

$$-u'' + u = 0$$

obtained by taking $p = q = 1$ in Example 1.4 (the vibrating spring). Writing as a first-order system we obtain

$$\mathbf{y}' = \begin{pmatrix} 0 & 1 \\ 1 & 0 \end{pmatrix} \mathbf{y}.$$

The eigenvalues of this matrix are $\lambda_1 = -1$ and $\lambda_2 = 1$. Hence this IVP is unstable. (Note that in Chapter 1 we considered this ODE in the context of a BVP. With appropriate boundary conditions the problem can become stable, as we'll see in Chapter 6.)

Returning to the experiment of Example 2.1, here we have one source of growing error and one source of decreasing error for the IVP. Obviously, after a sufficiently long time the growing perturbation error will dominate, even if it starts from a very small deviation δ. This is why one "bad" eigenvalue of A is sufficient for the onset of instability. ♦

Example 2.3
The general homogeneous, scalar ODE with constant coefficients,

$$a_k u + a_{k-1}u' + \cdots + a_0 u^{(k)} = 0 \tag{2.5}$$

(or $\sum_{j=0}^{k} a_j \frac{d^{k-j}u}{dt^{k-j}} = 0$), with $a_0 > 0$, can be converted as we saw in Chapter 1 to a first-order ODE system. This gives a special case of (2.3) with $m = k$, $y_1 = u$, and

$$A = \begin{pmatrix} 0 & 1 & & & \\ & 0 & 1 & & \\ & & \ddots & \ddots & \\ & & & 0 & 1 \\ -a_k/a_0 & -a_{k-1}/a_0 & & & -a_1/a_0 \end{pmatrix}.$$

It is easy to verify that the eigenvalues of this matrix are the roots of the *characteristic polynomial*

$$\phi(\zeta) = \sum_{j=0}^{k} a_j \zeta^{k-j}. \tag{2.6}$$

The solution of the higher-order ODE (2.5) is therefore

- stable iff all roots ζ of the characteristic polynomial satisfy either $\mathcal{R}\text{e}(\zeta) < 0$ or $\mathcal{R}\text{e}(\zeta) = 0$ and ζ is simple,
- asymptotically stable iff all roots ζ of the characteristic polynomial satisfy $\mathcal{R}\text{e}(\zeta) < 0$. ♦

2.3 Linear, Variable-Coefficient Systems

The general form of a linear ODE system is

$$\mathbf{y}' = A(t)\mathbf{y} + \mathbf{q}(t), \tag{2.7}$$

where the $m \times m$ matrix $A(t)$ and the m-vector inhomogeneity $\mathbf{q}(t)$ are given for each t, $0 \le t \le b$.

We briefly review elementary ODE theory. The *fundamental solution* $Y(t)$ is the $m \times m$ matrix function which satisfies

$$Y'(t) = A(t)Y(t), \qquad 0 \le t \le b, \tag{2.8a}$$

$$Y(0) = I; \tag{2.8b}$$

i.e., the jth column of $Y(t)$, often referred to as a *mode*, satisfies the homogeneous version of the ODE (2.7) with the jth unit vector as initial value. The solution of ODE (2.7) subject to given initial values

$$\mathbf{y}(0) = \mathbf{c}$$

is then

$$\mathbf{y}(t) = Y(t)\left[\mathbf{c} + \int_0^t Y^{-1}(s)\mathbf{q}(s)ds\right]. \tag{2.9}$$

Turning to stability, it is clear that for a linear problem the difference between two solutions $\mathbf{y}(t)$ and $\hat{\mathbf{y}}(t)$ can be directly substituted into (2.9) in place of $\mathbf{y}(t)$ with the corresponding differences in data substituted into the right-hand side (say, $\mathbf{c} - \hat{\mathbf{c}}$ in place of $\mathbf{c}$). So the question of stability relates to the boundedness of $\mathbf{y}(t)$ for a homogeneous problem (i.e., with $\mathbf{q} = \mathbf{0}$) as we let $b \to \infty$. Then the solution of the ODE is

- stable iff $\sup_{0 \le t < \infty} \|Y(t)\|$ is bounded,
- asymptotically stable iff in addition to being stable, $\|Y(t)\| \to 0$ as $t \to \infty$.

We can define the stability constant

$$\kappa = \sup_{0 \le t < \infty} \|Y(t)\|$$

in an attempt to get a somewhat more quantitative feeling. But an examination of (2.9) suggests that a more careful definition of the stability constant, also taking into account perturbations in the inhomogeneity, is

$$\kappa = \sup_{0 \le s \le t < \infty} \|Y(t)Y^{-1}(s)\|. \tag{2.10}$$

Example 2.4
The simple ODE

$$y' = (\cos t)y$$

has the eigenvalue $\lambda(t) = \cos t$ and the fundamental solution $Y(t) = e^{\sin t}$. This problem is stable, with a moderate stability constant $\kappa = e^2 < 8$, even though the eigenvalue does not always remain below 0. ♦

2.4 Nonlinear Problems

A full exposition of stability issues for nonlinear problems is well beyond the scope of this book. A fundamental difference from the linear case is that the stability depends on the particular solution trajectory considered.

For a given, isolated solution $\mathbf{y}(t)$ of (2.2)[3], a linear analysis can be applied *locally*, to consider trends of *small* perturbations. Thus, if $\hat{\mathbf{y}}(t)$ satisfies the same ODE with $\hat{\mathbf{y}}(0) = \hat{\mathbf{c}}$ not too far from $\mathbf{c}$, then (under certain conditions) we can ignore the higher-order term $\mathbf{r}(t, \mathbf{y}, \hat{\mathbf{y}})$ in the Taylor expansion

$$f(t, \hat{\mathbf{y}}) = f(t, \mathbf{y}) + \frac{\partial \mathbf{f}}{\partial \mathbf{y}}(\hat{\mathbf{y}} - \mathbf{y}) + \mathbf{r}(t, \mathbf{y}, \hat{\mathbf{y}})$$

and consider the linear, *variational equation*

$$\mathbf{z}' = A(t, \mathbf{y})\mathbf{z} \tag{2.11}$$

for $\mathbf{z}$ (not $\mathbf{y}$), with the Jacobian matrix $A = \frac{\partial \mathbf{f}}{\partial \mathbf{y}}$.

Example 2.5
Often, one is interested in steady state solutions, i.e., when $\mathbf{y}(t)$ becomes independent of t, hence $\mathbf{y}' = \mathbf{0} = \mathbf{f}(\mathbf{y})$. An example is

$$y' = y(1 - y),$$

which obviously has the steady state solutions $y = 0$ and $y = 1$. The Jacobian is $A = 1 - 2y$, hence $A > 0$ for the value $y = 0$ and $A < 0$ for $y = 1$. We conclude that the steady state solution $y = 0$ is unstable, whereas the steady state solution $y = 1$ is stable. Thus, even if we begin the integration of the ODE from an initial value close to the steady state $y = 0$, $0 < c \ll 1$, the solution $y(t)$ will be repelled from it and attracted to the stable steady state $y = 1$. ♦

Since the Jacobian matrix depends on the solution trajectory $\mathbf{y}(t)$, its eigenvalues do not necessarily retain the same sign throughout the integration interval. It is then possible to have a system with a bounded solution over an arbitrarily long integration interval, which contains time subintervals whose total length is also arbitrarily large, where the system behaves unstably. This is already possible for linear problems with variable coefficients, e.g., Example 2.4, but it is not possible for the constant-coefficient problems of Section 2.2. Through the periods of solution growth, perturbation errors grow as well. Then, unless the system is sufficiently simple

[3] In other words, there is some tube in which $\mathbf{y}(t)$ is the only solution of (2.2), but no global uniqueness is postulated.

so that these errors shrink again, they may remain bounded through stable periods only to grow even further when the system becomes unstable again. This generates an effect of *unpredictability*, where the effect of errors in data grows uncontrollably even if the solution remains bounded.

Note: In the following section we give some brief background on material of current research interest. But for those who operate on a need-to-know basis, we note that this material appears, in later chapters, only in the "Notes and References" sections and in selected exercises.

2.5 Hamiltonian Systems

A lot of attention has been devoted in recent years to *Hamiltonian systems*. A Hamiltonian system consists of $m = 2l$ differential equations,

$$q_i' = \frac{\partial H}{\partial p_i}, \tag{2.12a}$$

$$i = 1, \ldots, l,$$

$$p_i' = -\frac{\partial H}{\partial q_i}, \tag{2.12b}$$

or in vector notation (with $\nabla_{\mathbf{p}} H$ denoting the gradient of H with respect to $\mathbf{p}$, etc.),

$$\mathbf{q}' = \nabla_{\mathbf{p}} H(\mathbf{q}, \mathbf{p}), \quad \mathbf{p}' = -\nabla_{\mathbf{q}} H(\mathbf{q}, \mathbf{p}).$$

The scalar function $H(\mathbf{q}, \mathbf{p})$, assumed to have continuous second derivatives, is the *Hamiltonian*.[4]

Differentiating H with respect to time t and substituting (2.12) we get

$$H' = \nabla_{\mathbf{p}} H^T \mathbf{p}' + \nabla_{\mathbf{q}} H^T \mathbf{q}' = 0$$

so $H(\mathbf{q}, \mathbf{p})$ is constant for all t. A typical example to keep in mind is that of a conservative system of particles. Then the components of $\mathbf{q}(t)$ are the generalized positions of the particles, and those of $\mathbf{p}(t)$ are the generalized momenta. The Hamiltonian H in this case is the total energy (the sum of kinetic and potential energies), and the constancy of H is a statement of *conservation of energy*.

Next, consider an autonomous ODE system of order $m = 2$,

$$\mathbf{y}' = \mathbf{f}(\mathbf{y}),$$

[4]In Chapters 9 and 10 we use e instead of H to denote the Hamiltonian.

with $\mathbf{y}(0) = (y_1(0), y_2(0))^T \in B$, for some set B in the plane. Each initial value $\mathbf{y}(0) = \mathbf{c}$ from B spawns a trajectory $\mathbf{y}(t) = \mathbf{y}(t; \mathbf{c})$, and we can follow the evolution of the set B under this flow,

$$S(t)B = \{\mathbf{y}(t, \mathbf{c})\,;\, \mathbf{c} \in B\}\,.$$

We then ask how the area of $S(t)B$ compares to the initial area of B: does it grow or shrink in time? It is easy to see for linear problems that this area shrinks for asymptotically stable problems and grows for unstable problems (recall Example 2.1). It is less easy to see, but it can be shown that the area of $S(t)B$ remains constant, even for nonlinear problems, if the divergence of $\mathbf{f}$ vanishes,

$$\nabla \mathbf{f} = \frac{\partial f_1}{\partial y_1} + \frac{\partial f_2}{\partial y_2} = 0.$$

This remains valid for $m > 2$ provided that $\nabla \mathbf{f} = 0$, with an appropriate extension of the concept of volume in m dimensions.

Now, for a Hamiltonian system with $l = 1$,

$$q' = H_p, \quad p' = -H_q,$$

we have for $\nabla \mathbf{f}$

$$\nabla \mathbf{f} = \frac{\partial^2 H}{\partial p \partial q} - \frac{\partial^2 H}{\partial q \partial p} = 0;$$

hence the Hamiltonian flow preserves area. In more dimensions, $l > 1$, it turns out that the area of each projection of $S(t)B$ on a $q_i \times p_i$ plane, $i = 1, \ldots, l$, is preserved, and this property is referred to as a *symplectic map*.

Since a Hamiltonian system cannot be asymptotically stable, its stability (*if* it is stable, which is true when H can be considered a norm at each t, e.g., if H is the total energy of a friction-free multibody system) is in a sense marginal. The solution trajectories do not simply decay to a rest state, and their long-time behavior is therefore of interest. This leads to some serious numerical challenges.

We conclude this brief exposition with a simple example.

Example 2.6

The simplest Hamiltonian system is the linear harmonic oscillator. The quadratic Hamiltonian

$$H = \frac{\omega}{2}(p^2 + q^2)$$

yields the linear equations of motion

$$q' = \omega p, \quad p' = -\omega q$$

or

$$\begin{pmatrix} q \\ p \end{pmatrix}' = \omega J \begin{pmatrix} q \\ p \end{pmatrix}, \quad J = \begin{pmatrix} 0 & 1 \\ -1 & 0 \end{pmatrix}.$$

Here $\omega > 0$ is a known parameter. The general solution is

$$\begin{pmatrix} q(t) \\ p(t) \end{pmatrix} = \begin{pmatrix} \cos \omega t & \sin \omega t \\ -\sin \omega t & \cos \omega t \end{pmatrix} \begin{pmatrix} q(0) \\ p(0) \end{pmatrix}.$$

Hence, $S(t)B$ is just a rotation of the set B at a constant rate depending on ω. Clearly, this keeps the area of B unchanged.

Note that the eigenvalues of J are purely imaginary. Thus, a small "push" (i.e., a perturbation of the system) of these eigenvalues towards the positive half-plane can make the system unstable. ♦

2.6 Notes and References

There are many books and papers on the subject of this chapter. The books by Hairer, Norsett, and Wanner [50], Mattheij and Molnaar [67], and Stuart and Humphries [93] treat the theory carefully with computations in mind, so we recommend them in particular. See also [77, 47, 8, 55, 4]. For Hamiltonian systems, see [82, 93, 50].

2.7 Exercises

2.1. For each of the following constant-coefficient systems $\mathbf{y}' = A\mathbf{y}$, determine if the system is stable, asymptotically stable, or unstable.

(a) $A = \begin{pmatrix} -1 & 0 \\ 0 & -100 \end{pmatrix}$, (b) $A = \begin{pmatrix} -1 & 10 \\ 0 & -2 \end{pmatrix}$,

(c) $A = \begin{pmatrix} 1 & 3 \\ 3 & 1 \end{pmatrix}$, (d) $A = \begin{pmatrix} 0 & 1 \\ -1 & 0 \end{pmatrix}$.

2.2. (a) Compute the eigenvalues of the matrix

$$A(t) = \begin{pmatrix} -\frac{1}{4} + \frac{3}{4}\cos 2t & 1 - \frac{3}{4}\sin 2t \\ -1 - \frac{3}{4}\sin 2t & -\frac{1}{4} - \frac{3}{4}\cos 2t \end{pmatrix}.$$

(b) Determine whether the variable coefficient system $\mathbf{y}' = A(t)\mathbf{y}$ is stable, asymptotically stable, or unstable.

[You may want to use $T(t) = \begin{pmatrix} \cos t & \sin t \\ -\sin t & \cos t \end{pmatrix}$.]

2.3. The *Lyapunov function* is an important tool for analyzing stability of nonlinear problems. The scalar, C^1-function $V(\mathbf{y})$ is a Lyapunov function at $\bar{\mathbf{y}}$ if

$$\frac{d}{dt}V(\mathbf{y}(t)) \leq 0 \tag{2.13}$$

for all $\mathbf{y}$ in a neighborhood of $\bar{\mathbf{y}}$. If also $V(\bar{\mathbf{y}}) = 0$ and $V(\mathbf{y}) > 0$ in the neighborhood, then V is a *positive definite Lyapunov function* at $\bar{\mathbf{y}}$.

It can be shown that if $\bar{\mathbf{y}}$ is a steady state solution of (2.2) then $\bar{\mathbf{y}}$ is stable if there is a corresponding positive definite Lyapunov function. If the inequality in (2.13) is sharp (except at $\mathbf{y} = \bar{\mathbf{y}}$), then the steady state solution is asymptotically stable.

(a) Construct a suitable Lyapunov function to show that $\bar{y} = 1$ is stable in Example 2.5. (You should find it difficult to construct a similar function for the other steady state, $\bar{y} = 0$, for this example.)

(b) Let $U(\mathbf{y})$ be a smooth, scalar function with a minimum at $\bar{\mathbf{y}}$ (note that $\mathbf{y}$ is not necessarily scalar), and consider the system

$$\mathbf{y}' = -\boldsymbol{\nabla}_{\mathbf{y}} U = -\left(\frac{\partial U}{\partial \mathbf{y}}\right)^T.$$

Show that $\bar{\mathbf{y}}$ is a stable steady state solution of this nonlinear ODE system.

2.4. Consider a nonlinear ODE system (2.2) which has an invariant set $\mathcal{M}$ defined by the equations

$$\mathbf{h}(t, \mathbf{y}) = \mathbf{0}; \tag{2.14}$$

i.e., assuming that the initial conditions satisfy $\mathbf{h}(0, \mathbf{y}(0)) = 0$, the solution of the ODE satisfies $\mathbf{h}(t, \mathbf{y}(t)) = \mathbf{0}$ for all later times $t \geq 0$. Let us assume below, to save on notation, that $\mathbf{f}$ and $\mathbf{h}$ are autonomous. Define the Jacobian matrix

$$H(\mathbf{y}) = \frac{\partial \mathbf{h}}{\partial \mathbf{y}}$$

and assume that it has full row rank for all t (in particular, there are no more equations in (2.14) than in (2.2)).

Next we *stabilize* the vector field, replacing the autonomous (2.2) by

$$\mathbf{y}' = \mathbf{f}(\mathbf{y}) - \gamma H^T(HH^T)^{-1}\mathbf{h}(\mathbf{y}). \tag{2.15}$$

(a) Show that if $\mathbf{h}(\mathbf{y}(0)) = 0$, then the solution of (2.15) coincides with that of the original $\mathbf{y}' = \mathbf{f}(\mathbf{y})$.

(b) Show that if there is a constant γ_0 such that

$$|H\mathbf{f}(\mathbf{y})|_2 \leq \gamma_0 |\mathbf{h}(\mathbf{y})|_2$$

for all $\mathbf{y}$ in the neighborhood of the invariant set $\mathcal{M}$, then $\mathcal{M}$ becomes asymptotically stable; i.e., $|\mathbf{h}(\mathbf{y}(t))|$ decreases in t for trajectories of (2.15) starting near $\mathcal{M}$, provided that $\gamma \geq \gamma_0$.

Chapter 3

Basic Methods, Basic Concepts

We begin our discussion of numerical methods for initial value ordinary differential equations (ODEs) with an introduction of the most basic concepts involved. To illustrate these concepts, we use three simple discretization methods: *forward Euler*, *backward Euler* (also called *implicit Euler*), and *trapezoidal*. The problem to be solved is written, as before, in the general form

$$\mathbf{y}' = \mathbf{f}(t, \mathbf{y}), \qquad 0 \le t \le b, \tag{3.1}$$

with $\mathbf{y}(0) = \mathbf{c}$ given. You can think of this at first as a scalar ODE—most of what we are going to discuss generalizes to systems directly, and we will highlight occasions where the size of the system is important.

We will assume sufficient smoothness and boundedness on $\mathbf{f}(t, \mathbf{y})$ so as to guarantee a unique existence of a solution $\mathbf{y}(t)$ with as many bounded derivatives as referred to in the sequel. This assumption will be relaxed in Section 3.7.

3.1 A Simple Method: Forward Euler

To approximate (3.1), we first discretize the interval of integration by a mesh

$$0 = t_0 < t_1 < \cdots < t_{N-1} < t_N = b$$

and let $h_n = t_n - t_{n-1}$ be the nth *step size*. We then construct approximations

$$\mathbf{y}_0 (= \mathbf{c}), \mathbf{y}_1, \ldots, \mathbf{y}_{N-1}, \mathbf{y}_N$$

with $\mathbf{y}_n$ an intended approximation of $\mathbf{y}(t_n)$.

In the case of an initial value problem (IVP) we know $\mathbf{y}_0$ and may proceed to integrate the ODE in steps, where on each step n $(1 \le n \le N)$ we know an approximation $\mathbf{y}_{n-1}$ at t_{n-1} and we seek $\mathbf{y}_n$ at t_n. Thus, as we progress towards t_n we do not need advance knowledge of the entire mesh beyond it (or even of N, for that matter). Let us concentrate on one such step, n (≥ 1).

Review: Order notation. Throughout this book we consider various computational errors depending on a discretization step size $h > 0$, and ask how they decrease as h decreases. We denote for a vector $\mathbf{d}$ depending on h

$$\mathbf{d} = O(h^p)$$

if there are two positive constants p and C such that for all $h > 0$ small enough,

$$|\mathbf{d}| \le Ch^p.$$

For example, comparing (3.2) and (3.3) we see that in (3.3) the order notation involves a constant C which bounds $\frac{1}{2}\|\mathbf{y}''\|$.

In other instances, such as when estimating the efficiency of a particular algorithm, we are interested in a bound on the work estimate as a parameter N increases unboundedly (e.g., $N = 1/h$). For instance,

$$w = O(N \log N)$$

means that there is a constant C such that

$$w \le CN \log N$$

as $N \to \infty$. It will be easy to figure out from the context which of these two meanings is the relevant one.

To construct a discretization method consider Taylor's expansion

$$\mathbf{y}(t_n) = \mathbf{y}(t_{n-1}) + h_n\mathbf{y}'(t_{n-1}) + \frac{1}{2}h_n^2\mathbf{y}''(t_{n-1}) + \cdots, \tag{3.2}$$

which we can also write, using the order notation, as

$$\mathbf{y}(t_n) = \mathbf{y}(t_{n-1}) + h_n\mathbf{y}'(t_{n-1}) + O(h_n^2). \tag{3.3}$$

The forward Euler method can be derived by dropping the rightmost term in this Taylor expansion and replacing $\mathbf{y}'$ by $\mathbf{f}$, yielding the scheme

$$\mathbf{y}_n = \mathbf{y}_{n-1} + h_n\mathbf{f}(t_{n-1}, \mathbf{y}_{n-1}). \tag{3.4}$$

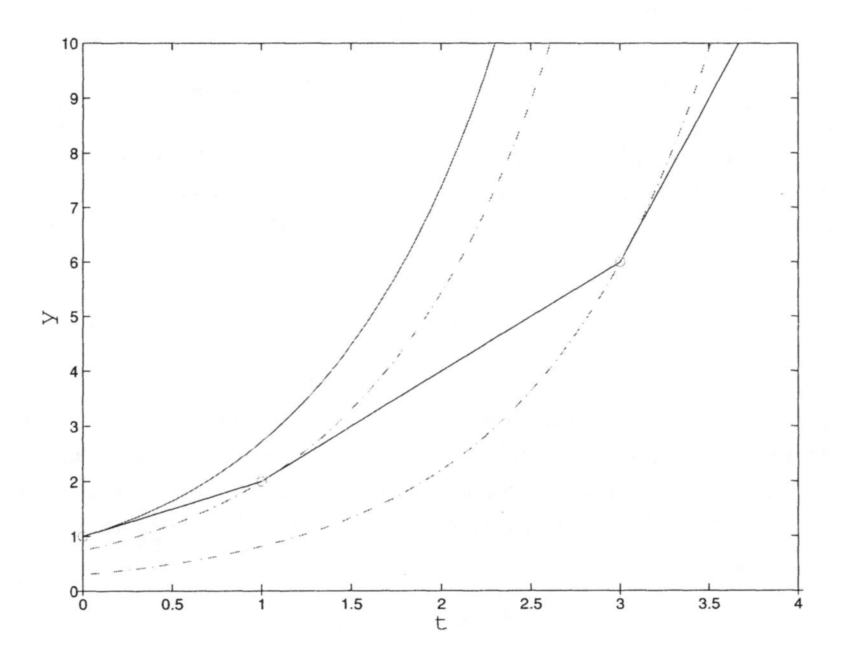

Figure 3.1: *The forward Euler method. The exact solution is the curved solid line. The numerical values are circled. The broken line interpolating them is tangential at the beginning of each step to the ODE trajectory passing through that point (dashed lines).*

This is a simple, *explicit* method; starting from $\mathbf{y}_0 = \mathbf{c}$ we apply (3.4) iteratively for $n = 1, 2, \ldots, N$. The effect of the approximation is depicted in Figure 3.1. The curved lines represent a family of solutions for the ODE with different initial values. At each step, the approximate solution $\mathbf{y}_{n-1}$ is on one of these curves at t_{n-1}. The forward Euler step amounts to taking a straight line in the tangential direction to the exact trajectory starting at $(t_{n-1}, \mathbf{y}_{n-1})$, continuing until the end of the step. (Recall Example 2.1.) One hopes that if h is small enough, then $\mathbf{y}_n$ is not too far from $\mathbf{y}(t_n)$. Let us assess this hope.

3.2 Convergence, Accuracy, Consistency, and 0-Stability

We now rewrite Euler's method (3.4) in a form compatible with the approximated ODE,

$$\frac{\mathbf{y}_n - \mathbf{y}_{n-1}}{h_n} - \mathbf{f}(t_{n-1}, \mathbf{y}_{n-1}) = \mathbf{0}.$$

To formalize a bit, let the difference operator

$$\mathcal{N}_h \mathbf{u}(t_n) \equiv \frac{\mathbf{u}(t_n) - \mathbf{u}(t_{n-1})}{h_n} - \mathbf{f}(t_{n-1}, \mathbf{u}(t_{n-1})) \tag{3.5}$$

be applied for $n = 1, 2, \ldots, N$ for any function $\mathbf{u}$ defined at mesh points with $\mathbf{u}(t_0)$ specified, and consider $\mathbf{y}_h$ to be a mesh function which takes on the value $\mathbf{y}_n$ at each t_n, $n = 0, 1, \ldots, N$. Then the numerical method is given by

$$\mathcal{N}_h \mathbf{y}_h(t_n) = \mathbf{0}$$

(with $\mathbf{y}_0 = \mathbf{c}$).

Much of the study of numerical ODEs is concerned with the errors on each step that are due to the difference approximation, and how they accumulate. One measure of the error made on each step is the *local truncation error*. It is the residual of the difference operator when it is applied to the exact solution,

$$\mathbf{d}_n = \mathcal{N}_h \mathbf{y}(t_n). \tag{3.6}$$

The local truncation error measures how closely the difference operator approximates the differential operator. This definition of the local truncation error applies to other discretization methods as well (they differ from one another in the definition of the difference operator). The difference method is said to be *consistent* (or *accurate*) *of order p* if

$$\mathbf{d}_n = O(h_n^p) \tag{3.7}$$

for a positive integer p.

For the forward Euler method (3.4), the Taylor expansion (3.2) yields

$$\mathbf{d}_n = \frac{h_n}{2}\mathbf{y}''(t_n) + O(h_n^2),$$

so the method is consistent of order 1.

A straightforward design of difference approximations to derivatives naturally leads to consistent approximations to differential equations. However, our real goal is not consistency but *convergence*. Let

$$h = \max_{1 \le n \le N} h_n$$

and assume Nh is bounded independent of N. The difference method is said to be *convergent of order p* if the *global error* $\mathbf{e}_n$, where $\mathbf{e}_n = \mathbf{y}_n - \mathbf{y}(t_n)$, $\mathbf{e}_0 = \mathbf{0}$, satisfies

$$\mathbf{e}_n = O(h^p) \tag{3.8}$$

for $n = 1, 2, \ldots, N$.

The positive integer p in (3.8) does not really have to be the same as the one in (3.7) for the definition to hold. But throughout this book we will consider methods where the order of convergence is inherited from the order of accuracy. For this we need 0-*stability.*

The difference method is 0-*stable* if there are positive constants h_0 and K such that for any mesh functions $\mathbf{x}_h$ and $\mathbf{z}_h$ with $h \leq h_0$,

$$|\mathbf{x}_n - \mathbf{z}_n| \leq K\{|\mathbf{x}_0 - \mathbf{z}_0| + \max_{1\leq j\leq N} |\mathcal{N}_h\mathbf{x}_h(t_j) - \mathcal{N}_h\mathbf{z}_h(t_j)|\}, \quad 1 \leq n \leq N. \tag{3.9}$$

What this bound says in effect is that the difference operator is invertible and that its inverse is bounded by K. Note the resemblance between (3.9) and the bound (1.6) which the differential operator satisfies. The bound in (3.9) measures the effect on the numerical solution of small perturbations in the data. The importance of this requirement lies in the following fundamental theorem.

Theorem 3.1

consistency + 0-*stability* $\Rightarrow$ convergence.

In fact, if the method is consistent of order p and 0-stable, then it is convergent of order p:

$$|\mathbf{e}_n| \leq K \max_j |\mathbf{d}_j| = O(h^p). \tag{3.10}$$

The proof of this fundamental theorem is immediate: simply let $\mathbf{x}_n \leftarrow \mathbf{y}_n$ and $\mathbf{z}_n \leftarrow \mathbf{y}(t_n)$ in the stability bound (3.9), and use the definitions of accuracy and local truncation error. ♦

Turning to the forward Euler method, by this fundamental convergence theorem we will obtain convergence of order 1 (assuming that a bounded $\mathbf{y}''$ exists) if we show that the 0-stability bound (3.9) holds. To see this, denote

$$\mathbf{s}_n = \mathbf{x}_n - \mathbf{z}_n, \qquad \theta = \max_{1\leq j\leq N} |\mathcal{N}_h\mathbf{x}_h(t_j) - \mathcal{N}_h\mathbf{z}_h(t_j)|.$$

Then for each n,

$$\begin{aligned} \theta &\geq \left| \frac{\mathbf{s}_n - \mathbf{s}_{n-1}}{h_n} - (\mathbf{f}(t_{n-1}, \mathbf{x}_{n-1}) - \mathbf{f}(t_{n-1}, \mathbf{z}_{n-1})) \right| \\ &\geq \frac{|\mathbf{s}_n|}{h_n} - \left| \frac{\mathbf{s}_{n-1}}{h_n} + (\mathbf{f}(t_{n-1}, \mathbf{x}_{n-1}) - \mathbf{f}(t_{n-1}, \mathbf{z}_{n-1})) \right|. \end{aligned}$$

Using Lipschitz continuity,

$$\begin{aligned}\left|\frac{\mathbf{s}_{n-1}}{h_n} + (\mathbf{f}(t_{n-1}, \mathbf{x}_{n-1}) - \mathbf{f}(t_{n-1}, \mathbf{z}_{n-1}))\right| &\leq \frac{|\mathbf{s}_{n-1}|}{h_n} + L|\mathbf{s}_{n-1}| \\ &= \left(\frac{1}{h_n} + L\right)|\mathbf{s}_{n-1}|\end{aligned}$$

so that

$$\begin{aligned}|\mathbf{s}_n| &\leq (1 + h_n L)|\mathbf{s}_{n-1}| + h_n\theta \\ &\leq (1 + h_n L)[(1 + h_{n-1}L)|\mathbf{s}_{n-2}| + h_{n-1}\theta] + h_n\theta \\ &\leq \dots \\ &\leq (1 + h_1 L)\cdots(1 + h_{n-1}L)(1 + h_n L)|\mathbf{s}_0| \\ &\quad + \theta\sum_{j=1}^{n} h_j(1 + h_{j+1}L)\cdots(1 + h_n L) \\ &\leq e^{Lt_n}|\mathbf{s}_0| + \frac{1}{L}(e^{Lt_n} - 1)\theta.\end{aligned}$$

The last inequality above is obtained by noting that $1 + hL \leq e^{Lh}$ implies

$$(1 + h_{j+1}L)\cdots(1 + h_n L) \leq e^{L(t_n - t_j)}, \qquad 0 \leq j \leq n,$$

and also,

$$\sum_{j=1}^{n} h_j e^{L(t_n - t_j)} \leq \sum_{j=1}^{n}\int_{t_{j-1}}^{t_j} e^{L(t_n - t)}dt = e^{Lt_n}\int_0^{t_n} e^{-Lt}dt = \frac{1}{L}(e^{Lt_n} - 1).$$

The stability bound is therefore satisfied, with $K = \max\left\{e^{Lb}, \frac{1}{L}(e^{Lb} - 1)\right\}$ in (3.9).

It is natural to ask next if the error bound (3.10) is useful in practice, i.e., if it can be used to reliably estimate the step size h needed to achieve a given accuracy. This is a tempting possibility. For instance, let M be an estimated bound on $\|\mathbf{y}''\|$. Then the error using forward Euler can be bounded by

$$|\mathbf{e}_n| \leq h\frac{M}{2L}(e^{Lt_n} - 1), \qquad 1 \leq n \leq N. \tag{3.11}$$

However, it turns out that this bound is too pessimistic in many applications, as the following example indicates.

Example 3.1

Consider the scalar problem

$$y' = -5ty^2 + \frac{5}{t} - \frac{1}{t^2}, \qquad y(1) = 1,$$

for $1 \le t \le 25$. (Note that the starting point of the integration is $t = 1$, not $t = 0$ as before. But this is of no significance: just change the independent variable to $\tau = t - 1$.) The exact solution is $y(t) = \frac{1}{t}$.

To estimate the Lipschitz constant L, note that near the exact solution,

$$f_y = -10ty \approx -10.$$

Similarly, use the exact solution to estimate $M = 2 \ge 2/t^3$. Substitution into (3.11) yields the bound

$$|e_n| \le \frac{h}{10} e^{10(t_n - 1)}$$

so $|e_N| \le \frac{h}{10} e^{240}$, not a very useful bound at all. ♦

We will be looking in later chapters into the question of realistic error estimation.

We close this section by mentioning another, important measure of the error made at each step, the *local error*. It is defined as the amount by which the numerical solution $\mathbf{y}_n$ at each step differs from the solution $\bar{\mathbf{y}}(t_n)$ to the IVP

$$\begin{aligned} \bar{\mathbf{y}}'(t) &= \mathbf{f}(t, \bar{\mathbf{y}}(t)), \\ \bar{\mathbf{y}}(t_{n-1}) &= \mathbf{y}_{n-1}\,. \end{aligned} \tag{3.12}$$

Thus the local error is given by

$$\mathbf{l}_n = \bar{\mathbf{y}}(t_n) - \mathbf{y}_n. \tag{3.13}$$

Under normal circumstances, it can be shown that the numerical solution exists and

$$|\mathbf{d}_n| = |\mathcal{N}_h \bar{\mathbf{y}}(t_n)| + O(h^{p+1}).$$

Moreover, it is easy to show, for all of the numerical ODE methods considered in this book, that[5]

$$h_n |\mathcal{N}_h \bar{\mathbf{y}}(t_n)| = |\mathbf{l}_n|(1 + O(h_n)). \tag{3.14}$$

The two local error indicators, $h_n \mathbf{d}_n$ and $\mathbf{l}_n$, are thus often closely related.

---

[5] We caution here that for stiff problems, to be discussed in Section 3.4, the constant implied in this $O(h_n)$ may be quite large.

3.3 Absolute Stability

Example 3.1 may make one wonder about the meaning of the fundamental convergence Theorem 3.1. The theorem is not violated: we still have that $|\mathbf{e}_n| \leq Ch$ for some constant C, even if large, so as $h \to 0$, $|\mathbf{e}_n| \to 0$. However, the theorem may not give a quantitative indication of what happens when we actually compute with a step size h, which is not very small. (The name "0-stability" now becomes more intuitive—this concept deals with the limit of $h \to 0$.) The basic reason why the constant in this example is so pessimistically large is that while $f_y \approx -10$, i.e., the exact solution mode is decaying, the stability bound uses the Lipschitz constant $L = 10$, and consequently is exponentially increasing in b.

For large step sizes, the difference equation should mimic the behavior of the differential equation in the sense that their stability properties should be similar.

What stability requirements arise for h which is not vanishingly small? Consider the scalar *test equation*

$$y' = \lambda y, \tag{3.15}$$

where λ is a complex constant (complex because later we will be looking at ODE systems, and there λ corresponds to an eigenvalue). If $y(0) = c$ (say $c > 0$ for notational convenience), then the exact solution is

$$y(t_n) = ce^{\lambda t_n},$$

whereas Euler's method, with a uniform step size $h_n = h$, gives

$$y_n = y_{n-1} + h\lambda y_{n-1} = (1 + h\lambda)y_{n-1} = \cdots = c(1 + h\lambda)^n.$$

Let us distinguish between three cases (cf. Example 2.1).

- If $\mathcal{R}\mathrm{e}(\lambda) > 0$, then $|y(t)| = ce^{\mathcal{R}\mathrm{e}(\lambda)t}$ grows exponentially with t. This is an unstable problem, although for $e^{\mathcal{R}\mathrm{e}(\lambda)b}$ not too large, one can still compute solutions which are meaningful in the *relative* sense. In this case, the error bound (3.10) is realistic. For unstable problems, the distance between solution curves increases in time.

- If $\mathcal{R}\mathrm{e}(\lambda) = 0$, the solution is oscillating (unless $\lambda = 0$) and the distance between solution curves stays the same.

- If $\mathcal{R}\mathrm{e}(\lambda) < 0$, then $|y(t)|$ decays exponentially. The distance between solution curves decreases. The problem is (asymptotically) stable, and we cannot tolerate growth in $|y_n|$. This is usually the interesting case, and it yields an additional *absolute stability* requirement,

$$|y_n| \leq |y_{n-1}|, \qquad n = 1, 2, \ldots . \tag{3.16}$$

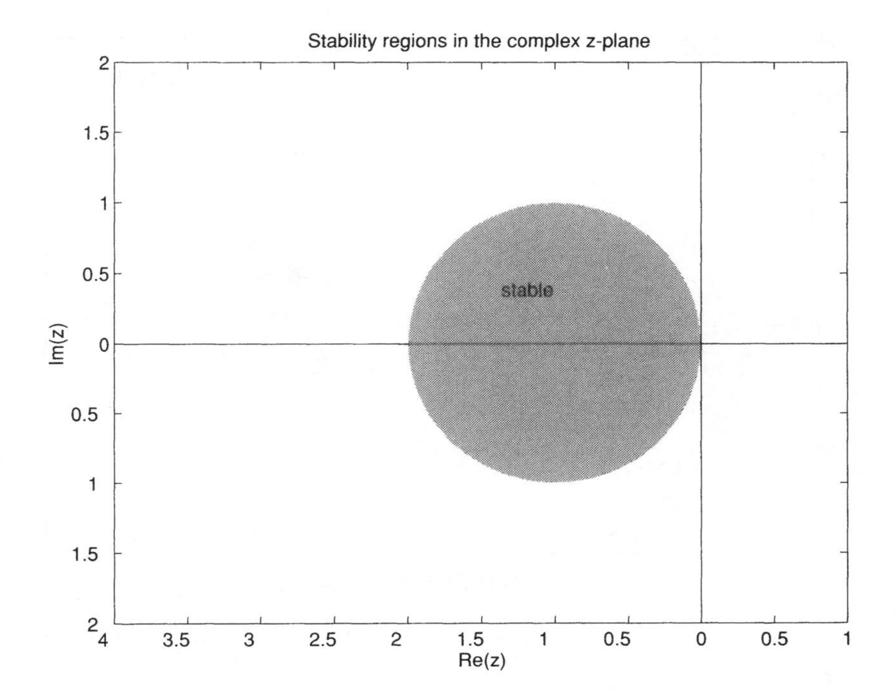

Figure 3.2: *Absolute stability region for the forward Euler method.*

For a given numerical method, the *region of absolute stability* is that region of the complex z-plane such that applying the method for the test equation (3.15), with $z = h\lambda$ from within this region, yields an approximate solution satisfying the absolute stability requirement (3.16).

For the forward Euler method we obtain the condition

$$|1 + h\lambda| \leq 1 \tag{3.17}$$

which yields the region of absolute stability depicted in Figure 3.2. For instance, if λ is negative, then h must be restricted to satisfy

$$h \leq \frac{2}{-\lambda}.$$

For Example 3.1 this gives $h < .2$. In this case the restriction is not practically unbearable. To see the effect of violating the absolute stability restriction, we plot in Figure 3.3 the approximate solutions obtained with uniform step sizes $h = .19$ and $h = .21$. For $h = .19$ the solution profile looks like the exact one ($y = 1/t$). The other, oscillatory profile is obtained for $h = .21$, which is outside the absolute stability region. When computing with $h = .4$ using the same forward Euler method and floating point arithmetic with a 14-hexadecimal-digit mantissa (this is the standard "double precision" in IEEE FORTRAN, for example), the computed solution oscillates and then blows up (i.e., overflow is detected) before reaching $t = 25$.

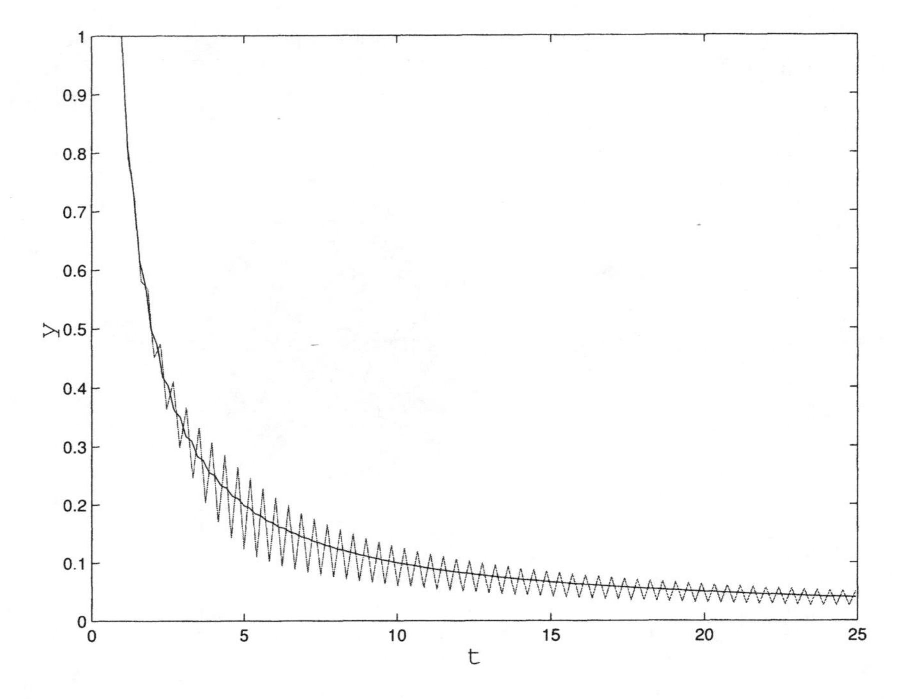

Figure 3.3: *Approximate solutions for Example* 3.1 *using the forward Euler method, with* $h = .19$ *and* $h = .21$. *The oscillatory profile corresponds to* $h = .21$; *for* $h = .19$ *the qualitative behavior of the exact solution is obtained.*

It is important to understand that the absolute stability restriction is indeed a stability, not accuracy, requirement. Consider the initial value $c = 10^{-15}$ for the test equation with $\mathcal{R}e(\lambda) < 0$, so the exact solution is approximated very well by the constant 0. Such an initial value corresponds to an unavoidable perturbation in the numerical method, due to roundoff errors. Now, the forward Euler solution corresponding to this initial perturbation of 0 remains very close to 0 for all $t_n > 0$, like the exact solution, when using any h from the absolute stability region, but it blows up as n increases if h is from outside that region, i.e., if $|1 + h\lambda| > 1$.

The concept of absolute stability was defined with respect to a very simple test equation (3.15), an ODE whose numerical solution is not a computationally challenging problem in itself. Nonetheless, it turns out that absolute stability gives useful information, at least qualitatively, in more general situations, where complicated systems of nonlinear ODEs are integrated numerically.

Note: Those readers who are prepared to trust us on the above statement may wish to skip the rest of this section, at least on first reading (especially if your linear algebra needs some dusting).

Systems with Constant Coefficients

We now consider the extension of the test equation analysis to a simple ODE system,

$$\mathbf{y}' = A\mathbf{y}, \tag{3.18}$$

where A is a constant, diagonalizable, $m \times m$ matrix.

Denote the eigenvalues of A by $\lambda_1, \lambda_2, \ldots, \lambda_m$, and let

$$\Lambda = \text{diag}\{\lambda_1, \lambda_2, \ldots, \lambda_m\}$$

be the diagonal $m \times m$ matrix composed of these eigenvalues. Again, the interesting case is when (3.18) is stable, i.e., $\mathcal{R}e(\lambda_j) \le 0$, $j = 1, \ldots, m$. The diagonalizability of A means that there is a nonsingular matrix T, consisting of the eigenvectors of A (scaled to have unit Euclidean norm, say), such that

$$T^{-1}AT = \Lambda.$$

Consider the following change of dependent variables,

$$\mathbf{w} = T^{-1}\mathbf{y}.$$

For $\mathbf{w}(t)$ we obtain, upon multiplying (3.18) by T^{-1} and noting that T is constant in t, the decoupled system

$$\mathbf{w}' = \Lambda\mathbf{w}. \tag{3.19}$$

The components of $\mathbf{w}$ are separated, and for each component we get a scalar ODE in the form of the test equation (3.15) with $\lambda = \lambda_j$, $j = 1, \ldots, m$. Moreover, since A and therefore T are constant, we can apply the same transformation to the discretization: let $\mathbf{w}_n = T^{-1}\mathbf{y}_n$, all n. Then the forward Euler method for (3.18),

$$\mathbf{y}_n = \mathbf{y}_{n-1} + h_n A\mathbf{y}_{n-1},$$

transforms into

$$\mathbf{w}_n = \mathbf{w}_{n-1} + h_n \Lambda\mathbf{w}_{n-1},$$

which is the forward Euler method for (3.19). The same commutativity of the discretization and the $\mathbf{w}$-transformation (in the case when T is constant!) holds for other discretization methods as well.

Now, for the decoupled system (3.19), where we can look at each scalar ODE separately, we obtain that if h_n are chosen such that $h\lambda_1, h\lambda_2, \ldots, h\lambda_m$ are all in the absolute stability region of the difference method (recall $h = \max_n h_n$), then

$$|\mathbf{w}_n| \le |\mathbf{w}_{n-1}| \le \cdots \le |\mathbf{w}_0|,$$

so

$$|\mathbf{y}_n| \le \|T\| |\mathbf{w}_n| \le \cdots \le \|T\| |\mathbf{w}_0| \le \|T\| \|T^{-1}\| |\mathbf{y}_0|.$$

Denoting by

$$\text{cond}(T) = \|T\| \|T^{-1}\| \tag{3.20}$$

the *condition number* of the eigenvector matrix T (measured in the norm induced by the vector norm used for $|\mathbf{y}_n|$), we obtain the stability bound

$$|\mathbf{y}_n| \le \text{cond}(T)|\mathbf{c}|, \qquad n = 0, 1, \ldots, N \tag{3.21}$$

(recall that $\mathbf{y}(0) = \mathbf{c}$).

Note that in general the stability constant cond(T) is not guaranteed to be of moderate size, although it is independent of n, and it may often depend on the size m of the ODE system. An additional complication arises when A is not diagonalizable. The considerations here are very similar to those arising in eigenvalue sensitivity analysis in linear algebra. Indeed, the essential question is similar too: how representative are the eigenvalues of the properties of the matrix A as a whole?

But there are important special cases where we encounter more favorable winds. If A is (real and) symmetric, then not only are its eigenvalues real, but its eigenvectors are orthogonal to one another as well. We may therefore choose T to be orthogonal, i.e.,

$$T^{-1} = T^T.$$

In this case it is advantageous to use the Euclidean norm l_2, because we get

$$\text{cond}(T) = 1$$

regardless of the size of the system. Thus, if $h(\min_{1\le j\le m} \lambda_j)$ is in the absolute stability region of the difference method, then (3.21) yields the stability bound in the l_2 norm,

$$\mathbf{y}_n{}^T\mathbf{y}_n \le \mathbf{c}^T\mathbf{c}, \qquad 0 \le n \le N. \tag{3.22}$$

The importance of obtaining a bound on cond(T) which is independent of m increases, of course, when m is large. Such is the case for the method of lines (Examples 1.3 and 1.7), where the ODE system arises from a spatially discretized time-dependent partial differential equation (PDE). In this case m is essentially the number of spatial grid points. This is worked out further for some instances in Exercises 3.6 and 3.7.

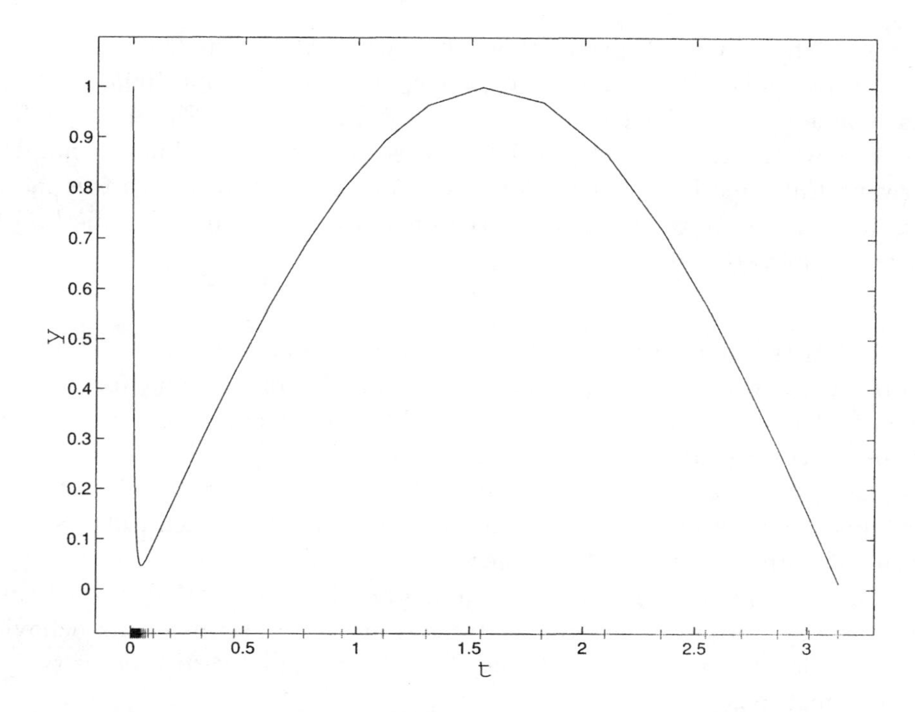

Figure 3.4: *Approximate solution and plausible mesh, Example* 3.2.

3.4 Stiffness: Backward Euler

Ideally, the choice of step size h_n should be dictated by approximation accuracy requirements. But we just saw that when using the forward Euler method (and, as it turns out, many other methods too), h_n must be chosen sufficiently small to obey an additional, absolute stability restriction, as well. Loosely speaking, the IVP is referred to as being *stiff* if this absolute stability requirement dictates a much smaller step size than is needed to satisfy approximation requirements alone. In this case other methods, which do not have such a restrictive absolute stability requirement, should be considered.

To illustrate this, consider a simple example.

Example 3.2
The scalar problem

$$y' = -100(y - \sin t), \quad t \geq 0, \quad y(0) = 1$$

has a solution which starts at the given initial value and varies rapidly. But after a short while, say, for $t \geq 0.03$, $y(t)$ varies much more slowly, satisfying $y(t) \approx \sin t$; see Figure 3.4. For the initial small interval of rapid change (commonly referred to as an initial *layer* or transient), we expect to use small step sizes, so that $100h_n \leq 1$, say. This is within the absolute stability region of the forward Euler method. But when $y(t) \approx \sin t$, accuracy considerations alone allow a much larger step size, so we want $100h_n \gg 2$. A reasonable mesh is plotted using markers on the t-axis in Figure 3.4. Obviously, however, the plotted solution in this figure was not found using the forward Euler method (but rather, using another method) with this mesh, because the absolute stability restriction of the forward Euler method is severely violated here. ♦

Scientists often describe stiffness in terms of *multiple time scales.* If the problem has widely varying time scales, and the phenomena (or *solution modes*) that change on fast scales are stable, then the problem is stiff. For example, controllers are often designed to bring a system rapidly back to a steady state and are thus a source of stiffness. In chemically reacting systems, stiffness often arises from the fact that some chemical reactions occur much more rapidly than others.

The concept of stiffness is best understood in qualitative, rather than quantitative, terms. In general, stiffness is defined in terms of the behavior of an explicit difference method, and the behavior of forward Euler is typical of such methods.

> An IVP is *stiff* in some interval $[0, b]$ if the step size needed to maintain stability of the forward Euler method is much smaller than the step size required to represent the solution accurately.

We note that stiffness depends, in addition to the differential equation itself, on

- the accuracy criterion,
- the length of the interval of integration, and
- the region of absolute stability of the method.

In Example 3.2, for a moderate error tolerance, the problem is stiff after about $t = 0.03$. If it were necessary to solve the problem to great accuracy, then it would not be stiff because the step size would need to be small in order to attain that accuracy, and hence it would not be restricted by stability.

For stable, homogeneous, linear systems, stiffness can be determined by the system's eigenvalues. For the test equation (3.15) on $[0, b]$, we say that the problem is *stiff* if

$$b\mathcal{R}e(\lambda) \ll -1. \tag{3.23}$$

Roughly, the general ODE system (3.1) is *stiff* in a neighborhood of the solution $\mathbf{y}(t)$ if there exists for some bounded data a component of $\mathbf{y}$ which decays rapidly on the scale of the interval length b. In the general case, stiffness can often be related to the eigenvalues λ_j of the local Jacobian matrix $\mathbf{f}_\mathbf{y}(t, \mathbf{y}(t))$, generalizing (3.23) to

$$b \min_j \mathcal{R}\text{e}(\lambda_j) \ll -1. \tag{3.24}$$

3.4.1 Backward Euler

Thus, we look for methods which do not violate the absolute stability requirement when applied to the test equation (3.15), even when $h\mathcal{R}\text{e}(\lambda) \ll -1$. Such a method is the backward Euler method. It is derived for the general ODE (3.1) just like the forward Euler method, except that everything is centered at t_n, rather than at t_{n-1}. This gives the first-order method

$$\mathbf{y}_n = \mathbf{y}_{n-1} + h_n \mathbf{f}(t_n, \mathbf{y}_n). \tag{3.25}$$

Geometrically, instead of using the tangent at $(t_{n-1}, \mathbf{y}_{n-1})$, as in the forward Euler method, the backward Euler method uses the tangent at the *future* point $(t_n, \mathbf{y}_n)$, thus enhancing the stability. The local truncation error of this method is similar in magnitude to that of the forward Euler method, and correspondingly, the convergence bound (3.10) is similar too (we leave the 0-stability proof in this case as an exercise). The two major differences between these simple methods are:

- While the forward Euler method is explicit, the backward Euler method is *implicit*: the unknown vector $\mathbf{y}_n$ at each step appears on both sides of the equation (3.25), generally in a nonlinear expression. Consequently, a nonlinear system of algebraic equations has to be (approximately) solved at each step. That's the bad news for backward Euler.

- The good news is the method's stability. Applying the backward Euler method (3.25) to the test equation, we obtain

  $$y_n = y_{n-1} + h\lambda y_n,$$

  i.e.,

  $$y_n = (1 - h\lambda)^{-1} y_{n-1}.$$

  The *amplification factor*, i.e., what multiplies $|y_{n-1}|$ to get $|y_n|$ in absolute value, satisfies

  $$\frac{1}{|1 - h\lambda|} \le 1$$

  for *all* values of $h > 0$ and λ satisfying $\mathcal{R}\text{e}(\lambda) \le 0$. In particular, there is no absolute stability prohibition from taking $h\mathcal{R}\text{e}(\lambda) \ll -1$, e.g., in Example 3.2.

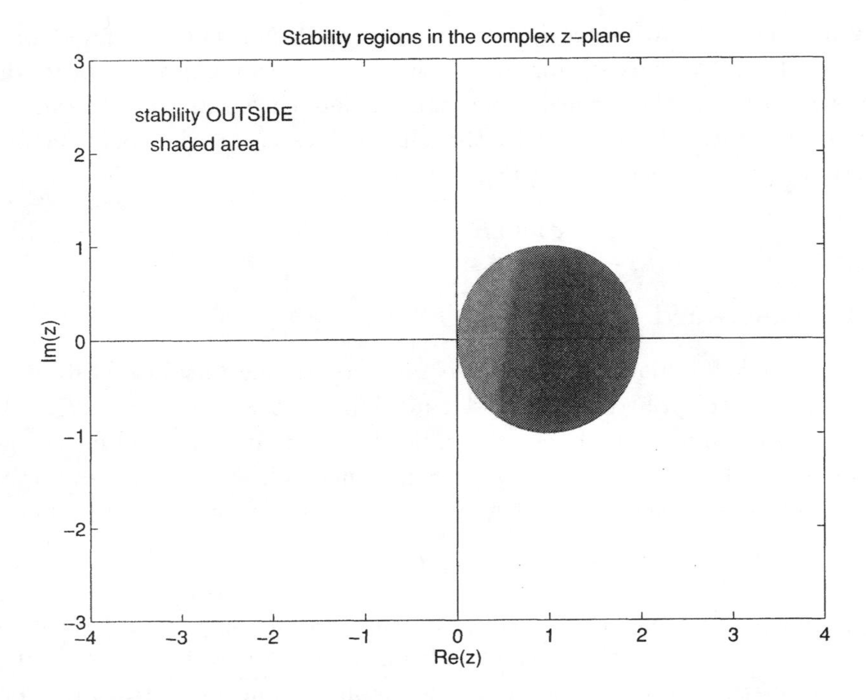

Figure 3.5: *Absolute stability region for the backward Euler method.*

The region of absolute stability of the backward Euler method is depicted in Figure 3.5. It contains, in addition to the entire left half-plane of $z = h\lambda$, a major part of the right half-plane as well. (The latter is a mixed blessing, though, as will be discussed later on.) For a given stiff problem, the backward Euler method needs fewer steps than the forward Euler method. In general, however, each backward Euler step may be more expensive in terms of computing time. Still, there are many applications where the overall computational expense using the implicit method is much less than with the explicit Euler method.

3.4.2 Solving Nonlinear Equations

For an implicit method like backward Euler, a nonlinear system of equations must be solved in each time step. For backward Euler, this nonlinear system is

$$\mathbf{g}(\mathbf{y}_n) = \mathbf{y}_n - \mathbf{y}_{n-1} - h\mathbf{f}(t_n, \mathbf{y}_n) = \mathbf{0}$$

(where $h = h_n$ for notational simplicity). There are a number of ways to solve this nonlinear system. We mention two basic ones.

Functional Iteration

Our first impulse might be to solve the nonlinear system by functional iteration. This yields

$$\mathbf{y}_n^{\nu+1} = \mathbf{y}_{n-1} + h\mathbf{f}(t_n, \mathbf{y}_n^\nu), \qquad \nu = 0, 1, \ldots,$$

where we can choose $\mathbf{y}_n^0 = \mathbf{y}_{n-1}$, for instance. (Note that ν is an iteration counter, not a power.)

The advantage here is simplicity. However, the convergence of this iteration requires $h\|\partial\mathbf{f}/\partial\mathbf{y}\| < 1$ in some norm.[6] For stiff systems, $\|\partial\mathbf{f}/\partial\mathbf{y}\|$ is large, so the step size h would need to be restricted and this would defeat the purpose of using the method.

Example 3.3

Let us generalize the ODE of Example 3.1 to

$$y' = \lambda(ty^2 - t^{-1}) - t^{-2}, \qquad t > 1,$$

where $\lambda < 0$ is a parameter. With $y(0) = 1$, the exact solution is still $y(t) = t^{-1}$. The backward Euler method gives a nonlinear equation for y_n,

$$y_n - y_{n-1} = h_n\lambda(t_n y_n^2 - t_n^{-1}) - h_n t_n^{-2}, \tag{3.26}$$

and functional iteration reads

$$y_n^{\nu+1} - y_{n-1} = h_n\lambda(t_n(y_n^\nu)^2 - t_n^{-1}) - h_n t_n^{-2}, \quad \nu = 0, 1, \ldots. \tag{3.27}$$

The question is, under what conditions does the iteration (3.27) converge rapidly?

Subtracting (3.27) from (3.26) and denoting $\varepsilon_n^\nu = y_n - y_n^\nu$, we get

$$\begin{aligned}\varepsilon_n^{\nu+1} &= h_n\lambda t_n(y_n^2 - (y_n^\nu)^2) \\ &= h_n\lambda t_n(y_n + y_n^\nu)\varepsilon_n^\nu \approx 2h_n\lambda\varepsilon_n^\nu, \quad \nu = 0, 1, \ldots.\end{aligned}$$

This iteration obviously converges iff $|\varepsilon_n^{\nu+1}| < |\varepsilon_n^\nu|$, and the approximate condition for this convergence is therefore

$$h_n < \frac{1}{2|\lambda|}.$$

The convergence is rapid if $h_n \ll \frac{1}{2|\lambda|}$. Now, if $\lambda = -5$, as in Example 3.1, then convergence of this nonlinear iteration is obtained with $h < 0.1$, and choosing $h = .01$ yields rapid convergence (roughly, one additional significant digit is gained at each iteration). But if $\lambda = -500$ then we must take $h < .001$ for convergence of the iteration, and this is a harsh restriction, given the smoothness and slow variation of the exact solution. Functional iteration is therefore seen to be effective only in the nonstiff case. ♦

[6]This would yield a *contraction* mapping, and therefore convergence as $\nu \to \infty$ to the fixed point $\mathbf{y}_n$.

Functional iteration is often used in combination with implicit methods for the solution of *nonstiff problems*, as we will see later, in Chapter 5.

Newton Iteration

Variants of Newton's method are used in virtually all modern stiff ODE codes.

Given the nonlinear system

$$\mathbf{g}(\mathbf{y}_n) = \mathbf{y}_n - \mathbf{y}_{n-1} - h\mathbf{f}(t_n, \mathbf{y}_n) = \mathbf{0},$$

Newton's method yields

$$\begin{aligned} \mathbf{y}_n^{\nu+1} &= \mathbf{y}_n^{\nu} - \left(\frac{\partial \mathbf{g}}{\partial \mathbf{y}}\right)^{-1} \mathbf{g}(\mathbf{y}_n^{\nu}) \\ &= \mathbf{y}_n^{\nu} - \left(I - h\frac{\partial \mathbf{f}}{\partial \mathbf{y}}\right)^{-1} (\mathbf{y}_n^{\nu} - \mathbf{y}_{n-1} - h\mathbf{f}(t_n, \mathbf{y}_n^{\nu})), \quad \nu = 0, 1, \ldots . \end{aligned}$$

The matrix $(I - h\partial \mathbf{f}/\partial \mathbf{y})$ is evaluated at the current iterate $\mathbf{y}_n^{\nu}$. This matrix is called the *iteration matrix*, and the costs of forming it and solving the linear system (for $\boldsymbol{\delta} = \mathbf{y}_n^{\nu+1} - \mathbf{y}_n^{\nu}$) often dominate the costs of solving the problem. We can take the initial guess

$$\mathbf{y}_n^0 = \mathbf{y}_{n-1},$$

although better ones are often available. Newton's method is iterated until an estimate of the error due to terminating the iteration is less than a user-specified tolerance, for example,

$$|\mathbf{y}_n^{\nu+1} - \mathbf{y}_n^{\nu}| \leq \text{NTOL}.$$

The tolerance NTOL is related to the local error bound that the user aims to achieve and is usually well above roundoff level. Because there is a very accurate initial guess, most ODE IVPs require no more than a few Newton iterations per time step. A strategy which iterates no more than, say, three times and, if there is no convergence, decreases the step size h_n (thus improving the initial guess) and repeats the process, can be easily conceived. We return to these ideas in the next two chapters.

Newton's method works well for Example 3.3, without a severe restriction on the time step.

Approximating the Jacobian Matrix

Newton's method requires the evaluation of the Jacobian matrix, $\frac{\partial \mathbf{f}}{\partial \mathbf{y}}$. This presents no difficulty for Example 3.3; however, in practical applications, specifying these partial derivatives analytically is often a difficult or cumbersome task. A convenient technique is to use *difference approximations*: at

Review: Newton's method. For a nonlinear equation

$$g(x) = 0,$$

we define a sequence of iterates as follows: x^0 is an initial guess. For a current iterate x^ν, we write

$$0 = g(x) = g(x^\nu) + g'(x^\nu)(x - x^\nu) + \cdots.$$

Approximating the solution x by neglecting the higher-order terms in this Taylor expansion, we define the next iterate $x^{\nu+1}$ by the linear equation

$$0 = g(x^\nu) + g'(x^\nu)(x^{\nu+1} - x^\nu).$$

We can generalize this directly to a system of m algebraic equations in m unknowns,

$$\mathbf{g}(\mathbf{x}) = \mathbf{0}.$$

Everything remains the same, except that the first derivative of g is replaced by the $m \times m$ Jacobian matrix $\frac{\partial \mathbf{g}}{\partial \mathbf{x}}$. We obtain the iteration

$$\mathbf{x}^{\nu+1} = \mathbf{x}^\nu - \left(\frac{\partial \mathbf{g}}{\partial \mathbf{x}}(\mathbf{x}^\nu)\right)^{-1} \mathbf{g}(\mathbf{x}^\nu), \quad \nu = 0, 1, \ldots.$$

We note that it is not good practice to compute a matrix inverse. Moreover, rather than computing $\mathbf{x}^{\nu+1}$ directly, it is better in certain situations (when ill-conditioning is encountered), and never worse in general, to solve the linear system for the difference $\boldsymbol{\delta}$ between $\mathbf{x}^{\nu+1}$ and $\mathbf{x}^\nu$, and then update. Thus, $\boldsymbol{\delta}$ is computed (for each ν) by solving the linear system

$$\left(\frac{\partial \mathbf{g}}{\partial \mathbf{x}}\right) \boldsymbol{\delta} = -\mathbf{g}(\mathbf{x}^\nu),$$

where the Jacobian matrix is evaluated at $\mathbf{x}^\nu$, and the next Newton iterate is obtained by

$$\mathbf{x}^{\nu+1} = \mathbf{x}^\nu + \boldsymbol{\delta}.$$

$\mathbf{y} = \mathbf{y}_n^\nu$, evaluate $\hat{\mathbf{f}} = \mathbf{f}(t_n, \hat{\mathbf{y}})$ and $\tilde{\mathbf{f}} = \mathbf{f}(t_n, \tilde{\mathbf{y}})$, where $\hat{\mathbf{y}}$ and $\tilde{\mathbf{y}}$ are perturbations of $\mathbf{y}$ in one coordinate, $\hat{y}_j = y_j + \epsilon$, $\tilde{y}_j = y_j - \epsilon$, and $\hat{y}_l = \tilde{y}_l = y_l$, $l \neq j$. Then the jth column of $\frac{\partial \mathbf{f}}{\partial \mathbf{y}}$ can be approximated by

$$\frac{\partial \mathbf{f}}{\partial y_j} \approx \frac{1}{2\epsilon}(\hat{\mathbf{f}} - \tilde{\mathbf{f}}),$$

where ϵ is a small positive parameter.

This simple trick is very easy to program and it does not affect the accuracy of the solution $\mathbf{y}_n$. It often works well in practice with the choice $\epsilon = 10^{-d}$, if floating point arithmetic with roughly $2d$ significant digits is being used (e.g., $d = 7$). The technique is useful also in the context of boundary value problems (BVPs); see, for instance, Sections 7.1 and 8.1.1. It does not *always* work well, though, and moreover, such an approximation of the Jacobian matrix may at times be relatively expensive, depending on the application. But it gives the user a simple technique for computing an approximate Jacobian matrix when it is needed. Most general-purpose codes provide a finite difference Jacobian as an option, using a somewhat more sophisticated algorithm to select the increment.

3.5 A-Stability, Stiff Decay

Ideally, one would desire that a numerical discretization method mimic all properties of the differential problem to be discretized, for all problems. This is not possible. One then lowers expectations, and designs discretization methods which capture the *essential* properties of a *class* of differential problems.

A first study of absolute stability suggests that, since for all stable test equations, $|y(t_n)| \leq |y(t_{n-1})|$, a good discretization method for stiff problems should do the same, i.e., satisfy $|y_n| \leq |y_{n-1}|$.

This gives the concept of A-stability: a difference method is *A-stable* if its region of absolute stability contains the entire left half-plane of $z = h\lambda$. A glance at Figures 3.2 and 3.5 indicates that the backward Euler method is A-stable, whereas the forward Euler method is not.

But a further probe into A-stability reveals two deficiencies. The first is that it does not distinguish between the cases

$$\mathcal{R}\mathrm{e}(\lambda) \ll -1$$

and

$$-1 \ll \mathcal{R}\mathrm{e}(\lambda) \leq 0, \quad |\mathcal{I}\mathrm{m}(\lambda)| \gg 1\,.$$

The latter case gives rise to a highly oscillatory exact solution, which does not decay much. The difficulties arising are of a different type, so when addressing stiffness of the type that we have been studying, it is not essential

Review: Matrix decompositions. Consider a linear system of m equations

$$A\mathbf{x} = \mathbf{b},$$

where A is real, square, and nonsingular, $\mathbf{b}$ is given, and $\mathbf{x}$ is a solution vector to be found. The solution is given by

$$\mathbf{x} = A^{-1}\mathbf{b}.$$

However, it is usually bad practice to attempt to form A^{-1}.

The well-known algorithm of *Gaussian elimination* (without pivoting) is equivalent to forming an *LU decomposition* of A:

$$A = LU,$$

where L is a unit lower triangular matrix (i.e., $l_{ij} = 0,\ i < j$, and $l_{ii} = 1$) and U is upper triangular (i.e., $u_{ij} = 0,\ i > j$). Note that this decomposition is independent of the right-hand side $\mathbf{b}$. It can be done without knowing $\mathbf{b}$ and it can be used for more than one right-hand side. The LU decomposition requires $\frac{1}{3}m^3 + O(m^2)$ flops (i.e., elementary floating point operations).

Given a data vector $\mathbf{b}$ we can now find $\mathbf{x}$ by writing

$$L(U\mathbf{x}) = A\mathbf{x} = \mathbf{b}\,.$$

Solving $L\mathbf{z} = \mathbf{b}$ for $\mathbf{z}$ involves *forward substitution* and costs $O(m^2)$ flops. Subsequently solving $U\mathbf{x} = \mathbf{z}$ completes the solution process using a *back substitution* and another $O(m^2)$ flops. The solution process is therefore much cheaper, when m is large, than the cost of the decomposition.

Not every nonsingular matrix has an LU decomposition, and even if there exists such a decomposition the numerical process may become unstable. Thus, partial pivoting must be applied (unless the matrix has some special properties, e.g., it is symmetric positive definite). A row-partial pivoting involves permuting rows of A to enhance stability and results in the decomposition

$$A = PLU,$$

where P is a permutation matrix (i.e., the columns of P are the m unit vectors, in some permuted order). We will refer to an LU decomposition, assuming that partial pivoting has been applied as necessary.

Review: Matrix decompositions, continued. Another important matrix decomposition is the *QR decomposition*

$$A = QR,$$

where R is upper triangular (like U) and Q is orthogonal: $Q^T Q = I$. This decomposition costs twice as much as the LU decomposition, but it has somewhat better stability properties, because $\|Q\|_2 = \|Q^{-1}\|_2 = 1$, which implies an ideal conditioning, $\mathrm{cond}(Q) = 1$ (see (3.20)). This is useful also for finding least squares solutions to overdetermined linear systems and as a building block in algorithms for finding matrix eigenvalues.

If A is large and *sparse* (i.e., most of its elements are zero) then the LU and the QR decompositions may or may not remain suitable. For instance, if all the nonzero elements of A are contained in a narrow band, i.e., in a few diagonals along the main diagonal (whence A is called *banded*), then both the LU and the QR algorithms can be easily adjusted to not do any work outside the band. For boundary value ODEs, this typically leads to a reduction in the algorithm's complexity from cubic to linear in the matrix dimension. But inside the band the sparsity is usually lost, and other, iterative algorithms become more attractive. The latter is typically the case for elliptic PDEs and is outside the scope of our book.

to include points near the imaginary axis in the absolute stability region of the difference method.

The second possible weakness of the A-stability definition arises from its exclusive use of absolute stability. In the very stiff limit, $h_n \mathcal{R}\mathrm{e}(\lambda) \ll -1$, the exact solution of the test equation satisfies $|y(t_n)| = |y(t_{n-1})| e^{h_n \mathcal{R}\mathrm{e}(\lambda)} \ll |y(t_{n-1})|$. The corresponding absolute stability requirement, $|y_n| \le |y_{n-1}|$, seems anemic in comparison, since it does not exclude $|y_n| \approx |y_{n-1}|$.

Let us generalize the test equation a bit, to include an inhomogeneity,

$$y' = \lambda(y - g(t)), \tag{3.28}$$

where $g(t)$ is a bounded, but otherwise arbitrary, function. We can rewrite (3.28) as

$$\varepsilon y' = \hat{\lambda}(y - g(t)),$$

where $\varepsilon = \frac{1}{|\mathcal{R}\mathrm{e}(\lambda)|}$, $\hat{\lambda} = \varepsilon\lambda$, and note that the *reduced solution*, obtained for $\varepsilon = 0$, is $y(t) = g(t)$. This motivates us to say that the discretization method has *stiff decay* if for $t_n > 0$ fixed,

$$|y_n - g(t_n)| \to 0 \qquad \text{as} \qquad h_n \mathcal{R}\mathrm{e}(\lambda) \to -\infty. \tag{3.29}$$

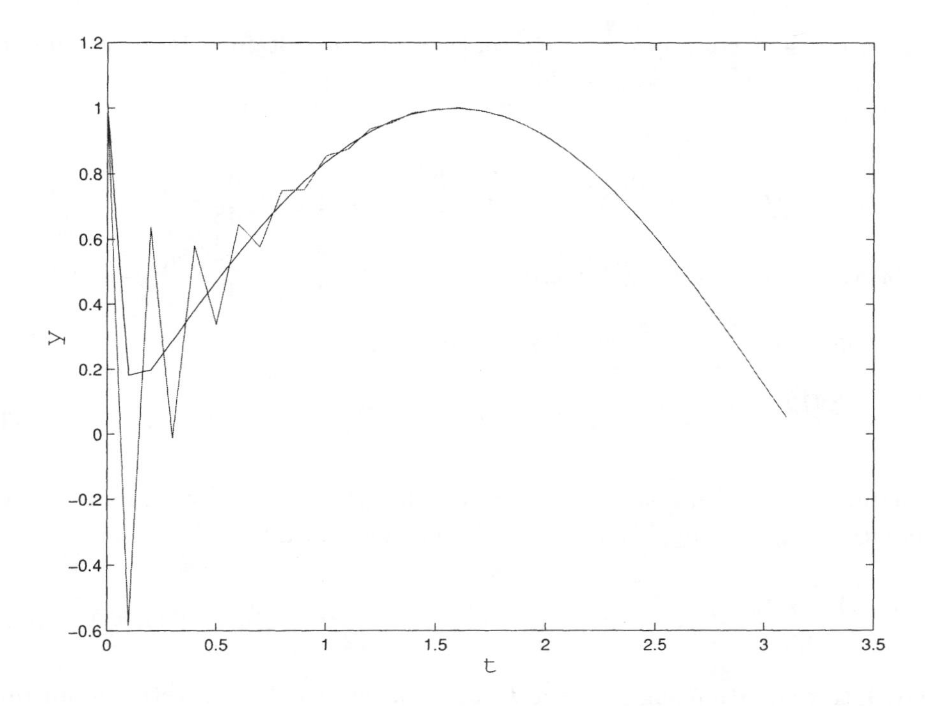

Figure 3.6: *Approximate solution on a coarse uniform mesh for Example* 3.2, *using backward Euler (the smoother curve) and trapezoidal methods.*

This is a stronger requirement than absolute stability in the very stiff limit, and it does not relate to what happens elsewhere in the $h\lambda$-plane. The backward Euler method has stiff decay, because when applied to (3.28) it yields

$$y_n - g(t_n) = (1 - h_n\lambda)^{-1}(y_{n-1} - g(t_n)).$$

The forward Euler method, of course, does not have stiff decay.

The practical advantage of methods with stiff decay lies in their ability to skip fine-level (i.e., rapidly varying) solution details and still maintain a decent description of the solution on a coarse level in the very stiff (not the highly oscillatory!) case. For instance, using backward Euler with a fixed step $h = .1$ to integrate the problem of Example 3.2, the initial layer is poorly approximated, and still the solution is qualitatively recovered where it varies slowly; see Figure 3.6. Herein lies a great potential for efficient use, as well as a great danger of misuse, of such discretization methods.

3.6 Symmetry: Trapezoidal Method

The forward Euler method was derived using a Taylor expansion centered at t_{n-1}. The backward Euler method was likewise derived, centered at t_n instead. Both methods are first-order accurate, which is often insufficient for an efficient computation. Better accuracy is obtained by centering the expansions at $t_{n-1/2} = t_n - \frac{1}{2}h_n$.

Writing

$$\mathbf{y}(t_n) = \mathbf{y}(t_{n-1/2}) + \frac{h_n}{2}\mathbf{y}'(t_{n-1/2}) + \frac{h_n^2}{8}\mathbf{y}''(t_{n-1/2}) + \frac{h_n^3}{48}\mathbf{y}'''(t_{n-1/2}) + \dots,$$
$$\mathbf{y}(t_{n-1}) = \mathbf{y}(t_{n-1/2}) - \frac{h_n}{2}\mathbf{y}'(t_{n-1/2}) + \frac{h_n^2}{8}\mathbf{y}''(t_{n-1/2}) - \frac{h_n^3}{48}\mathbf{y}'''(t_{n-1/2}) + \dots,$$

dividing by h_n, and subtracting, we obtain

$$\frac{\mathbf{y}(t_n) - \mathbf{y}(t_{n-1})}{h_n} = \mathbf{y}'(t_{n-1/2}) + \frac{h_n^2}{24}\mathbf{y}'''(t_{n-1/2}) + O(h_n^4). \tag{3.30}$$

Furthermore, writing similar expansions for $\mathbf{y}'$ instead of $\mathbf{y}$ and adding, we replace $\mathbf{y}'(t_{n-1/2})$ by $\frac{1}{2}(\mathbf{y}'(t_n) + \mathbf{y}'(t_{n-1}))$ and obtain

$$\frac{\mathbf{y}(t_n) - \mathbf{y}(t_{n-1})}{h_n} = \frac{1}{2}(\mathbf{y}'(t_n) + \mathbf{y}'(t_{n-1})) - \frac{h_n^2}{12}\mathbf{y}'''(t_{n-1/2}) + O(h_n^4). \tag{3.31}$$

The latter equation suggests the *trapezoidal method* for discretizing our prototype ODE system (3.1),

$$\mathbf{y}_n = \mathbf{y}_{n-1} + \frac{h_n}{2}(\mathbf{f}(t_n, \mathbf{y}_n) + \mathbf{f}(t_{n-1}, \mathbf{y}_{n-1})). \tag{3.32}$$

The local truncation error can be read off of (3.31): this method is second-order accurate.

The trapezoidal method is *symmetric*: a change of variable $\tau = -t$ on $[t_{n-1}, t_n]$ (i.e., integrating from right to left) leaves it unchanged.[7] Like the backward Euler method, it is implicit: the cost per step of these two methods is similar. But the trapezoidal method is more accurate, so perhaps fewer integration steps are needed to satisfy a given error tolerance. Before being able to conclude that, however, we must check stability.

Both the trapezoidal method and the backward Euler method are 0-stable (Exercise 3.1). To check absolute stability, apply the method with a step size h to the test equation. This gives

$$y_n = \frac{2 + h\lambda}{2 - h\lambda} y_{n-1}.$$

The region of absolute stability is precisely the left half-plane of $h\lambda$, so this method is A-stable. Moreover, the approximate solution is not dampened when $\mathcal{R}e(\lambda) > 0$, which is qualitatively correct since the exact solution grows in that case.

[7]Consider for notational simplicity the ODE $y' = f(y)$. A discretization method given by $y_n = y_{n-1} + h_n\psi(y_{n-1}, y_n, h)$ is *symmetric* if $\psi(u, v, h) = \psi(v, u, -h)$, because then, by letting $z_n \leftarrow y_{n-1}, z_{n-1} \leftarrow y_n$, and $h \leftarrow -h$, we get the same method for z_n as for y_n.

Method	h	Max error
Forward Euler	.1	.91e-2
Forward Euler	.05	.34e-2
Forward Euler	.025	.16e-2
Backward Euler	.1	.52e-2
Backward Euler	.05	.28e-2
Backward Euler	.025	.14e-2
Trapezoidal	.1	.42e-3
Trapezoidal	.05	.14e-3
Trapezoidal	.025	.45e-4

Table 3.1: *Maximum errors for Example* 3.1.

On the other hand, we cannot expect stiff decay with the trapezoidal method, because its amplification factor satisfies $|\frac{2+h\lambda}{2-h\lambda}| \to 1$ in the very stiff limit. This is typical for symmetric methods. Precisely, for the trapezoidal method,

$$\frac{2+h\lambda}{2-h\lambda} \to -1 \qquad \text{as} \qquad h\mathcal{R}\mathrm{e}(\lambda) \to -\infty.$$

The practical implication of this is that any solution detail must be resolved even if only a coarse picture of the solution is desired, because the fast mode components of local errors (for which h is "large") get propagated, almost undamped, throughout the integration interval $[0, b]$. This is evident in Figure 3.6, where we contrast integrations using the trapezoidal and the backward Euler methods for Example 3.2 with a uniform step size $h = .1$. To apply the trapezoidal method intelligently for this example, we must use a small step size through the initial layer, as in Figure 3.4. Then the step size can become larger. The indicated mesh in Figure 3.4 yields the solution profile shown when using the trapezoidal method.

Finally, in Table 3.1 we display the maximum error at mesh points for Example 3.1, when using each of the three methods introduced hitherto, with uniform step sizes $h = .1, h = .05$, and $h = .025$. Note that the error in the two Euler methods is not only larger in magnitude than the error obtained using the trapezoidal method; it also decreases linearly with h, while the trapezoidal error decreases more favorably, like h^2.

3.7 Rough Problems

In the beginning of this chapter we assumed that the given ODE (3.1) is "sufficiently smooth," in the sense that all derivatives mentioned in the sequel are bounded by a constant of moderate size. This is often the case in practice. Still, there are many important instances where the problem is not very smooth. In this section we discuss some such situations and their implication on the choice of discretization method.

In general, if $\mathbf{f}(t, \mathbf{y})$ has k bounded derivatives at the solution $\mathbf{y}(t)$,

$$\sup_{0 \le t \le b} \left| \frac{d^j}{dt^j} \mathbf{f}(t, \mathbf{y}(t)) \right| \le M, \qquad j = 0, 1, \ldots, k,$$

then by (3.1), $\mathbf{y}(t)$ has $k+1$ bounded derivatives, and in particular,

$$\|\mathbf{y}^{(j)}\| \le M, \qquad j = 1, \ldots, k+1.$$

So, if $\mathbf{f}$ is discontinuous but bounded, then $\mathbf{y}$ has a bounded, discontinuous first derivative. But the higher derivatives of $\mathbf{y}$ appearing in the Taylor expansion (3.2) (and hence in the expression for the local truncation error) are not bounded, so a discretization across such a discontinuity may yield rather inaccurate approximations.

Suppose first that there is one point, $\bar{t}$, $0 < \bar{t} < b$, where $\mathbf{f}(\bar{t}, \mathbf{y}(\bar{t}))$ is discontinuous, and everywhere else $\mathbf{f}$ is smooth and bounded. Note that the conditions of Theorem 1.1 do not hold on the interval $[0, b]$, but they do hold on each of the intervals $[0, \bar{t}]$ and $[\bar{t}, b]$. Thus, we may consider integrating the problem

$$\mathbf{y}' = \mathbf{f}(t, \mathbf{y}), \;\; 0 < t < \bar{t}; \qquad \mathbf{y}(0) = \mathbf{c}$$

followed by the problem

$$\mathbf{z}' = \mathbf{f}(t, \mathbf{z}), \;\; \bar{t} < t < b; \qquad \mathbf{z}(\bar{t}) = \mathbf{y}(\bar{t}).$$

For each of these subproblems we can discretize using one of the methods described in this chapter and the next one, and expect to realize the full accuracy order of the method. Now, the algorithm does not "know" that we have switched problems at $t = \bar{t}$. The integration can therefore proceed as before from 0 to b, provided that *$\bar{t}$ coincides with one of the mesh points, or step ends, $t_{\bar{n}}$*. On the other hand, if $\bar{t}$ is in the interior of a step, i.e., for some $\bar{n}$,

$$t_{\bar{n}-1} < \bar{t} < t_{\bar{n}},$$

then an $O(h_{\bar{n}})$ error results, regardless of the order of the (consistent) discretization method applied.

Example 3.4
Consider the function

$$f(t, y) = t - j\tau, \qquad j\tau \le t < (j+1)\tau, \quad j = 0, 1, \ldots, J,$$

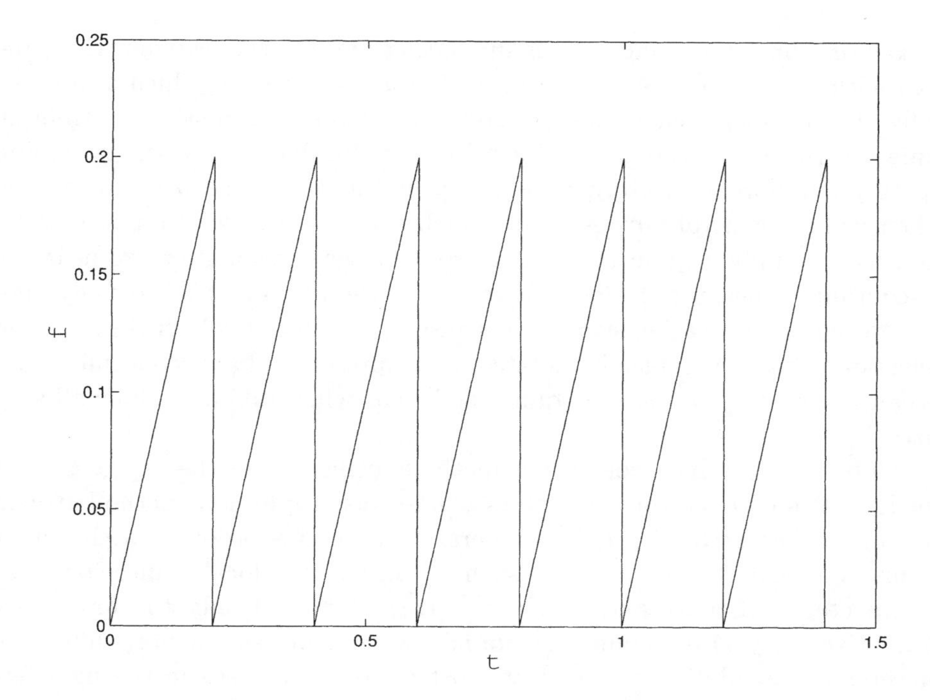

Figure 3.7: *Sawtooth function for* $\tau = 0.2$.

where $\tau > 0$ is a parameter. The ODE $y' = f$ is therefore a quadrature problem for a sawtooth function; see Figure 3.7. With $y(0) = 0$, the solution is

$$y(t) = j\tau^2/2 + (t - j\tau)^2/2, \qquad j\tau \le t < (j+1)\tau, \quad j = 0, 1, \ldots, J.$$

We also calculate that away from the discontinuity points $j\tau$,

$$y''(t) = 1, \qquad y'''(t) = 0.$$

For the trapezoidal method, the local truncation error therefore vanishes on a step $[t_{n-1}, t_n]$ that does not contain a point of discontinuity. For this special case, the trapezoidal method using a constant step size h reproduces $y(t_n)$ exactly if $\tau = lh$ for some positive integer l. If, on the other hand, $\tau = (l + r)h$ for some fraction r, then an $O(h\tau)$ error results, up to J times, for a combined $O(h)$ error. A worse, $O(1)$ error, may result if the "teeth are sharper," e.g., $f(t) = t/\tau - j$, $j\tau \le t < (j+1)\tau$, and $\tau = O(h)$. ♦

If $\mathbf{f}(t, \mathbf{y})$ has a discontinuity, then it is important to locate it and place a mesh point as close to it as possible. Unlike in the simple Example 3.4, however, the precise location of such a discontinuity may not be known. In many examples, the discontinuity is defined by a *switching function*, e.g.,

$$\mathbf{f}(t, \mathbf{y}) = \begin{cases} \mathbf{f}_I(t, \mathbf{y}(t)) & \text{if } g(t, \mathbf{y}(t)) < 0, \\ \mathbf{f}_{II}(t, \mathbf{y}(t)) & \text{if } g(t, \mathbf{y}(t)) > 0. \end{cases}$$

This situation often occurs when simulating mechanical systems (Example 1.6) with dry friction. A simple *event location* algorithm, which automatically detects when g changes sign and then solves a local nonlinear algebraic equation (using interpolation of nearby $\mathbf{y}_n$-values) to locate the switching point $g = 0$ more accurately, proves very useful in practice. An alternative, when using a general-purpose code which features adaptive step-size selection, is to simply rely on the code to select a very small step size near the discontinuity (because the local error is large there). But this is usually inferior to the event location strategy because the theory on which the step-size selection is based is typically violated in the presence of a discontinuity, and because the code can become quite inefficient when taking such small step sizes.

Note that the first-order Euler methods utilize $\mathbf{y}''$ in the expression for the local truncation error, while the second-order trapezoidal method utilizes the higher derivative $\mathbf{y}'''$. This is general: a method of order p matches $p+1$ terms in a local Taylor expansion such as (3.2), so the local truncation error is expected to behave like $O(h_n^p \mathbf{y}^{(p+1)}(t_n))$. Thus, if only the first $l+1$ derivatives of $\mathbf{y}(t)$ exist and are bounded, then, in general, *any* difference method will exhibit an order of at most l. As we will see in the next few chapters, higher-order methods cost more per step, so if the problem is rough (i.e., l is low), then lower-order methods get the nod.

Example 3.5
The harmonic oscillator

$$\begin{aligned} u'' + \omega^2 u = 0, \qquad 0 < t < b, \\ u(0) = 1, \qquad u'(0) = 0 \end{aligned}$$

has the solution

$$u(t) = \cos \omega t.$$

If the frequency ω is high, $\omega \gg 1$, then the derivatives grow larger and larger, because

$$\|u^{(p)}\| = \omega^p.$$

The local error of a discretization method of order p is

$$O(h^{p+1}\omega^{p+1}).$$

This means that to recover the highly oscillatory solution $u(t)$ accurately, we must restrict

$$h < 1/\omega$$

regardless of the order of the method. In fact, for $h > 1/\omega$, increasing the order of the method as such is useless. ♦

3.8 Software, Notes, and References

3.8.1 Notes

The basic questions of a numerical method's order, consistency, 0-stability, and convergence are discussed, in one form or another, in most books that are concerned with the numerical solution of differential equations. But there is a surprisingly wide variety of expositions; see, e.g., [54, 43, 62, 85, 50, 26, 8, 93, 67]. We have chosen to expose concepts in a way which highlights how they mimic properties of the differential equation. One benefit of this is that the concepts and the treatment naturally extend to BVPs and to PDEs (which we do not pursue directly in this book). Omitted is therefore a general derivation of the global error as an accumulation of local errors, such as what leads to (3.11).

For this reason we also chose to define the local truncation error as in (3.5)–(3.6). Some other authors have chosen to define this quantity multiplied by h_n, making $\mathbf{d}_n$ and $\mathbf{l}_n$ generally of the same order. We also chose not to use the local error (whose definition depends on the existence of the approximate solution: see, e.g., Exercise 3.9 to note that existence of the numerical solution is not always a foregone conclusion) as a tool in our exposition of fundamental concepts, despite its practical usefulness as discussed in the next chapter.

Another decision we made which deviates from most texts on numerical methods for initial value ODEs was to introduce the concept of stiffness at an early stage. This not only highlights the basic importance of the topic, but it is also a natural approach if one has the numerical solution of PDEs in mind, and moreover it allows a natural treatment of the concept of absolute stability. The extension of ODE stability restrictions to time-dependent PDEs is somewhat facilitated in Section 3.3, but it remains a nontrivial issue in general, because the ODE system size m is very large for the corresponding method of lines. Some cases do extend directly, though; see Exercises 3.6 and 3.7. For more challenging cases, see, e.g., Reddy and Trefethen [78] and references therein.

While the concept of stiffness is intuitively simple, its precise definition and detection have proved to be elusive. Our discussion has excluded, for instance, the influence of rough forcing terms and of small but positive eigenvalues. These issues are treated at length in many texts, e.g., Hairer and Wanner [52], Shampine [85], and Butcher [26].

The concept of absolute stability is due to Dahlquist [33]. But the search for a good stability concept to capture the essence of stiffness in numerical methods has brought about an often confusing plethora of definitions (for an extensive exposition, see, e.g., [52]). We feel that an appropriate concept should reflect what one aims to capture: a simple, intuitive phenomenon which is independent of the discretization approach. Hence we have used the somewhat less well known terminology and definition of stiff decay. This term was introduced by J. Varah in 1980, following Prothero and Robinson [73].

A more reliable, and in some cases also more efficient, means than finite differences for obtaining a Jacobian matrix without user intervention is to use *automatic differentiation* software [16]. This software takes as input a user-provided routine that computes **f** and produces another routine which efficiently computes the Jacobian matrix. At present, this requires an initial time investment, to install and learn to use the automatic differentiation software. However, we expect that simpler interfaces will soon be available.

The stiff problems dealt with in Chapters 3–5 have eigenvalues with large, negative real parts. Another type of "stiffness" is when the problem has (nearly) purely imaginary large eigenvalues. This yields highly (i.e., rapidly) oscillatory problems. See Examples 3.5 and 9.8. A recent survey on the numerical solution of such problems is given by Petzold, Jay, and Yen [71].

3.8.2 Software

In each of the following chapters which deals with numerical methods, there is a section (just before the "Exercises" section) which briefly describes some available software packages where corresponding methods and techniques are implemented. The methods described in the current chapter are too basic to be implemented alone in quality general-purpose codes. Here we quickly mention instead some outlets through which such software is available. There are three types.

1. A complete environment is made available in which users can do their own code development interactively, having direct access to various software tools. These tools include certain ODE packages (and much more). Examples of such programming environments are MATLAB and MATHEMATICA. We have used MATLAB in most of the coding developed for the examples and exercises in this book, and we strongly recommend it. An interactive environment such as MATLAB does not replace procedural languages like C or FORTRAN, though, especially for production runs and large-scale computing. The tools provided with MATLAB currently do not cover all the problem classes of this book, nor are they always the best available for a given application. MATLAB does allow interfacing with external software written in C or FORTRAN.

2. Collected programs are available for a fee through software libraries such as NAG and IMSL. These programs are written in FORTRAN or C. The range of software available in this way in some areas is considerably more extensive than what is available as part of the integrated programming environments, and it is more suitable for production codes. The advantage here, compared to the next alternative, is that there is a measure of quality control and support of the software avail-

able in this way. This occasionally also implies some limitations on the range and richness of the software available.

3. A large collection of codes is available electronically through NETLIB. The web page is at

   http://netlib.bell-labs.com/netlib/

   It is also possible to e-mail

   netlib@research.att.com

   with a request such as

   ```
   send codename from ode
   ```

   which causes the (hypothetical) ODE code CODENAME to be e-mailed back. NETLIB is a software repository: the software is available for free and comes with no guarantee.

The codes COLSYS, DASSL, and their derivatives, which solve stiff IVPs, BVPs, and differential-algebraic problems, and are distinguished by the fact that one of the present authors took part in writing them, are available through NETLIB.

Most software for scientific computation to date is written in FORTRAN, but software is available to convert to, or interface with, C and C++ programs. The user therefore does not need to be fluent in FORTRAN in order to use these codes. Such porting programs are available through NETLIB.

3.9 Exercises

3.1. Show that the backward Euler method and the trapezoidal method are 0-stable.

3.2. To draw a circle of radius r on a graphics screen, one may proceed to evaluate pairs of values $x = r\cos\theta$, $y = r\sin\theta$ for a succession of values θ. But this is expensive. A cheaper method may be obtained by considering the ODE

$$\begin{aligned} \dot{x} &= -y, \qquad x(0) = r, \\ \dot{y} &= x, \qquad y(0) = 0, \end{aligned}$$

where $\dot{x} = \frac{dx}{d\theta}$, and approximating this using a simple discretization method. However, care must be taken so as to ensure that the obtained approximate solution looks right, i.e., that the approximate curve closes rather than spirals.

For each of the three discretization methods introduced in this chapter, namely, forward Euler, backward Euler, and trapezoidal methods, carry out this integration using a uniform step size $h = .02$ for $0 \le \theta \le 120$. Determine if the solution spirals in, spirals out, or forms an approximate circle as desired. Explain the observed results. [Hint: This has to do with a certain invariant function of x and y, rather than with the order of the methods.]

3.3. The following ODE system:

$$y_1' = \alpha - y_1 - \frac{4y_1y_2}{1+y_1^2},$$

$$y_2' = \beta y_1 \left(1 - \frac{y_2}{1+y_1^2}\right),$$

where α and β are parameters, represents a simplified approximation to a chemical reaction [92]. There is a parameter value $\beta_c = \frac{3\alpha}{5} - \frac{25}{\alpha}$ such that for $\beta > \beta_c$ solution trajectories decay in amplitude and spiral in phase space into a stable fixed point, whereas for $\beta < \beta_c$ trajectories oscillate without damping and are attracted to a stable limit cycle. [This is called a *Hopf bifurcation*.]

(a) Set $\alpha = 10$ and use any of the discretization methods introduced in this chapter with a fixed step size $h = .01$ to approximate the solution starting at $y_1(0) = 0$, $y_2(0) = 2$, for $0 \le t \le 20$. Do this for the parameter values $\beta = 2$ and $\beta = 4$. For each case plot y_1 vs. t and y_2 vs. y_1. Describe your observations.

(b) Investigate the situation closer to the critical value $\beta_c = 3.5$. [You may have to increase the length of the integration interval b to get a better look.]

3.4. When deriving the trapezoidal method, we proceeded to replace $\mathbf{y}'(t_{n-1/2})$ in (3.30) by an average and then use the ODE (3.1). If instead we first use the ODE, replacing $\mathbf{y}'(t_{n-1/2})$ by $\mathbf{f}(t_{n-1/2}, \mathbf{y}(t_{n-1/2}))$, and then average $\mathbf{y}$, we obtain the implicit *midpoint method*,

$$\mathbf{y}_n = \mathbf{y}_{n-1} + h_n \mathbf{f}\left(t_{n-1/2}, \frac{1}{2}(\mathbf{y}_n + \mathbf{y}_{n-1})\right). \tag{3.33}$$

(a) Show that this method is symmetric, second-order, and A-stable. How does it relate to the trapezoidal method for the constant-coefficient ODE (3.18)?

(b) Show that even if we allow λ to vary in t, i.e., we consider the scalar ODE

$$y' = \lambda(t)y$$

in place of the test equation, what corresponds to A-stability holds; namely, using the midpoint method,

$$|y_n| \le |y_{n-1}| \qquad \text{if} \qquad \mathcal{R}e(\lambda) \le 0$$

(this property is called *AN-stability* [24]). Show that the same cannot be said about the trapezoidal method: the latter is not AN-stable.

3.5. (a) Show that the trapezoidal step (3.32) can be viewed as a half-step of forward Euler followed by a half-step of backward Euler.

(b) Show that the midpoint step (3.33) can be viewed as a half-step of backward Euler followed by a half-step of forward Euler.

(c) Consider an autonomous system $\mathbf{y}' = \mathbf{f}(\mathbf{y})$ and a fixed step size, $h_n = h$, $n = 1, \ldots, N$. Show that the trapezoidal method applied N times is equivalent to applying first a half-step of forward Euler (i.e., forward Euler with step size $h/2$), followed by $N-1$ midpoint steps, finishing off with a half-step of backward Euler.

Conclude that these two symmetric methods are *dynamically equivalent* [34]; i.e., for h small enough their performance is very similar independently of N, even over a very long time: $b = Nh \gg 1$.

(d) However, if h is not small enough (compared to the problem's small parameter, say λ^{-1}) then these methods do not necessarily perform similarly. Construct an example where one of these methods blows up (error $> 10^5$, say) while the other yields an error below 10^{-5}. [Do not program anything: this is a (nontrivial!) pen-and-paper question.]

3.6. Consider the method of lines applied to the simple heat equation in one space dimension,

$$u_t = au_{xx},$$

with $a > 0$ a constant, $u = 0$ at $x = 0$, $x = 1$ for $t \geq 0$, and $u(x, 0) = g(x)$ given as well. Formulate the method of lines, as in Example 1.3, to arrive at a system of the form (3.18) with A symmetric. Find the eigenvalues of A and show that, when using the forward Euler discretization for the time variable, the resulting method is stable if

$$h \leq \frac{1}{2a}\Delta x^2.$$

(This is a rather restrictive condition on the time step.) On the other hand, if we discretize in time using the trapezoidal method (the resulting method, second-order in both space and time, is called Crank–Nicolson) or the backward Euler method, then no stability restriction for the time step arises. [Hint: To find the eigenvalues, try eigenvectors $\mathbf{v}^k$ in the form $v_i^k = \sin(ik\pi\Delta x)$, $i = 1, \ldots, m$, for $1 \leq k \leq m$.]

3.7. Consider the same question as the previous one, but this time the heat equation is in two space variables on a unit square,

$$u_t = a(u_{xx} + u_{yy}), \qquad 0 \le x, y \le 1, \quad t \ge 0.$$

The boundary conditions are $u = 0$ around the square, and $u(x, y, 0) = g(x, y)$ is given as well.

Formulate a system (3.18) using a uniform grid with spacing Δx on the unit square. Conclude again that no restrictions on the time step arise when using the implicit methods which we have presented for time discretization. What happens with the forward Euler method? [Hint: Don't try this exercise before you have done the previous one.]

3.8. Consider the ODE

$$\frac{dy}{dt} = f(t, y), \qquad 0 \le t \le b,$$

where $b \gg 1$.

(a) Apply the *stretching* transformation $t = \tau b$ to obtain the equivalent ODE

$$\frac{dy}{d\tau} = b\ f(\tau b, y), \qquad 0 \le \tau \le 1.$$

(Strictly speaking, y in these two ODEs is not quite the same function. Rather, it stands in each case for the unknown function.)

(b) Show that applying any of the discretization methods in this chapter to the ODE in t with step size $h = \Delta t$ is equivalent to applying the same method to the ODE in τ with step size $\Delta\tau$ satisfying $\Delta t = b\Delta\tau$. In other words, the same stretching transformation can be equivalently applied to the discretized problem.

3.9. Write a short program which uses the forward Euler, the backward Euler, and the trapezoidal *or* midpoint methods to integrate a linear, scalar ODE with a known solution, using a fixed step size $h = b/N$, and which finds the maximum error. Apply your program to the following problem:

$$\frac{dy}{dt} = (\cos t)y, \qquad 0 \le t \le b,$$

$y(0) = 1$. The exact solution is

$$y(t) = e^{\sin t}.$$

Verify those entries given in Table 3.2 and complete the missing ones. Make as many (useful) observations as you can on the results in the complete table. Attempt to provide explanations. [Hint: Plotting these solution curves for $b = 20$, $N = 10b$, say, may help.]

b	N	Forward Euler	Backward Euler	Trapezoidal	Midpoint
1	10	.35e-1	.36e-1	.29e-2	.22e-2
	20	.18e-1	.18e-1	.61e-3	.51e-3
10	100				
	200				
100	1000	2.46	25.90	.42e-2	.26e-2
	2000				
1000	1000				
	10000	2.72	1.79e+11	.42e-2	.26e-2
	20000				
	100000	2.49	29.77	.42e-4	.26e-4

Table 3.2: *Maximum errors for long interval integration of* $y' = (\cos t)y$.

3.10. Consider two linear harmonic oscillators (recall Example 2.6), one fast and one slow, $u_1'' = -\varepsilon^{-2}(u_1 - \bar{u}_1)$ and $u_2'' = -(u_2 - \bar{u}_2)$. The parameter is small: $0 < \varepsilon \ll 1$. We write this as a first-order system

$$\mathbf{u}' = \begin{pmatrix} \varepsilon^{-1} & 0 \\ 0 & 1 \end{pmatrix} \mathbf{v},$$

$$\mathbf{v}' = -\begin{pmatrix} \varepsilon^{-1} & 0 \\ 0 & 1 \end{pmatrix} (\mathbf{u} - \bar{\mathbf{u}}),$$

where $\mathbf{u}(t), \mathbf{v}(t)$, and the given constant vector $\bar{\mathbf{u}}$ each have two components. It is easy to see that $E_F = \frac{1}{2\varepsilon}(v_1^2 + (u_1 - \bar{u}_1)^2)$ and $E_S = \frac{1}{2}(v_2^2 + (u_2 - \bar{u}_2)^2)$ remain constant for all t (see Section 2.5).

Next, we apply the following time-dependent linear transformation:

$$\mathbf{u} = Q\mathbf{x}, \quad \mathbf{v} = Q\mathbf{z}, \quad Q(t) = \begin{pmatrix} \cos \omega t & \sin \omega t \\ -\sin \omega t & \cos \omega t \end{pmatrix},$$

$$K = \dot{Q}^T Q = \begin{pmatrix} 0 & -1 \\ 1 & 0 \end{pmatrix},$$

where $\omega \geq 0$ is another parameter. This yields the coupled system

$$\mathbf{x}' = Q^T \begin{pmatrix} \varepsilon^{-1} & 0 \\ 0 & 1 \end{pmatrix} Q\mathbf{z} + \omega K\mathbf{x}, \tag{3.34a}$$

$$\mathbf{z}' = -Q^T \begin{pmatrix} \varepsilon^{-1} & 0 \\ 0 & 1 \end{pmatrix} Q(\mathbf{x} - \bar{\mathbf{x}}) + \omega K\mathbf{z}\,, \tag{3.34b}$$

where $\bar{\mathbf{x}} = Q^T\bar{\mathbf{u}}$. We can write the latter system in our usual notation as a system of order 4,

$$\mathbf{y}' = A(t)\mathbf{y} + \mathbf{q}(t)\,.$$

(a) Show that the eigenvalues of the matrix A are all purely imaginary for all ω. [Hint: Show that $A^T = -A$.]

(b) Using the values $\bar{\mathbf{u}} = (1, \pi/4)^T$, $\mathbf{u}(0) = \bar{\mathbf{u}}$, $\mathbf{v}(0)^T = (1, -1)/\sqrt{2}$, and $b = 20$, apply the midpoint method with a constant step size h to the system (3.34) for the following parameter combinations: $\varepsilon = 0.001$, $k = 0.1, 0.05, 0.001$, and $\omega = 0, 1, 10$ (a total of nine runs). Compute the error indicators $\max_t |E_F(t) - E_F(0)|$ and $\max_t |E_S(t) - E_S(0)|$. Discuss your observations.

(c) Attempt to show that the midpoint method is unstable for this problem if $h > 2\sqrt{\varepsilon/\omega}$ (see [10]). Conclude that A-stability and AN-stability do not automatically extend to ODE systems.

3.11. Consider the implicit ODE

$$M(\mathbf{y})\mathbf{y}' = \mathbf{f}(t, \mathbf{y}),$$

where $M(\mathbf{y})$ is nonsingular for all $\mathbf{y}$. The need to integrate IVPs of this type typically arises in robotics. When the system size m is large, the cost of inverting M may dominate the entire solution cost. Also, $\frac{\partial M}{\partial \mathbf{y}}$ is complicated to evaluate, but it is given that its norm is not large, say $O(1)$.

(a) Extend the forward Euler and the backward Euler discretizations for this case (without inverting M). Justify.

(b) Propose a method for solving the nonlinear system of equations resulting at each time step when using backward Euler, for the case where $\|\partial\mathbf{f}/\partial\mathbf{y}\|$ is very large.

Chapter 4

One-Step Methods

The basic methods developed in Chapter 3 can be adequate for computing approximate solutions of a relatively low accuracy (as we will see, for instance, in Example 4.1) or if the problem being solved is rough in a certain sense (see Section 3.7). But often in practice a quality solution of high accuracy to a relatively smooth problem is sought, and then using a basic, low-order method necessitates taking very small steps in the discretization. This makes the integration process inefficient. Substantially fewer steps are needed when using a higher-order method in such circumstances.

In order to develop efficient, highly accurate approximation algorithms, we therefore design higher-order difference methods. The higher-order methods we consider in this book are of two types: one step and linear multistep. In each of these classes of methods it will be useful to distinguish further between methods for stiff problems and methods for nonstiff problems. The big picture is depicted in Figure 4.1.

In this chapter we will explore higher-order *one-step methods*. These are methods which do not use any information from previous steps (in contrast to linear multistep methods, which will be taken up in the next chapter). Thus, in a typical step of size $h = h_n = t_n - t_{n-1}$, we seek an approximation y_n to $y(t_n)$ given the previous step's end result, y_{n-1}.

Taylor Series Method

The conceptually simplest approach for achieving higher order is to use the differential equation to construct the Taylor series for the solution. For a scalar ordinary differential equation (ODE) $y' = f(t, y)$, the Taylor series method is given by replacing the higher derivatives in a truncated Taylor expansion, yielding the formula

$$y_n = y_{n-1} + hy'_{n-1} + \frac{h^2}{2}y''_{n-1} + \cdots \frac{h^p}{p!}y^{(p)}_{n-1},$$

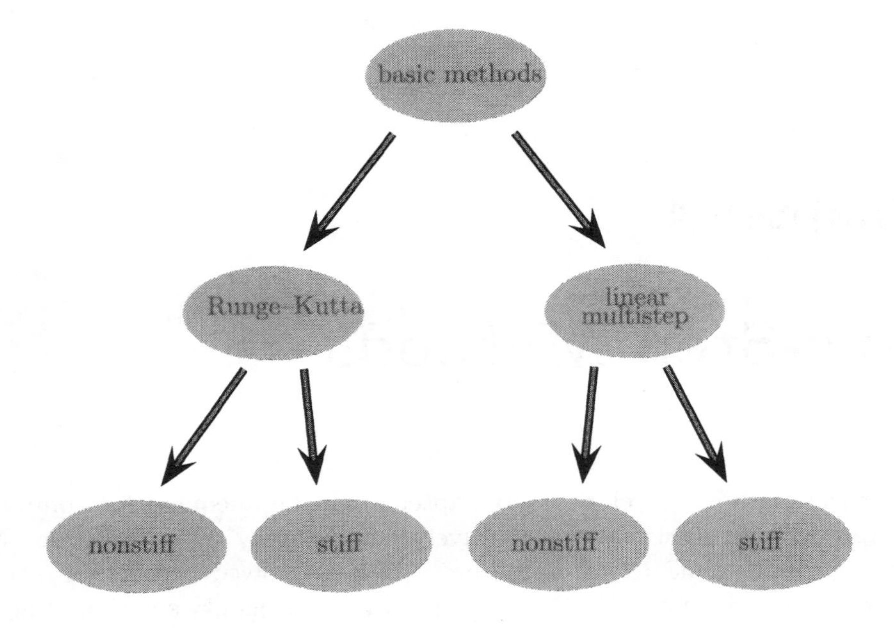

Figure 4.1: *Classes of higher-order methods.*

with $f(t, y)$ and its derivatives evaluated at (t_{n-1}, y_{n-1}),

$$\begin{aligned} y'_{n-1} &= f, \\ y''_{n-1} &= f_t + f_y f, \\ y'''_{n-1} &= f_{tt} + 2f_{ty}f + f_y f_t + f_{yy}f^2 + f_y^2 f, \end{aligned} \tag{4.1}$$

etc. The local truncation error is $h^p y^{(p+1)}(t_n)/(p+1)! + O(h^{p+1})$. For a system of differential equations, the derivatives are defined similarly.

A problem with this method is that it requires analytic expressions for derivatives which in a practical application can be quite complicated. On the other hand, advances in compiler technology have enabled much more robust programs for symbolic and automatic differentiation in recent years, which may make this method more attractive for some applications.

We thus seek one-step methods that achieve a higher accuracy order without forming the symbolic derivatives of the Taylor series method. This leads to *Runge–Kutta methods*, to which the rest of this chapter is devoted.

4.1 The First Runge–Kutta Methods

We stay with a scalar ODE for some of this exposition. The extension to ODE systems is straightforward, and will be picked up later.

Many Runge–Kutta methods are based on quadrature schemes. In fact, the reader may want to quickly review basic quadrature rules at this point.

Review: Recall from Advanced Calculus that **Taylor's theorem for a function of several variables** gives

$$\begin{aligned} F(x,y) &= F + \left[\frac{\partial F}{\partial x}(x-\hat{x}) + \frac{\partial F}{\partial y}(y-\hat{y})\right] \\ &+ \frac{1}{2!}\left[\frac{\partial^2 F}{\partial x^2}(x-\hat{x})^2 + 2\frac{\partial^2 F}{\partial x \partial y}(x-\hat{x})(y-\hat{y}) + \frac{\partial^2 F}{\partial y^2}(y-\hat{y})^2\right] \\ &+ \cdots + \frac{1}{n!}\ell^n F + \cdots, \end{aligned}$$

where the functions on the right-hand side are evaluated at $(\hat{x}, \hat{y})$ and

$$\ell^n F = \sum_{j=0}^{n} \binom{n}{j} \left(\frac{\partial^n F}{\partial x^j \partial y^{n-j}}(\hat{x}, \hat{y}) \right) (x-\hat{x})^j (y-\hat{y})^{n-j}.$$

Let's reconsider the methods we have already seen in the previous chapter in the context of quadrature. Writing

$$y(t_n) - y(t_{n-1}) = \int_{t_{n-1}}^{t_n} y'(t)dt, \tag{4.2}$$

we can approximate the area under the curve $y'(t)$ (see Figure 4.2) using either the lower sum based on $y'(t_{n-1})$ (forward Euler) or the upper sum based on $y'(t_n)$ (backward Euler). These are first-order methods.

For a better approximation, we can use the height at the midpoint of the interval, i.e., $y'(t_{n-1/2})$, where $t_{n-1/2} = t_n - h/2$; see Figure 4.3. This leads to the *midpoint method* (recall (3.33))

$$y_n = y_{n-1} + hf\left(t_{n-1/2}, \frac{y_{n-1} + y_n}{2}\right).$$

This is an implicit Runge–Kutta method. We can construct an explicit method based on the same idea by first approximating $y(t_{n-1/2})$ by the forward Euler method and then substituting into the midpoint method to obtain

$$\hat{y}_{n-1/2} = y_{n-1} + \frac{h}{2} f(t_{n-1}, y_{n-1}), \tag{4.3a}$$

$$y_n = y_{n-1} + hf(t_{n-1/2}, \hat{y}_{n-1/2}). \tag{4.3b}$$

The obtained *explicit midpoint* method (4.3) gives us a first real taste of the original Runge–Kutta idea: a higher order is achieved by repeated

Review: Basic quadrature rules. Given the task of evaluating an integral

$$\int_a^b f(t)dt$$

for some function $f(t)$ on an interval $[a, b]$, basic quadrature rules are derived by replacing $f(t)$ with an interpolating polynomial $\phi(t)$ and integrating the latter exactly. If there are s distinct interpolation points $c_1, \ldots, c_s$, then we can write the interpolating polynomial of degree $< s$ in Lagrange form

$$\phi(t) = \sum_{j=1}^{s} f(c_j)L_j(t),$$

where

$$L_j(t) = \Pi_{i=1, i \neq j}^{s} \frac{(t - c_i)}{(c_j - c_i)}.$$

Then

$$\int_a^b f(t)dt \approx \sum_{j=1}^{s} w_j f(c_j),$$

where the weights w_j are given by

$$w_j = \int_a^b L_j(t)dt.$$

The *precision* of the quadrature rule is p if the rule is exact for all polynomials of degree $< p$, i.e., if for any polynomial f of degree $< p$,

$$\int_a^b f(t)dt = \sum_{j=1}^{s} w_j f(c_j).$$

If $b - a = O(h)$ then the error in a quadrature rule of precision p is $O(h^{p+1})$. Obviously, $p \geq s$, but p may be significantly larger than s if the points c_j are chosen carefully.

The midpoint and trapezoidal rules have precision $p = 2$. Simpson's rule has precision $p = 4$. Gaussian quadrature at s points has the highest precision possible at $p = 2s$.

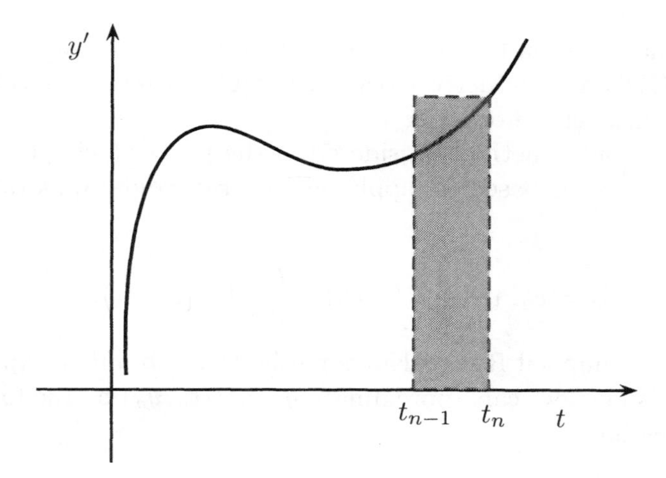

Figure 4.2: *Approximate area under curve.*

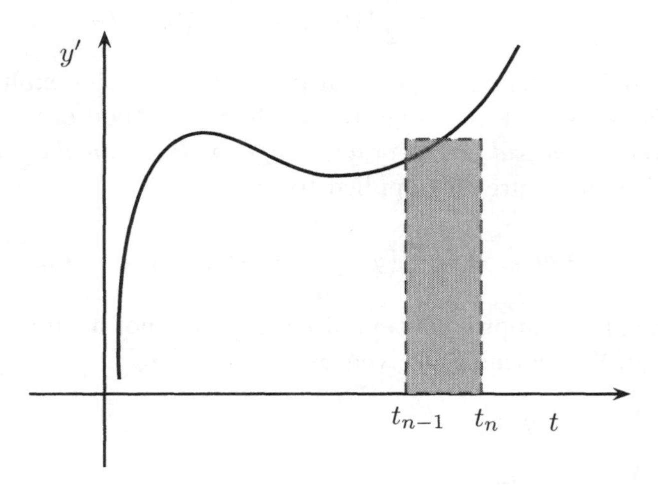

Figure 4.3: *Midpoint quadrature.*

function evaluations of f within the interval $[t_{n-1}, t_n]$. Note that this method is not linear in f anymore (substitute (4.3a) into (4.3b) to see this). At first glance, it might seem that the order would be limited to 1, because the first stage (4.3a) uses the forward Euler method, which is first order. However, note that the term involving $\hat{y}_{n-1/2}$ enters into (4.3b) multiplied by h, and therefore its error becomes less important.

Indeed, the local truncation error of (4.3) is given by

$$\begin{aligned} d_n &= \frac{y(t_n) - y(t_{n-1})}{h} - f\left(t_{n-1/2}, y(t_{n-1}) + \frac{h}{2} f(t_{n-1}, y(t_{n-1}))\right) \\ &= y' + \frac{h}{2} y'' + \frac{h^2}{6} y''' \qquad (4.4) \\ &- \left(f + \frac{h}{2}(f_t + f_y f) + \frac{h^2}{8}(f_{tt} + 2f_{ty} f + f_{yy} f^2)\right) + O(h^3), \end{aligned}$$

where all quantities on the right-hand side are evaluated at $(t_{n-1}, y(t_{n-1}))$. Using the ODE and its derivatives, all but $O(h^2)$ terms cancel. Thus the method is consistent of order 2.

The trapezoidal method considered in the previous chapter is obtained in a similar manner based on applying the trapezoidal quadrature rule to (4.2):

$$y_n = y_{n-1} + \frac{h}{2} f(t_n, y_n) + \frac{h}{2} f(t_{n-1}, y_{n-1}).$$

This is another implicit Runge–Kutta method. To obtain an explicit method based on this idea, we can approximate y_n in $f(t_n, y_n)$ by the forward Euler method, yielding

$$\hat{y}_n = y_{n-1} + hf(t_{n-1}, y_{n-1}), \qquad (4.5a)$$

$$y_n = y_{n-1} + \frac{h}{2} f(t_n, \hat{y}_n) + \frac{h}{2} f(t_{n-1}, y_{n-1}). \qquad (4.5b)$$

This is called the *explicit trapezoidal* method. Like the explicit midpoint method it is an explicit two-stage Runge–Kutta method of order 2.

The famous *classical fourth-order Runge–Kutta method* is closely related to Simpson's quadrature rule applied to (4.2),

$$y(t_n) - y(t_{n-1}) \approx \frac{h}{6}\left(y'(t_{n-1}) + 4y'(t_{n-1/2}) + y'(t_n)\right). \qquad (4.6)$$

To build an explicit approximation of $y'(t_{n-1/2})$ is not a simple matter anymore, though. The formula is given by

$$\begin{aligned}
Y_1 &= y_{n-1}, \\
Y_2 &= y_{n-1} + \frac{h}{2} f(t_{n-1}, Y_1), \\
Y_3 &= y_{n-1} + \frac{h}{2} f(t_{n-1/2}, Y_2), \\
Y_4 &= y_{n-1} + hf(t_{n-1/2}, Y_3), \\
y_n &= y_{n-1} + \frac{h}{6}\left(f(t_{n-1}, Y_1) + 2f(t_{n-1/2}, Y_2)\right. \\
&\quad \left. + 2f(t_{n-1/2}, Y_3) + f(t_n, Y_4)\right).
\end{aligned} \qquad (4.7)$$

It has order 4.

Example 4.1
We compute the solution of the simple Example 3.1

$$y' = -5ty^2 + \frac{5}{t} - \frac{1}{t^2}, \qquad y(1) = 1$$

using three explicit Runge–Kutta methods: forward Euler, explicit midpoint (RK2), and the classical fourth-order method (RK4). We use various fixed

Step h	Euler error	Rate	RK2 error	Rate	RK4 error	Rate
0.2	.40e-2		.71e-3		.66e-6	
0.1	.65e-6	12.59	.33e-6	11.08	.22e-7	4.93
0.05	.32e-6	1.00	.54e-7	2.60	.11e-8	4.34
0.02	.13e-6	1.00	.72e-8	2.20	.24e-10	4.16
0.01	.65e-7	1.00	.17e-8	2.08	.14e-11	4.07
0.005	.32e-7	1.00	.42e-9	2.04	.89e-13	3.98
0.002	.13e-7	1.00	.66e-10	2.02	.13e-13	2.13

Table 4.1: *Errors and calculated convergence rates for the forward Euler, the explicit midpoint (RK2), and the classical Runge–Kutta (RK4) methods.*

step sizes to integrate up to $t = 25$ and record the absolute errors at the end of the interval in Table 4.1 (the exact solution, to recall, is $y(t) = 1/t$). We also record for each method a calculated convergence rate. This "rate" is calculated as follows: if the error at step n behaves like $e_n(h) = y_n - y(t_n) \approx ch^p$ for some unknown constant c and rate p, then the error with half the step size should satisfy $e_{2n}(h/2) \approx c\left(\frac{h}{2}\right)^p$. Thus $p \approx rate := \log_2\left(\frac{e_n(h)}{e_{2n}(h/2)}\right)$.

A number of general observations can be deduced already from this very simple example.

- The error for a given step size is much smaller for the higher-order methods. That is the basic reason for embarking on the search for higher-order methods in this chapter and the next. Of course, the cost of each step is also higher for a higher-order method. Roughly, if the cost is measured simply by the number of evaluations of f (which in complex applications is usually the determining cost factor), then the cost of an RK4 step is double that of the explicit midpoint method, which in turn is double that of the forward Euler method.

- Thus, the choice of method depends on the accuracy requirements. Generally, the smaller the error tolerance and the smoother the problem and its solution, the more advantageous it becomes to use higher-order methods (see Section 3.7). Here, if the maximum error tolerance is 10^{-4}, then the best choice would be forward Euler. But for an error tolerance 10^{-12}, the fourth-order method is best.

- The error is polluted, as evidenced by the deviations of the computed rates from their predicted values of 1, 2, and 4, for both very large and very small step sizes. For $h = .2$ an error due to partial violation of absolute stability is observed. For $h = .002$ the truncation error in the classical fourth-order method is so small that the total error begins to be dominated by roundoff error (we have been using floating point arithmetic with 14 hexadecimal digits). Roundoff error generally increases as h decreases, because more steps are required to cover the integration interval. The assumption $e_n(h) \approx ch^p$ presupposes that the roundoff error is dominated for this h by the truncation error, which is often the case in practice. ♦

4.2 General Formulation of Runge–Kutta Methods

In general, an s-stage Runge–Kutta method for the ODE system

$$\mathbf{y}' = \mathbf{f}(t, \mathbf{y})$$

can be written in the form

$$\mathbf{Y}_i = \mathbf{y}_{n-1} + h \sum_{j=1}^{s} a_{ij} \mathbf{f}(t_{n-1} + c_j h, \mathbf{Y}_j), \qquad 1 \le i \le s, \tag{4.8a}$$

$$\mathbf{y}_n = \mathbf{y}_{n-1} + h \sum_{i=1}^{s} b_i \mathbf{f}(t_{n-1} + c_i h, \mathbf{Y}_i). \tag{4.8b}$$

The $\mathbf{Y}_i$'s are intermediate approximations to the solution at times $t_{n-1} + c_i h$, which may be correct to a lower order of accuracy than the solution $\mathbf{y}_n$ at the end of the step. Note that $\mathbf{Y}_i$ are local to the step from t_{n-1} to t_n, and the only approximation that the next step "sees" is $\mathbf{y}_n$. The coefficients of the method are chosen in part so that error terms cancel in such a way that $\mathbf{y}_n$ is more accurate.

The method can be represented conveniently in a shorthand notation

$$\begin{array}{c|cccc} c_1 & a_{11} & a_{12} & \cdots & a_{1s} \\ c_2 & a_{21} & a_{22} & \cdots & a_{2s} \\ \vdots & \vdots & \vdots & \ddots & \vdots \\ c_s & a_{s1} & a_{s2} & \cdots & a_{ss} \\ \hline & b_1 & b_2 & \cdots & b_s \end{array}$$

We will always choose

$$c_i = \sum_{j=1}^{s} a_{ij}, \qquad i = 1, \ldots, s. \tag{4.9}$$

The Runge–Kutta method is *explicit* iff $a_{ij} = 0$ for $j \ge i$, because then each $\mathbf{Y}_i$ in (4.8a) is given in terms of known quantities. Historically, the first Runge–Kutta methods were explicit. However, implicit Runge–Kutta methods are useful for the solution of stiff systems, as well as for boundary value problems (BVPs; see Chapter 8).

Some examples of explicit Runge–Kutta methods are given below:
Forward Euler:

$$\begin{array}{c|c} 0 & 0 \\ \hline & 1 \end{array}$$

One-parameter family of second-order methods:

$$\begin{array}{c|cc} 0 & 0 & 0 \\ \alpha & \alpha & 0 \\ \hline & 1-\frac{1}{2\alpha} & \frac{1}{2\alpha} \end{array}$$

For $\alpha = 1$, we have the explicit trapezoidal method, and for $\alpha = 1/2$ it is the explicit midpoint method.

There are three one-parameter families of *third-order three-stage methods*. One such family is

$$\begin{array}{c|ccc} 0 & 0 & 0 & 0 \\ \frac{2}{3} & \frac{2}{3} & 0 & 0 \\ \frac{2}{3} & \frac{2}{3}-\frac{1}{4\alpha} & \frac{1}{4\alpha} & 0 \\ \hline & \frac{1}{4} & \frac{3}{4}-\alpha & \alpha \end{array}$$

where α is a parameter.

Finally, the classical fourth-order method is written using this notation as

$$\begin{array}{c|cccc} 0 & 0 & 0 & 0 & 0 \\ \frac{1}{2} & \frac{1}{2} & 0 & 0 & 0 \\ \frac{1}{2} & 0 & \frac{1}{2} & 0 & 0 \\ 1 & 0 & 0 & 1 & 0 \\ \hline & \frac{1}{6} & \frac{1}{3} & \frac{1}{3} & \frac{1}{6} \end{array}$$

We see that there are s-stage explicit Runge–Kutta methods of order $p = s$, at least for $p \leq 4$. One may wonder if it is possible to obtain order $p > s$ and if it is possible to always maintain at least $p = s$. The answers are both negative. There will be more on this in the next section.

The choice of intermediate variables $\mathbf{Y}_i$ to describe the Runge–Kutta method (4.8) is not the only natural one. Sometimes it is more natural to use intermediate approximations to $\mathbf{f}$ rather than $\mathbf{y}$ at the interior stages. We leave it to the reader to verify that the general s-stage Runge–Kutta method (4.8) can be written as

$$\mathbf{K}_i = \mathbf{f}\left(t_{n-1} + c_i h, \mathbf{y}_{n-1} + h\sum_{j=1}^{s} a_{ij}\mathbf{K}_j\right), \qquad 1 \le i \le s, \tag{4.10a}$$

$$\mathbf{y}_n = \mathbf{y}_{n-1} + h\sum_{i=1}^{s} b_i\mathbf{K}_i. \tag{4.10b}$$

4.3 Convergence, 0-Stability, and Order for Runge–Kutta Methods

The basic convergence of one-step methods is essentially automatic. All of the methods we have seen so far, and any that we will see, are accurate to at least first order; i.e., they are consistent. The fundamental Theorem 3.1 tells us that convergence (to the order of accuracy, as in (3.10)) is achieved, provided only that the method is 0-stable. We can write any reasonable one-step method in the form

$$\mathbf{y}_n = \mathbf{y}_{n-1} + h\boldsymbol{\psi}(t_{n-1}, \mathbf{y}_{n-1}, h), \tag{4.11}$$

where $\boldsymbol{\psi}$ satisfies a Lipschitz condition in $\mathbf{y}$. (This is obvious for explicit methods. For implicit methods the Implicit Function Theorem is applied.) In the previous chapter we showed, following Theorem 3.1, that the forward Euler method is 0-stable. Replacing $\mathbf{f}$ in that proof by $\boldsymbol{\psi}$ yields the same conclusion for the general one-step method (4.11). We leave the details to the exercises.

We next consider the question of verifying the order of a given Runge–Kutta method. Order conditions for general Runge–Kutta methods are obtained by expanding the numerical solution in Taylor series, as we did for second-order methods. For an autonomous ODE system[8]

$$\mathbf{y}' = \mathbf{f}(\mathbf{y}),$$

the exact solution satisfying $\mathbf{y}(t_{n-1}) = \mathbf{y}_{n-1}$ (recall that we are interested

[8]Without loss of generality, we can consider systems of autonomous differential equations only. This is because the ODE $y' = f(t, y)$, can be transformed to autonomous form by adding t to the dependent variables as follows: $t' = 1$, $y' = f(t, y)$.

here in what happens in just one step) has at t_{n-1} the derivatives

$$\begin{aligned}
\mathbf{y}' &= \mathbf{f} =: \mathbf{f}^0, \\
\mathbf{y}'' &= \mathbf{f}' = \left(\frac{\partial \mathbf{f}}{\partial \mathbf{y}}\right)\mathbf{f} =: \mathbf{f}^1, \\
\mathbf{y}''' &= \left(\frac{\partial \mathbf{f}^1}{\partial \mathbf{y}}\right)\mathbf{f} =: \mathbf{f}^2, \\
&\vdots \\
\mathbf{y}^{(k)} &= \left(\frac{\partial \mathbf{f}^{k-2}}{\partial \mathbf{y}}\right)\mathbf{f} =: \mathbf{f}^{k-1}.
\end{aligned}$$

(Note: $\mathbf{f}^j$ is *not* the jth power of $\mathbf{f}$.) By Taylor's expansion at $\mathbf{y} = \mathbf{y}_{n-1}$,

$$\begin{aligned}
\mathbf{y}(t_n) &= \mathbf{y} + h\mathbf{y}' + \frac{h^2}{2}\mathbf{y}'' + \cdots + \frac{h^{p+1}}{(p+1)!}\mathbf{y}^{(p+1)} + \cdots \\
&= \mathbf{y} + h\mathbf{f} + \frac{h^2}{2}\mathbf{f}^1 + \cdots + \frac{h^{p+1}}{(p+1)!}\mathbf{f}^p + \cdots .
\end{aligned}$$

For an s-stage Runge–Kutta method (4.8), substituting $\mathbf{y}(t)$ into the difference equations gives that in order to obtain a method of order p we must have

$$\sum_{i=1}^{s} b_i \mathbf{f}(\mathbf{Y}_i) = \mathbf{f} + \frac{h}{2}\mathbf{f}^1 + \cdots + \frac{h^{p-1}}{p!}\mathbf{f}^{p-1} + O(h^p).$$

A Taylor expansion of the $\mathbf{f}(\mathbf{Y}_i)$ therefore follows.

Although this is conceptually simple, there is an explosion of terms to match which is severe for higher-order methods. An elegant theory involving *trees* for enumerating these terms was developed by J. C. Butcher, in a long series of papers starting in the mid-1960s. The details are complex, though, and do not yield a methodology for designing a method of a desirable order, only for checking the order of a given one.

Instead, we proceed to derive simple, necessary order conditions, to get the hang of it. The idea is that the essence of a method's accuracy is often captured by applying our analysis to very simple equations.

Necessary Order Conditions

Consider the scalar ODE

$$y' = y + t^{l-1}, \qquad t \geq 0, \tag{4.12}$$

l a positive integer, with $y(0) = 0$, for the first step (i.e., $t_{n-1} = y_{n-1} = 0$). Then

$$Y_i = h \sum_{j=1}^{s} a_{ij}(Y_j + (hc_j)^{l-1}).$$

We can write this in matrix form

$$(I - hA)\mathbf{Y} = h^l AC^{l-1}\mathbf{1},$$

where A is the $s \times s$ coefficient matrix from the tableau defining the Runge–Kutta method, $\mathbf{Y} = (Y_1, \ldots, Y_s)^T$, $C = \mathrm{diag}\{c_1, \ldots, c_s\}$ is the diagonal matrix with the coefficients c_j on its diagonal, and $\mathbf{1} = (1, 1, \ldots, 1)^T$. It follows that $\mathbf{Y} = h^l (I - hA)^{-1} AC^{l-1}\mathbf{1}$ and

$$\begin{aligned} y_n &= h \sum_{i=1}^{s} b_i (Y_i + (hc_i)^{l-1}) \\ &= h^l \mathbf{b}^T [I + hA + \cdots + h^k A^k + \cdots] C^{l-1}\mathbf{1}, \end{aligned}$$

where $\mathbf{b}^T = (b_1, b_2, \ldots, b_s)$. Now we compare the two expansions for y_n and for $y(t_n)$ and equate equal powers of h. For the exact solution $y(t) = \int_0^t e^{t-s} s^{l-1} ds$, we have

$$y(0) = \cdots = y^{(l-1)}(0) = 0, \quad y^{(l+j)}(0) = (l-1)!, \quad j \geq 0.$$

This yields that for the method to be of order p the following order conditions must be satisfied:

$$\mathbf{b}^T A^k C^{l-1}\mathbf{1} = \frac{(l-1)!}{(l+k)!} = \frac{1}{l(l+1)\cdots(l+k)}, \qquad 1 \leq l + k \leq p. \tag{4.13}$$

(The indices run as follows: for each l, $1 \leq l \leq p$, we have order conditions for $k = 0, 1, \ldots, p - l$.)

In component form the order conditions (4.13) read

$$\sum_{i, j_1, \ldots, j_k} b_i a_{i,j_1} a_{j_1,j_2} \cdots a_{j_{k-1},j_k} c_{j_k}^{l-1} = \frac{(l-1)!}{(l+k)!}.$$

The vector form is not only more compact, though; it is also easy to program.

We next consider two simple subsets of these order conditions. Setting $k = 0$ in (4.13), we obtain the pure quadrature order conditions

$$\mathbf{b}^T C^{l-1}\mathbf{1} = \sum_{i=1}^{s} b_i c_i^{l-1} = \frac{1}{l}, \qquad l = 1, 2, \ldots, p. \tag{4.14}$$

Note that the coefficients a_{ij} of the Runge–Kutta method do not appear in (4.14). Next, setting $l = 1$ in (4.12) and $k \leftarrow k + 1$ in (4.13), we obtain that for the method to be of order p the following order conditions must be satisfied:

$$\mathbf{b}^T A^{k-1}\mathbf{1} = \frac{1}{k!}, \qquad k = 1, 2, \ldots, p. \tag{4.15}$$

These conditions are really additional to (4.14) only when $k \geq 3$, because $A\mathbf{1} = \mathbf{c}$, so $A^{k-1}\mathbf{1} = A^{k-2}\mathbf{c}$.

The leading term of the local truncation error for the ODE (4.12) with $l = 1$ is

$$d_n \approx h^p \left(\mathbf{b}^T A^p \mathbf{1} - \frac{1}{(p+1)!} \right).$$

For explicit Runge–Kutta methods, A is strictly lower triangular, hence $A^j = 0$ for all $j \geq s$. This immediately leads to the following conclusions.

- An explicit Runge–Kutta method can have at most order s; i.e., $p \leq s$.
- If $p = s$ then the leading term of the local truncation error for the test equation (4.12) with $l = 1$ cannot be reduced by any choice of coefficients of the explicit Runge–Kutta method.

Example 4.2
Consider all explicit two-stage Runge–Kutta methods

$$\begin{array}{c|cc} 0 & 0 & 0 \\ c_2 & c_2 & 0 \\ \hline & b_1 & b_2 \end{array}$$

For $l = 1$, $k = 0$, condition (4.13) reads $b_1 + b_2 = 1$. For $l = 1$, $k = 1$, we have $b_2 c_2 = 1/2$. The condition for $l = 2$, $k = 0$ is the same. Denoting $\alpha = c_2$ results in the family of two-stage, second-order methods displayed in the previous section. For the choice of the parameter α, we can minimize the local truncation error for the quadrature test equation $y' = t^p$. It is

$$d_n \approx h^p \left(\sum_{i=1}^{s} b_i c_i^p - \frac{1}{p+1} \right).$$

Trying to achieve $b_2 c_2^2 = 1/3$ gives the choice $\alpha = 2/3, b_2 = 3/4$. But this choice does nothing special for the ODE $y' = y$, for instance. ♦

The obtained set of order conditions (4.13) (recall also (4.9)) is certainly necessary for the method to have order p. These order conditions are **not sufficient** in general! Still, they can be used to find a simple upper bound on the order of a given method (Exercise 4.4), and also for the purpose of designing new Runge–Kutta methods. In fact, often the order is already determined by the conditions (4.14) plus (4.15) alone.

Example 4.3
We can now view the classical Runge–Kutta method as a result of a methodical design process. The starting point is an attempt to extend Simpson's quadrature rule, which is fourth order. Although Simpson's rule has only three abscissae, $0, 1/2$, and 1, we know already that a method of order 4 will not result from only three stages, so we settle for four stages and choose abscissae $c_i = 0, 1/2, 1/2, 1$. Next, we must have $\mathbf{b}^T A^3 \mathbf{1} = b_4 a_{21} a_{32} a_{43} = 1/24$ from (4.15). In particular, we must choose $a_{i+1,i} \neq 0$. The simplest choice then is to set the rest of the a_{ij} to 0, yielding $c_{i+1} = a_{i+1,i}$, $i = 1, 2, 3$. The choice $b_1 = b_4 = 1/6$ is as in Simpson's rule and results from the quadrature conditions (4.14) alone. The final choice $b_2 = b_3$ is determined by the condition (4.15) with $k = 3$. This completes the definition of the method. Its order does turn out to be $p = 4$. ♦

Example 4.4
The simple necessary order conditions (4.14), (4.15) give an upper bound on a method's order which turns out to agree with the order for many if not all the methods in actual use. However, counterexamples where these conditions are not sufficient can be constructed. For explicit Runge–Kutta methods of order p with p stages, these conditions are sufficient for $p = 1$, 2, and 3. One is tempted to conclude that this is always true, as the famous joke goes, "by induction." For $p = 4$, however, there are two additional conditions: $b_3 a_{32} c_2^2 + b_4(a_{42} c_2^2 + a_{43} c_3^2) = 1/12$ and $b_3 c_3 a_{32} c_2 + b_4 c_4 (a_{42} c_2 + a_{43} c_3) = 1/8$. The first of these is covered by (4.13), but the second, which is in the form $\mathbf{b}^T CAC\mathbf{1} = 2/4!$, is not. Together with (4.14) and (4.15) these conditions imply in particular that we must choose $c_4 = 1$ (Exercise 4.5). But the conditions (4.13) alone do not imply this. A particular example where these conditions are not sufficient is

$$
\begin{array}{c|cccc}
0 & 0 & 0 & 0 & 0 \\
\frac{1}{4} & \frac{1}{4} & 0 & 0 & 0 \\
\frac{1}{2} & 0 & \frac{1}{2} & 0 & 0 \\
\frac{3}{4} & 0 & \frac{1}{4} & \frac{1}{2} & 0 \\
\hline
 & 0 & \frac{2}{3} & -\frac{1}{3} & \frac{2}{3}
\end{array}
$$

♦

Example 4.5
For the sake of completeness, we list below the full set of conditions that must hold for a method to have order at least 5, in addition to those conditions already necessary for order 4. (The order-4 conditions are discussed in the previous Example 4.4. For a higher order we recommend that the reader

consult a more in-depth book.) The first four of these are included in (4.13).

$$\begin{aligned}
\mathbf{b}^T C^4 \mathbf{1} &= \frac{1}{5}, \\
\mathbf{b}^T A^4 \mathbf{1} &= \frac{1}{120}, \\
\mathbf{b}^T A^2 C^2 \mathbf{1} &= \frac{1}{60}, \\
\mathbf{b}^T A C^3 \mathbf{1} &= \frac{1}{20}, \\
\mathbf{b}^T C^2 A C \mathbf{1} &= \frac{1}{10}, \\
\mathbf{b}^T C A C^2 \mathbf{1} &= \frac{1}{15}, \\
\mathbf{b}^T C A^2 C \mathbf{1} &= \frac{1}{30}, \\
\mathbf{b}^T A C A C \mathbf{1} &= \frac{1}{40}, \\
\sum_{i,j,k} b_i a_{ij} c_j a_{ik} c_k &= \frac{1}{20}.
\end{aligned}$$

♦

Finally, we return to the question: what is the maximal attainable order p by an *explicit* s-stage Runge–Kutta method? This question turns out to have a complicated answer. Given (4.9), the number of coefficients in an explicit s-stage method is $s(s+1)/2$, but it does not bear a simple relationship to the number of independent order conditions. Indeed, one often encounters families of "eligible" methods for a given order and number of stages in practice. Still, we have the following limitations on the attainable order as a function of the number of stages:

number of stages	1	2	3	4	5	6	7	8	9	10
attainable order	1	2	3	4	4	5	6	6	7	7

This explains in part why the fourth-order explicit Runge–Kutta method is so popular (especially when no adaptive error control is contemplated).

4.4 Regions of Absolute Stability for Explicit Runge–Kutta Methods

In this section we investigate the regions of absolute stability for explicit Runge–Kutta methods. To recall, this region is obtained for a given method by determining for what values of $z = h\lambda$ we get $|y_n| \le |y_{n-1}|$ when applying the method to the test equation[9]

$$y' = \lambda y. \tag{4.16}$$

This test equation is an obvious generalization of (4.12) without the inhomogeneity. Repeating the same arguments here, we obtain

$$\begin{aligned} y_n &= [1 + z\mathbf{b}^T(I - zA)^{-1}\mathbf{1}]y_{n-1} \\ &= [1 + z\mathbf{b}^T(I + zA + \cdots + z^k A^k + \cdots)\mathbf{1}]y_{n-1}. \end{aligned} \tag{4.17}$$

Substituting (4.15) into this expression and writing (4.17) as

$$y_n = R(z)y_{n-1},$$

we get for a Runge–Kutta method of order p

$$R(z) = 1 + z + \frac{z^2}{2} + \cdots + \frac{z^p}{p!} + \sum_{j>p} z^j \mathbf{b}^T A^{j-1}\mathbf{1}. \tag{4.18}$$

For an s-stage explicit method of order p, since $A^{j-1} = 0$, $j > s$, we get

$$y_n = \left[1 + z + \frac{z^2}{2} + \cdots + \frac{z^p}{p!} + \sum_{j=p+1}^{s} z^j \mathbf{b}^T A^{j-1}\mathbf{1}\right] y_{n-1}.$$

In particular, the region of absolute stability of an explicit pth-order Runge–Kutta method for $s = p \le 4$ is given by

$$\left|1 + h\lambda + \frac{(h\lambda)^2}{2} + \cdots + \frac{(h\lambda)^p}{p!}\right| \le 1. \tag{4.19}$$

Thus we note that all p-stage explicit Runge–Kutta methods of order p have the same region of absolute stability. For an s-stage method with order $p < s$, the absolute stability region is seen to depend somewhat on the method's coefficients.[10]

The stability regions for the explicit p-stage pth-order Runge–Kutta methods, $1 \le p \le 4$, are shown in Figure 4.4.

How do you plot a region of absolute stability? Recall that the numbers of modulus 1 in the complex plane are represented by $e^{i\theta}$, for $0 \le \theta \le 2\pi$. The stability condition is given by $|R(z)| \le 1$, where $R(z)$ is given by (4.18).

[9] We must consider λ, hence z, to be a complex number, because it represents an eigenvalue of a matrix in general.

[10] For the fourth- (and to a lesser extent, the third-) order methods depicted in Figure 4.4 there is a stretch along the imaginary axis of z where $|R(z)| < 1$. This translates to dissipativity when such methods are used to construct finite difference approximations to hyperbolic PDEs, and it facilitates using such methods as smoothers when designing multigrid solvers for certain PDEs. No such effect occurs for lower-order discretizations with $p = s$. A full discussion of this is well beyond the scope of this book.

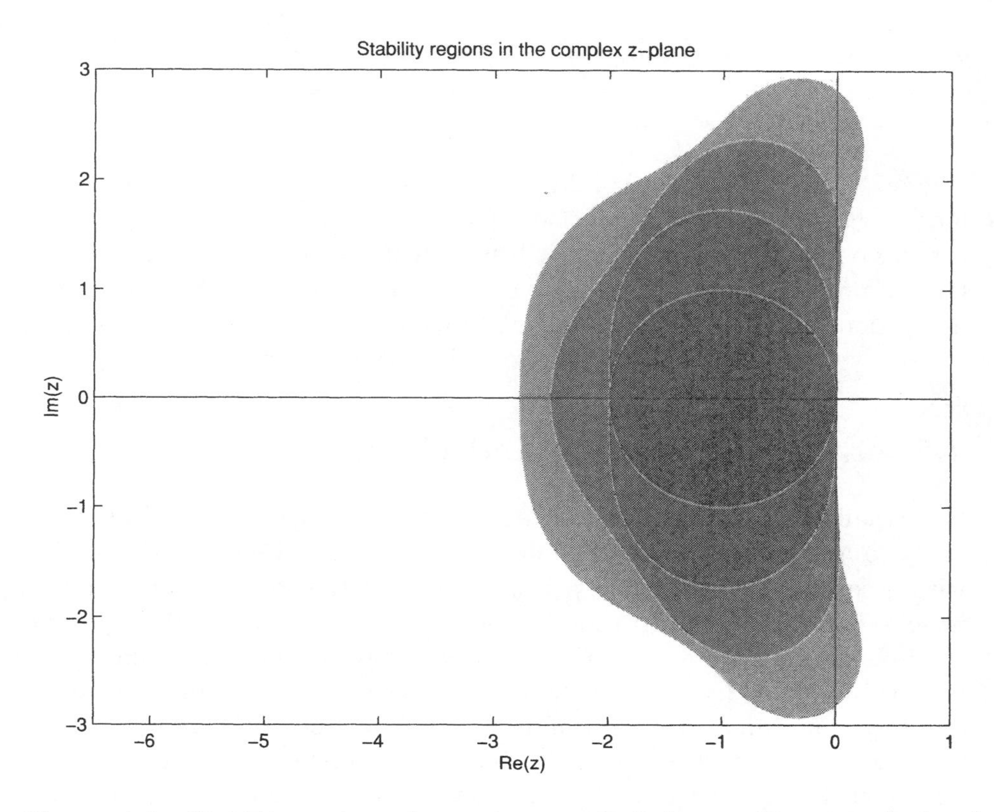

Figure 4.4: *Stability regions for p-stage explicit Runge–Kutta methods of order p, $p = 1, 2, 3, 4$. The inner circle corresponds to forward Euler, $p = 1$. The larger p is, the larger the stability region. Note the "ear lobes" of the fourth-order method protruding into the right half-plane.*

For explicit Runge–Kutta methods $R(z)$ is a polynomial in $z = h\lambda$ given, e.g., by the expression whose magnitude appears in (4.19). Thus, to find the boundary of the region of absolute stability, we find the roots $z(\theta)$ of

$$R(z) = e^{i\theta}$$

for a sequence of θ values. Starting with $\theta = 0$, for which $z = 0$, we repeatedly increase θ by a small increment, each time applying a root finder to find the corresponding z, starting from z of the previous θ as a first guess,[11] until the stability boundary curve returns to the origin.

It is also possible to compute the region of absolute stability via a brute force approach. To do this, we first form a grid over a large part of the complex plane including the origin. Then at each mesh point z_{ij}, if $|R(z_{ij})| < 1$, we mark z_{ij} as being inside the stability region.

Finally, we note that no explicit Runge–Kutta method can have an unbounded region of absolute stability. This is because all Runge–Kutta meth-

[11]This is an elementary example of a *continuation* method.

ods applied to the test equation $y' = \lambda y$ yield

$$y_n = R(z)y_{n-1}, \quad z = h\lambda,$$

where $R(z)$ is a polynomial of degree s. Since $|R(z)| \to \infty$ as $|z| \to \infty$, very large negative values of z cannot be in the region of absolute stability. In fact, it turns out that all known explicit Runge–Kutta methods are inappropriate for stiff problems, and we are led to consider implicit Runge–Kutta methods in Section 4.7. Before that, though, we discuss some of the ingredients necessary to write robust software for initial value ODEs.

4.5 Error Estimation and Control

In virtually all modern codes for ODEs, the step size is selected automatically to achieve both reliability and efficiency. Any discretization method with a constant step size will perform poorly if the solution varies rapidly in some parts of the integration interval and slowly in other, large parts of the integration interval, and if it is to be resolved well everywhere by the numerical method (see Exercise 4.12). In this section we will investigate several ways to estimate the error and select the next step $h = h_n = t_n - t_{n-1}$. Since we strive to keep the entire integration process local in time (i.e., we march in time with all the information locally known) we attempt to control the local error or the local truncation error, rather than the global error. Basically, by specifying an error tolerance ETOL a user can require a more accurate (and more expensive) approximate solution or a less accurate (and cheaper) one. Our step-size selection strategy may attempt to roughly equate the errors made at each step, e.g.,

$$|\mathbf{l}_n| \approx \text{ETOL},$$

where $\mathbf{l}_n$ is the local error.[12] (The vector $\mathbf{l}_n$ has m components for a system of m first order ODEs.) This makes the step size as large as possible, but to achieve a higher success rate in such step-size predictions, we typically use some fraction of ETOL for safety. The global error also relates to the tolerance in case it can be obtained as a simple sum of local errors.

If the components of the solution $\mathbf{y}$ are very different in magnitude, then we are better off to consider an array of tolerances. In fact, for each component j of $\mathbf{y}$ $(1 \le j \le m)$, it may be necessary to specify an absolute error tolerance ATOL_j, in addition to a common relative error tolerance RTOL. One then wants to choose h so that for each j, $1 \le j \le m$,

$$|(l_j)_n| \le \text{frac}\,[\text{ATOL}_j + |(y_j)_n|\,\text{RTOL}],$$

[12]Recall from (3.14) that the local truncation error in the nth time step is related to the local error by $h_n(|\mathbf{d}_n| + O(h^{p+1})) = |\mathbf{l}_n|(1 + O(h_n))$. Thus, local error control and step-size selection are sometimes viewed as controlling the local truncation error.

where frac is a safety fraction (say, frac = .9). Good codes allow the specification of $m+1$ tolerances as well as a default option of specifying only one or two.

Let us next assume again a scalar ODE, for notational simplicity. A basic problem for estimating the step size in Runge–Kutta methods is that the expression for the local truncation error is so complicated. For example, the local truncation error of the two-stage family of explicit second-order methods derived earlier for a scalar ODE is given by

$$hd_n = \frac{h^3}{6}\left[\frac{3}{4\gamma}\left(f_{yy}f^2 + 2f_{ty}f + f_{tt}\right) - y'''\right] + O(h^4).$$

Since the whole purpose of Runge–Kutta methods was to eliminate the need to calculate the symbolic partial derivatives explicitly, we will look for methods to estimate the error at each step which do not use the partial derivatives directly. For this reason, it is convenient in the case of Runge–Kutta methods to estimate the local error, rather than the local truncation error.

The essential idea of the methods described below is to calculate two approximate solutions y_n and $\hat{y}_n$ at t_n, such that $\hat{y}_n - y_n$ gives an estimate of the local error of the *less accurate* of the two approximate solutions, y_n. We can then check if $|\hat{y}_n - y_n| \le$ ETOL. If this inequality is not satisfied then the step h is rejected and another step $\tilde{h}$ is selected instead. If the method for finding y_n has order p, then $l_n(\tilde{h}) \approx c\tilde{h}^{p+1}$, so we choose $\tilde{h}$ to satisfy

$$\left(\frac{\tilde{h}}{h}\right)^{p+1} |\hat{y}_n - y_n| \approx \text{frac ETOL}$$

and repeat the process until an acceptable step size is found.[13] If the step is accepted then the same formula can be used to predict a larger step size $h_{n+1} = \tilde{h}$ for the next time step.

Embedded Methods

We are searching, then, for methods which deliver two approximations, y_n and $\hat{y}_n$, at t_n. A pair of Runge–Kutta methods of orders p and $p+1$, respectively, will do the job. The key idea of embedded methods is that such a pair will share stage computations. Thus we seek to derive an s-stage formula of order $p+1$ such that there is another formula of order p embedded inside it (therefore, using the same function evaluations).

If the original method is given by

$$\begin{array}{c|c} \mathbf{c} & A \\ \hline & \mathbf{b}^T \end{array}$$

[13]Another safety factor, ensuring that $\tilde{h}/h$ is neither too large nor too small, is used in practice, because we are using a simplified model for the error which does not take into account large h, roundoff error, and absolute stability effects.

then the embedded method is given by

$$\begin{array}{c|c} \mathbf{c} & A \\ \hline & \hat{\mathbf{b}}^T \end{array}$$

We therefore use a combined notation for an embedded method:

$$\begin{array}{c|c} \mathbf{c} & A \\ \hline & \mathbf{b}^T \\ \hline & \hat{\mathbf{b}}^T \end{array}$$

The simplest example is forward Euler embedded in a modified trapezoid:

$$\begin{array}{c|cc} 0 & 0 & 0 \\ 1 & 1 & 0 \\ \hline & 1 & 0 \\ \hline & \frac{1}{2} & \frac{1}{2} \end{array}$$

Probably the most famous embedded formula is the Fehlberg 4(5) pair. It has six stages and delivers a method of order 4 with an error estimate (or a method of order 5 without):

$$\begin{array}{c|cccccc} 0 & & & & & & \\ \frac{1}{4} & \frac{1}{4} & & & & & \\ \frac{3}{8} & \frac{3}{32} & \frac{9}{32} & & & & \\ \frac{12}{13} & \frac{1932}{2197} & -\frac{7200}{2197} & \frac{7296}{2197} & & & \\ 1 & \frac{439}{216} & -8 & \frac{3680}{513} & -\frac{845}{4104} & & \\ \frac{1}{2} & -\frac{8}{27} & 2 & -\frac{3544}{2565} & \frac{1859}{4104} & -\frac{11}{40} & \\ \hline & \frac{25}{216} & 0 & \frac{1408}{2565} & \frac{2197}{4104} & -\frac{1}{5} & 0 \\ \hline & \frac{16}{135} & 0 & \frac{6656}{12825} & \frac{28561}{56430} & -\frac{9}{50} & \frac{2}{55} \end{array}$$

The Fehlberg 4(5) pair.

Note that we have omitted obvious 0's in the tableau. The somewhat unintuitive coefficients of the Fehlberg pair arise not only from satisfying the order conditions but also from an attempt to minimize the local error in y_n.

With any error estimate, a question arises as to whether one should add the estimate to the solution to produce a more accurate method (but now with no close error estimate). Here this would simply mean using $\hat{y}_n$ rather than y_n for the start of the next step. Of course, this casts doubt on the quality of the error estimation, but users rarely complain when a code provides more accuracy than requested. Besides, the quality of ETOL as an actual error estimate is questionable in any case, because it does not directly relate to the actual, global error (more on this later). This strategy, called *local extrapolation*, has proven to be successful for some methods for nonstiff problems, and all quality explicit Runge–Kutta codes use it, but it is not as common in the solution of stiff problems.

The methods of Dormand and Prince bite the bullet. They are designed to minimize the local error in $\hat{y}_n$, in anticipation that the latter will be used for the next step. The 4(5) pair given below has seven stages, but the last stage is the same as the first stage for the next step ($\hat{y}_n = Y_7$, and in the next step $n - 1 \leftarrow n$, $Y_1 = y_{n-1}$, so Y_7 at the current step and Y_1 at the next step are the same), so this method has the cost of a six-stage method.

$$
\begin{array}{c|ccccccc}
0 & & & & & & & \\
\frac{1}{5} & \frac{1}{5} & & & & & & \\
\frac{3}{10} & \frac{3}{40} & \frac{9}{40} & & & & & \\
\frac{4}{5} & \frac{44}{45} & -\frac{56}{15} & \frac{32}{9} & & & & \\
\frac{8}{9} & \frac{19372}{6561} & -\frac{25360}{2187} & \frac{64448}{6561} & -\frac{212}{729} & & & \\
1 & \frac{9017}{3168} & -\frac{355}{33} & \frac{46732}{5247} & \frac{49}{176} & -\frac{5103}{18656} & & \\
1 & \frac{35}{384} & 0 & \frac{500}{1113} & \frac{125}{192} & -\frac{2187}{6784} & \frac{11}{84} & \\
\hline
 & \frac{5179}{57600} & 0 & \frac{7571}{16695} & \frac{393}{640} & -\frac{92097}{339200} & \frac{187}{2100} & \frac{1}{40} \\
\hline
 & \frac{35}{384} & 0 & \frac{500}{1113} & \frac{125}{192} & -\frac{2187}{6784} & \frac{11}{84} & 0
\end{array}
$$

The Dormand–Prince 4(5) *pair.*

Note: for stiff problems, the stability properties of a method and its embedded pair should be similar; see Exercise 5.7.

Step Doubling

The idea behind step doubling is simple. By subtracting the solution obtained with two steps of size $h_n = h$ from the solution obtained using one step of size $2h$, we obtain an estimate of the local error. Since we know the form of the local error as $h \to 0$, we can estimate it well. To make this more precise, write the local error (recall (3.12)–(3.13)) as

$$l_n = \psi(t_n, y(t_n))h^{p+1} + O(h^{p+2}).$$

The function ψ is called the *principal error function*. Now, let y_n be the solution using two steps of size h starting from y_{n-2}, and let $\tilde{y}_n$ be the solution taking one step of size $2h$ from y_{n-2}. Then the two local errors satisfy

$$\begin{aligned} 2l_n(h) &= 2h^{p+1}\psi + O(h^{p+2}), \\ l_n(2h) &= (2h)^{p+1}\psi + O(h^{p+2}), \end{aligned}$$

where we have assumed that the local error after two steps is twice the local error after one step. (This is true in the limit as $h \to 0$.) Then

$$|\tilde{y}_n - y_n| \approx 2h^{p+1}(2^p - 1)|\psi(t_n, y_n)| + O(h^{p+2}).$$

Thus,

$$2|l_n| \approx 2h^{p+1}|\psi(t_n, y(t_n))| \approx |\tilde{y}_n - y_n|/(2^p - 1).$$

Although step doubling gives an accurate local error estimate, it is more expensive per step, especially for stiff problems. The embedded error estimates are cheaper, especially if the importance of an accurate local error estimate is discounted. The step doubling procedure is general, though, and works without inventing special embedded pair formulae.

Global Error

A similar step doubling procedure can be applied to estimate the *global error*. Here, after a sequence of steps for integrating the ODE over a given interval has been chosen and the integration carried out (say, by an embedded Runge–Kutta pair), the integration is repeated with each of these steps halved using the same discretization method. The above step doubling error estimation procedure is repeated, this time for all n, to obtain a global error estimate.

There are other procedures for estimating the global error, but like the one just described, they are all nonlocal, and as such are much more cumbersome than the local procedures used above. When solving BVPs one is naturally dealing with the global error anyway, so typical practical procedures estimate it. For initial value problems (IVPs), one would like to avoid estimating the global error if possible, for the reasons indicated before. However, there are applications where an estimate of the global error is required.

Example 4.6
While the global error often behaves like $|\mathbf{e}_n| \approx \max_j |\mathbf{d}_j|$, there are exceptions. Consider

$$y' = \lambda(y - \sin t) + \cos t, \qquad t \geq 0,$$

with $y(0) = 0$ and $\lambda = 50$, say. Here the exact solution $y(t) = \sin t$ is smooth and nicely bounded, and this is what local errors and locally based step-size selection relate to. But globally, the error accumulates roughly like $|e_n| \approx e^{50t_n} \max_j |d_j|$. Therefore, the actual, global error at $t = 1$, say, will be much poorer than the local error, and will not relate well to a user-given local error tolerance.

Fortunately, examples like this are rare in practice, or so one hopes (see Exercise 4.18). ♦

While local error estimates are more easily obtained and they allow a more robust, dynamical error control and step-size selection, satisfying a global error tolerance is typically closer to what the user (i.e., the person looking for a solution to a particular application) may want. But how does a user go about specifying a global error tolerance? And how accurately need the error tolerance(s) be specified and satisfied?

These questions arise in modeling and depend strongly on the application. Here we just note that a precise error bound is often unknown and not really needed. When a ballpark value for the global error would do, the stock value of a local error tolerance goes up.

4.6 Sensitivity to Data Perturbations

One important factor in assessing the choice of error tolerance for a given application is the accuracy expected of the *exact solution*. Real-world models often involve various parameters and initial data whose values are determined by inaccurate means. The exact solution of the given initial value ODE system may therefore be viewed as a sample out of a cluster of trajectories. It makes no sense, then (it is a waste of resources), to impose an error tolerance so strict that the computed solution is much closer to the exact solution than this exact solution trajectory is to its equally valid neighbor trajectories.

So, to assess solution accuracy (and worth) in practice a user often needs a *sensitivity analysis*; i.e., we ask by how much the exact solution changes when the data are perturbed. Below we consider the relatively simple case of *small* perturbations.

To be specific, let us consider an IVP depending on parameters

$$\begin{aligned} \mathbf{y}' &= \mathbf{f}(t, \mathbf{y}, \mathbf{p}), \qquad 0 < t < b, \\ \mathbf{y}(0) &= \mathbf{c}. \end{aligned} \tag{4.20}$$

The l parameters $\mathbf{p}$ can be functions of t, but for simplicity we assume they are all given constants. Denote the exact solution of (4.20) by $\mathbf{y}(t)$. We next consider a perturbation vector $\boldsymbol{\phi}$, i.e.,

$$\bar{\mathbf{p}} = \mathbf{p} + \boldsymbol{\phi}.$$

Call the resulting solution of (4.20) (i.e., with $\bar{\mathbf{p}}$ replacing $\mathbf{p}$) $\bar{\mathbf{y}}(t)$. We seek a bound on $\|\mathbf{y} - \bar{\mathbf{y}}\|$ in terms of $|\boldsymbol{\phi}|$ when $|\boldsymbol{\phi}|$ is so small that $O(|\boldsymbol{\phi}|^2)$ terms can be considered negligible.

Thus we write

$$\bar{\mathbf{y}}(t) = \bar{\mathbf{y}}(t; \mathbf{p} + \boldsymbol{\phi}) \approx \mathbf{y}(t; \mathbf{p}) + \left(\frac{\partial \mathbf{y}(t; \mathbf{p})}{\partial \mathbf{p}}\right) \boldsymbol{\phi}$$

and obtain

$$|\bar{\mathbf{y}}(t) - \mathbf{y}(t)| \leq |P(t)\boldsymbol{\phi}| + O(|\boldsymbol{\phi}|^2), \qquad 0 \leq t \leq b, \tag{4.21a}$$

where

$$P = \frac{\partial \mathbf{y}}{\partial \mathbf{p}} \tag{4.21b}$$

is an $m \times l$ matrix function. The simplest form of sensitivity analysis therefore consists of approximately calculating $P(t)$. Then, given bounds on the parameter variation

$$|\phi_j| \leq \phi_j^U, \qquad 1 \leq j \leq l,$$

we can determine for each t

$$\eta_i = \max \left| \sum_{j=1}^{l} P_{i,j} \phi_j \right| \leq \sum_{j=1}^{l} |P_{i,j}| \phi_j^U ,$$

giving the approximate bound

$$|\bar{\mathbf{y}}(t) - \mathbf{y}(t)| \leq \boldsymbol{\eta}(t) + O(|\boldsymbol{\phi}^U|^2), \qquad 0 \leq t \leq b. \tag{4.22}$$

The behavior of the perturbation matrix function $P(t)$ is governed by a linear initial value ODE. Differentiating (4.20) with respect to $\mathbf{p}$ and noting that the initial conditions are assumed to be independent of $\mathbf{p}$, we obtain

$$\begin{aligned} P' &= \left(\frac{\partial \mathbf{f}}{\partial \mathbf{y}}\right) P + \frac{\partial \mathbf{f}}{\partial \mathbf{p}}, \qquad 0 < t < b, \\ P(0) &= 0. \end{aligned} \tag{4.23}$$

For each column of P (corresponding to one parameter in $\mathbf{p}$), we therefore have an IVP which depends on $\mathbf{y}(t)$ but is linear in P. Thus, in order to estimate the perturbation function, in practice we may solve (4.20) with a relatively permissive error tolerance, and compute P by integrating (4.23) as well, using the same time step sequence. The combined system can be solved efficiently, noting that the sensitivity system (4.23) is linear and shares the iteration matrix of the original system (4.20), and exploiting this structure.

Before turning to an example, we remark that a similar treatment can be applied to assess solution sensitivity with respect to perturbations in the initial data $\mathbf{c}$. This is left to Exercise 6.4.

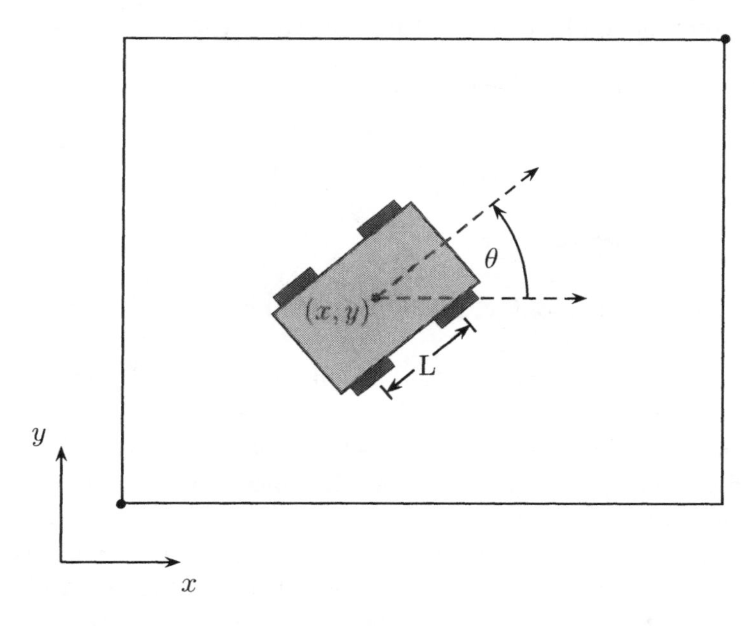

Figure 4.5: *Schematic of a toy car (not to scale).*

Example 4.7
A simplified description of the motion of a car in an arena is given by the equations corresponding to the ODE in (4.20),

$$
\begin{aligned}
x' &= v\cos\theta, && (4.24a)\\
y' &= v\sin\theta, && (4.24b)\\
\theta' &= v\frac{\tan\psi}{L}, && (4.24c)\\
v' &= a - \gamma v, && (4.24d)
\end{aligned}
$$

where x and y are the Cartesian coordinates of the car's center, θ is the angle the car makes with the x-axis (see Figure 4.5), and v is the velocity with which the car is moving. Denote $\mathbf{y} = (x, y, \theta, v)^T$.

The damping (friction) factor γ and the car's length L are given: we take $\gamma = 1/6$ and $L = 11.5$ cm (it's a toy car, representative of a mobile robot). The acceleration a and the steering angle ψ that the front wheels make with the car's body are two functions which one normally controls in order to drive a car. Here we take them, for simplicity, as constant parameters: we set $a = 100$ cm s^{-2}, $\psi = 1$. We are interested in the sensitivity of the car's position (x, y) with respect to a constant change in ψ.

Since we are checking sensitivity with respect to only one parameter, $P(t)$ is a vector of length 4. The differential equations (4.23) for $P(t)$ are found by differentiating (4.24) with respect to the parameter ψ, to obtain

$$
\begin{aligned}
P_1' &= -vP_3 \sin\theta + P_4 \cos\theta, \\
P_2' &= vP_3 \cos\theta + P_4 \sin\theta, \\
P_3' &= \left[P_4 \tan\psi + \frac{v}{\cos^2\psi} \right] / L, \\
P_4' &= -\gamma P_4 .
\end{aligned}
$$

Note that P depends on $\mathbf{y}$ but $\mathbf{y}$ does not depend on P, and that the ODE for P is linear, given $\mathbf{y}$.

We use MATLAB to compute the solutions for $\mathbf{y}(t)$ and $P(t)$, starting with $\mathbf{y}(0) = (10, 10, 1, 0)^T$, $P(0) = \mathbf{0}$. We evaluate $\bar{\mathbf{y}}_{\pm}(t) = \mathbf{y}(t) \pm \phi P(t)$, and we also numerically solve (4.20) directly for the perturbed problems, where ψ is replaced by $\psi \pm \phi$. The resulting plots for $\phi = 0.01$ and for $\phi = 0.05$ are given in Figure 4.6.

We see that for $\phi = 0.01$ the linear sensitivity analysis captures the trajectory perturbation rather well. Also, not surprisingly for anyone who drives, the distance between $\mathbf{y}(t)$ and the perturbed trajectories increases with t. As the size of the perturbation is increased to $\phi = 0.05$, the linear approximation becomes less valid. ♦

Sensitivity analysis plays an important role in a number of situations, in addition to assessing the accuracy of a model. In model development and model reduction, the sensitivity of the solution with respect to perturbations in the parameters is often used to help make decisions on which parts of the model are actively contributing in a given setting. Partial derivative matrices similar to those occurring in sensitivity analysis also arise in the shooting and multiple shooting methods for BVPs (see Chapter 7), as well as in parameter estimation, design optimization, and optimal control.

4.7 Implicit Runge–Kutta and Collocation Methods

Compared to explicit Runge–Kutta methods, for implicit Runge–Kutta methods there are many more parameters to choose in (4.8) or (4.10). Thus, we might expect to be able to attain a higher order for a given number of stages. This turns out to be the case, as we have already seen in the implicit midpoint method, which is a one-stage method of order 2. Moreover, the amplification function $R(z)$ is no longer a polynomial. This enables the construction of implicit Runge–Kutta methods which are appropriate for the solution of stiff systems.

Many of the most commonly used implicit Runge–Kutta methods are based on quadrature methods; that is, the points at which the intermediate stage approximations are taken are the same points used in certain classes

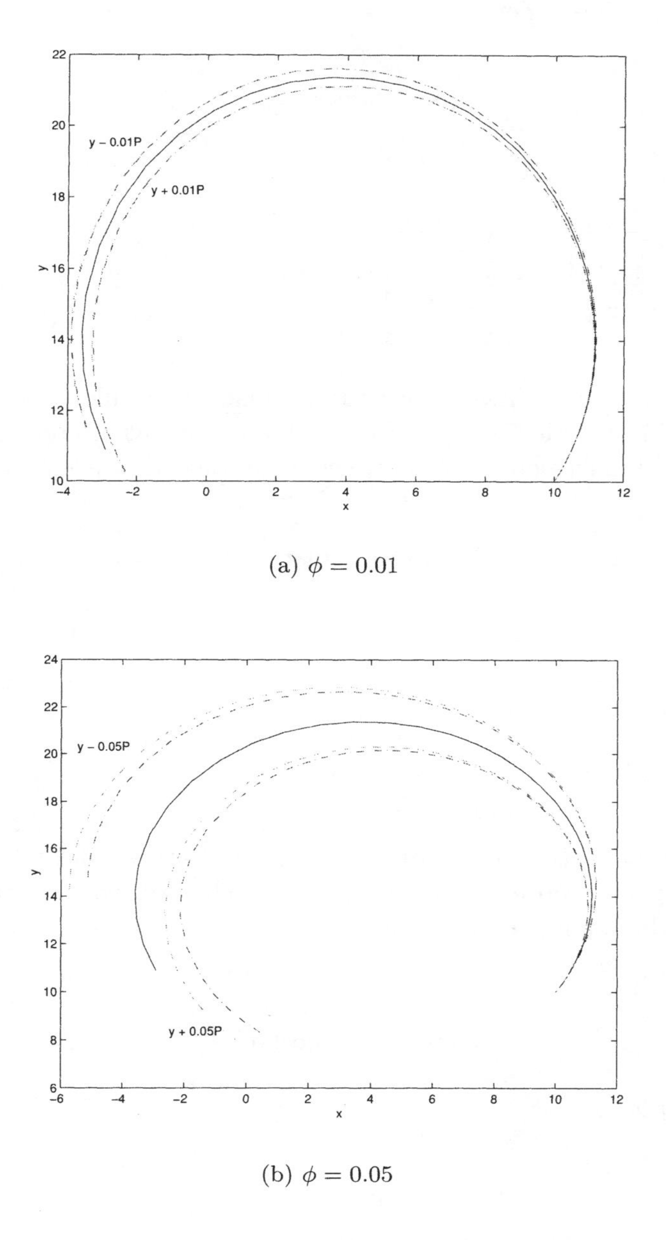

(a) $\phi = 0.01$

(b) $\phi = 0.05$

Figure 4.6: *Toy car routes under constant steering: unperturbed (solid line), steering perturbed by* $\pm\phi$ *(dash-dot lines), and corresponding trajectories computed by the linear sensitivity analysis (dashed lines).*

of quadrature formulas. There are several classes of these methods, of which we give some examples with the first two instances for each.

Gauss methods: These are the maximum order methods; an s-stage Gauss method has order $2s$:

$$\begin{array}{c|c} \frac{1}{2} & \frac{1}{2} \\ \hline & 1 \end{array} \qquad \text{implicit midpoint,} \qquad s=1, \quad p=2,$$

$$\begin{array}{c|cc} \frac{3-\sqrt{3}}{6} & \frac{1}{4} & \frac{3-2\sqrt{3}}{12} \\ \frac{3+\sqrt{3}}{6} & \frac{3+2\sqrt{3}}{12} & \frac{1}{4} \\ \hline & \frac{1}{2} & \frac{1}{2} \end{array} \qquad s=2, \quad p=4.$$

Radau methods: These correspond to quadrature rules where one end of the interval is included ($c_1 = 0$ or $c_s = 1$), and they attain order $2s-1$. The choice $c_1 = 0$ makes no sense, so we consider only the case $c_s = 1$:

$$\begin{array}{c|c} 1 & 1 \\ \hline & 1 \end{array} \qquad \text{backward Euler,} \qquad s=1, \quad p=1,$$

$$\begin{array}{c|cc} \frac{1}{3} & \frac{5}{12} & -\frac{1}{12} \\ 1 & \frac{3}{4} & \frac{1}{4} \\ \hline & \frac{3}{4} & \frac{1}{4} \end{array} \qquad s=2, \quad p=3.$$

Lobatto methods: These correspond to quadrature rules where the function is sampled at both ends of the interval. The order of accuracy is $2s-2$. There are three families. One such is:

$$\begin{array}{c|cc} 0 & 0 & 0 \\ 1 & \frac{1}{2} & \frac{1}{2} \\ \hline & \frac{1}{2} & \frac{1}{2} \end{array} \qquad \text{trapezoidal method,} \qquad s=2, \quad p=2,$$

$$\begin{array}{c|ccc} 0 & 0 & 0 & 0 \\ \frac{1}{2} & \frac{5}{24} & \frac{1}{3} & -\frac{1}{24} \\ 1 & \frac{1}{6} & \frac{2}{3} & \frac{1}{6} \\ \hline & \frac{1}{6} & \frac{2}{3} & \frac{1}{6} \end{array} \qquad s=3, \quad p=4.$$

Note that, in constructing a Runge–Kutta method, common sense should prevail. For example, while there is no analytical reason why we should choose $0 \le c_i \le 1$, in physical applications it sometimes does not make sense to evaluate the function outside the interval.

A Runge–Kutta method with a nonsingular coefficient matrix A which satisfies $a_{sj} = b_j, j = 1, \ldots, s$, is called *stiffly accurate.* This gives stiff decay (Exercise 4.7).

4.7.1 Implicit Runge–Kutta Methods Based on Collocation

Collocation is an idea which runs throughout numerical analysis. The basic idea is to choose a function from a simple space (usually a polynomial) and a set of collocation points, and require that the function satisfy the given problem equations at the collocation points.

Starting with a set of s distinct points $0 \le c_1 < c_2 < \cdots < c_s \le 1$, and considering for simplicity a scalar ODE $y' = f(t, y)$ at first, we seek the polynomial $\phi(t)$ of degree at most s which *collocates* the ODE as follows:

$$\begin{aligned} \phi(t_{n-1}) &= y_{n-1}, \\ \phi'(t_i) &= f(t_i, \phi(t_i)), \quad i = 1, 2, \ldots, s, \end{aligned}$$

where $t_i = t_{n-1} + c_i h$ are the collocation points. This defines $\phi(t)$ uniquely.[14] Now, take

$$y_n = \phi(t_n).$$

This gives an s-stage implicit Runge–Kutta method. Why? Observe that ϕ' is a polynomial of degree at most $s - 1$ which interpolates s data points $f(t_i, \phi(t_i))$. Define $K_i = \phi'(t_i)$. Now, write ϕ' as a Lagrange interpolation formula

$$\phi'(t_{n-1} + \tau h) = \sum_{j=1}^{s} L_j(t_{n-1} + \tau h) K_j,$$

where $L_j(t_{n-1} + \tau h) = \Pi_{i=1, i\ne j}^{s} \frac{(\tau - c_i)}{(c_j - c_i)}$. (Because ϕ' is a polynomial of degree $< s$, it agrees with its s-points interpolant identically.) Integrating ϕ' with respect to t from t_{n-1} to t_i, $i = 1, 2, \ldots, s$, and from t_{n-1} to t_n, we get

$$\begin{aligned} \phi(t_i) - \phi(t_{n-1}) &= h \sum_{j=1}^{s} \left(\int_0^{c_i} L_j(r) dr \right) K_j, \\ \phi(t_n) - \phi(t_{n-1}) &= h \sum_{j=1}^{s} \left(\int_0^{1} L_j(r) dr \right) K_j. \end{aligned}$$

[14] Note that if we collect the polynomial pieces defined in this way on each step interval $[t_{n-1}, t_n]$ into one function defined on $[0, b]$, then we get a continuous, piecewise polynomial approximation of the solution $y(t)$.

(Recall again our brief review of basic quadrature.) Now define

$$
\begin{aligned}
a_{ij} &= \int_0^{c_i} L_j(r)dr, \\
b_j &= \int_0^1 L_j(r)dr.
\end{aligned}
\tag{4.25}
$$

Thus, $K_i = f(t_i, \phi(t_i)) = f(t_i, y_{n-1} + h\sum_{j=1}^s a_{ij}K_j)$ and $y_n = y_{n-1} + h\sum_{i=1}^s b_i K_i$. The obtained formula is therefore a Runge–Kutta method in the form (4.10). Finally, note that for the general ODE system

$$\mathbf{y}' = \mathbf{f}(t, \mathbf{y}),$$

precisely the same argument can be repeated, where now we have a vector of m collocation polynomials, $\boldsymbol{\phi}(t)$.

The Gauss, Radau, and Lobatto methods introduced above are collocation methods. That is, given the quadrature points c_i, in each case all the other methods' coefficients are determined by (4.25). We note that

- Runge–Kutta methods which are also collocation methods are easy to derive;
- the order of such a collocation Runge–Kutta method is at least s and is determined only by the quadrature order condition (4.14) (i.e., the order limitation is a result from quadrature theory);
- the maximum order of an s-stage Runge–Kutta method is $2s$.

The last two conclusions require a proof which is left for the exercises. We note here that the order is restricted to be at most $2s$ by the quadrature order condition (4.14) and that a simple collocation analysis reveals that this order $2s$ is attained by collocation at Gaussian points.

With regard to absolute stability, we have already seen in (4.17) that for the test equation $y' = \lambda y$ a Runge–Kutta method reads

$$y_n = R(z)y_{n-1},$$

where $z = h\lambda$ and

$$R(z) = 1 + z\mathbf{b}^T(I - zA)^{-1}\mathbf{1}. \tag{4.26}$$

The region of absolute stability is given by the set of values z such that $|R(z)| \leq 1$. For an explicit method, we saw that $R(z)$ is a polynomial, and hence the method cannot be A-stable. For implicit Runge–Kutta methods, in contrast, $R(z)$ is a rational function; i.e., it is a quotient of two polynomials

$$R(z) = \frac{P(z)}{Q(z)},$$

and A-stable methods are abundant. *All* of the implicit Runge–Kutta methods which we have seen so far turn out to be A-stable.

When $\mathcal{R}e(z) \to -\infty$ we would also like a method to have stiff decay. For this we must have $R(-\infty) = 0$ in (4.26), which is achieved if $P(z)$ has a lower degree than $Q(z)$. Note that by (4.26), if A is nonsingular, then $R(-\infty) = 1 - \mathbf{b}^T A^{-1}\mathbf{1}$, so $R(-\infty) = 0$ if the last row of A coincides with $\mathbf{b}^T$. For a collocation Runge–Kutta method, this happens when $c_1 > 0$ and $c_s = 1$. In particular,

- the Radau methods, extending backward Euler, have stiff decay;
- the Gauss and Lobatto methods, which extend midpoint and trapezoid, do not have stiff decay, although they are A-stable.

These innocent-looking conclusions in fact have far-reaching importance. The Gauss and Lobatto methods are families of *symmetric* methods—this is important particularly in the context of BVPs. Symmetric methods can work for stiff problems, but for very stiff problems they do not approximate the exponential function well. The arguments of the previous chapter extend here directly. The Radau methods, on the other hand, are particularly suitable for the solution of stiff initial value ODEs.

Note: The technical level of the rest of Section 4.7 is higher than that of the material we've already presented, and although it is of practical interest, skipping it should not impede the reading of the next section nor of subsequent chapters.

4.7.2 Implementation and Diagonally Implicit Methods

One of the challenges for implicit Runge–Kutta methods is the development of efficient implementations. To see why efficiency could be a problem, we consider again the general Runge–Kutta method

$$\begin{aligned}\mathbf{Y}_i &= \mathbf{y}_{n-1} + h\sum_{j=1}^{s} a_{ij}\mathbf{f}(t_{n-1}+c_jh, \mathbf{Y}_j), \qquad 1 \le i \le s,\\ \mathbf{y}_n &= \mathbf{y}_{n-1} + h\sum_{i=1}^{s} b_i\mathbf{f}(t_{n-1}+c_ih, \mathbf{Y}_i).\end{aligned}$$

For the νth Newton iterate, let $\boldsymbol{\delta}_i = \mathbf{Y}_i^{\nu+1} - \mathbf{Y}_i^{\nu}$ and $\mathbf{r}_i = \mathbf{Y}_i^{\nu} - \mathbf{y}_{n-1} - h\sum_{j=1}^{s} a_{ij}\mathbf{f}(\mathbf{Y}_j^{\nu})$. Then the Newton iteration takes the form

$$\begin{pmatrix} I - ha_{11}J_1 & -ha_{12}J_2 & \cdots & -ha_{1s}J_s \\ -ha_{21}J_1 & I - ha_{22}J_2 & \cdots & -ha_{2s}J_s \\ \vdots & \vdots & \ddots & \vdots \\ -ha_{s1}J_1 & -ha_{s2}J_2 & \cdots & I - ha_{ss}J_s \end{pmatrix} \begin{pmatrix} \boldsymbol{\delta}_1 \\ \boldsymbol{\delta}_2 \\ \vdots \\ \boldsymbol{\delta}_s \end{pmatrix} = - \begin{pmatrix} \mathbf{r}_1 \\ \mathbf{r}_2 \\ \vdots \\ \mathbf{r}_s, \end{pmatrix},$$

where $J_i = (\partial\mathbf{f}/\partial\mathbf{y})$ is evaluated at $\mathbf{Y}_i^{\nu}$, $i = 1, 2, \dots, s$. We note that for a system of m differential equations, this is an $sm \times sm$ system of equations to be solved at each time step. This is usually not competitive with the multistep methods to come in Chapter 5, which require only the solution of an $m \times m$ nonlinear system at each time step. Thus, it is important to look for ways to make the iteration process less expensive.

Review: The **Kronecker product**, or *direct product* of two matrices A and B is given by

$$A \otimes B = \begin{pmatrix} a_{11}B & a_{12}B & \cdots & a_{1s}B \\ a_{21}B & a_{22}B & \cdots & a_{1s}B \\ \vdots & \vdots & \ddots & \vdots \\ a_{s1}B & a_{s2}B & \cdots & a_{ss}B \end{pmatrix}.$$

There are two important properties of the Kronecker product that we will need:

1. $(A \otimes B)(C \otimes D) = AC \otimes BD$,
2. $(A \otimes B)^{-1} = A^{-1} \otimes B^{-1}$.

First, we simplify the Newton iteration by taking $J_1 = J_2 = \cdots = J_s = J = (\partial\mathbf{f}/\partial\mathbf{y})$, evaluated at $\mathbf{y}_{n-1}$. Using the approximate Jacobian does not reduce the method's accuracy, provided that the Newton iteration can still converge. Using the Kronecker product notation, the simplified Newton method can now be written as

$$(I - hA \otimes J)\boldsymbol{\delta} = -\mathbf{r}. \tag{4.27}$$

Note that while $\boldsymbol{\delta}$ and $\mathbf{r}$ depend on the iteration ν, the matrix in (4.27) is the same for all iterations, and it depends only on the step counter n. So, at most one LU decomposition is needed per time step (at most, because we may hold this matrix fixed over a few time steps).

Diagonally Implicit Runge–Kutta

Unfortunately, collocation tends to yield Runge–Kutta methods where the coefficient matrix $A = (a_{ij})_{i,j=1}^{s}$ has few zeros. In particular, there are no zeros in the coefficient matrices of Radau collocation methods. For efficiency reasons, one can therefore also consider noncollocation implicit Runge–Kutta methods for which A is a lower triangular matrix.

One such family of methods is the diagonally implicit Runge–Kutta (DIRK) methods. These are implicit Runge–Kutta methods for which the coefficient matrix A is lower triangular, with equal coefficients a along the diagonal.[15] Thus, the stages in the Runge–Kutta method are defined by

$$\begin{aligned} \mathbf{Y}_i - ha\mathbf{f}(t_{n-1} + c_i h, \mathbf{Y}_i) &= \mathbf{y}_{n-1} + h\sum_{j=1}^{i-1} a_{i,j}\mathbf{f}(t_{n-1} + c_j h, \mathbf{Y}_j), \\ & i = 1, \ldots, s. \end{aligned} \tag{4.28}$$

Instead of having to solve an $sm \times sm$ system of linear equations, we now have to solve s systems of size $m \times m$ each, all with the same matrix $I - haJ$. Hence the nonlinear system can be solved by block back substitution. Only one evaluation of the local Jacobian J and one LU decomposition of the $m \times m$ submatrix $I - haJ$ are needed on each time step.

DIRK methods have become popular for the numerical solution of time-dependent partial differential equations (PDEs) via the method of lines (recall Examples 1.3 and 1.7). Here the Jacobian J is very large and sparse, and iterative methods are often used for the linear algebra. A fully implicit Runge–Kutta method makes things cumbersome, but DIRK methods offer a Runge–Kutta alternative to the backward differentiation formula methods of the next chapter in this situation. When iterative methods are used it becomes less important to insist that the diagonal elements of the coefficient matrix A all be the same; however, it turns out that this extra freedom in designing the DIRK method does not buy much.

---

[15]Strictly speaking, DIRK methods do not require equal coefficients along the diagonal. However, the subset of DIRK methods with equal diagonal coefficients, which is called singly diagonally implicit Runge–Kutta (SDIRK), are the methods most commonly used because of the possibility for a more efficient implementation. So, we refer only to the SDIRK methods, and we call them DIRK to distinguish from the rather different SIRK methods which arise in Section 4.7.4.

Because so many of the coefficients of DIRK methods have been specified by construction to be zero, it is not surprising that the maximum attainable order is much less than for general implicit Runge–Kutta methods. In fact, it has been shown that the maximum order of an s-stage DIRK method cannot exceed $s+1$. Some such methods are the midpoint method ($s=1, p=2$) and

$$\begin{array}{c|cc} \gamma & \gamma & 0 \\ 1-\gamma & 1-2\gamma & \gamma \\ \hline & 1/2 & 1/2 \end{array} \qquad s=2, \quad p=3$$

with $\gamma = \frac{3+\sqrt{3}}{6}$. This latter method satisfies $R(-\infty) = 1-\sqrt{3} \approx -0.7321$. Thus $R(-\infty) < 1$, a marked improvement over the midpoint method which has no attenuation ($R(-\infty) = -1$) at the stiffness limit $\mathcal{R}\text{e}(z) = \mathcal{R}\text{e}(\lambda)h = -\infty$.

If the method is to have stiff decay as well, then the order is further restricted to s. Examples are backward Euler ($s=p=1$),

$$\begin{array}{c|cc} \gamma & \gamma & 0 \\ 1 & 1-\gamma & \gamma \\ \hline & 1-\gamma & \gamma \end{array} \qquad s=2, \quad p=2,$$

where $\gamma = \frac{2-\sqrt{2}}{2}$, and

$$\begin{array}{c|ccc} .4358665215 & .4358665215 & 0 & 0 \\ .7179332608 & .2820667392 & .4358665215 & 0 \\ 1 & 1.208496649 & -.644363171 & .4358665215 \\ \hline & 1.208496649 & -.644363171 & .4358665215 \end{array} \qquad s=3, \quad p=3.$$

4.7.3 Order Reduction

When speaking of the order of accuracy of a method and making statements like

$$\|\mathbf{e}\| = \max_{0 \le n \le N} |\mathbf{e}_n| = O(h^p)$$

regarding the error in a method of order p, we mean that as the maximum step size $h \to 0$ (and $N \to \infty$), the error also shrinks so fast that $h^{-p}\|\mathbf{e}\|$ remains bounded in the limit. In a finite world we must of course think of h as small but finite, and we normally think of it as being much smaller than the smallest scale of the ODE being approximated.

This changes in the very stiff limit. We have already considered, in Chapter 3, problems like

$$y' = \lambda(y - q(t)), \qquad 0 < t < 1, \tag{4.29}$$

where $0 < \frac{1}{-\mathcal{R}e(\lambda)} \ll h \ll 1$. Here there are two small parameters, one the method's step size, and the other the problem's. We consider the limit process in which $\frac{1}{-\mathcal{R}e(\lambda)} \to 0$ faster than $h \to 0$. In this case our statements about the method's order may have to be revised.

Indeed, some Runge–Kutta methods suffer from a reduction in their order of convergence in the very stiff limit. The essential reason is that these methods are based on quadrature, and in the very stiff case the integration effect is very weak, at least for some solution components. For example, assume that λ is real in (4.29). Upon dividing (4.29) by $-\lambda$ we see that y' is multiplied by a constant which shrinks to 0 while the right-hand side is scaled to 1, so there is almost no integration in determining $y(t)$ from $q(t)$. Upon applying a Runge–Kutta method in this case we obtain almost an interpolation problem at the internal stages of a Runge–Kutta method. The accuracy order of this interpolation, and not the quadrature precision, then takes center stage.

The phenomenon of order reduction is important particularly for differential-algebraic equations, so we leave a fuller discussion to Chapter 10. Here we note that some methods are affected by this more than others. Unfortunately, DIRK methods are only first-order accurate at the very stiff limit, which suggests that they should not be used for very stiff problems. Fortunately, many time-dependent PDEs are stiff but not very stiff. Collocation at Radau points retains its full usual order in the very stiff limit. That is one reason why these methods are so popular in practice despite the more expensive linear algebra necessary for their implementation.

4.7.4 More on Implementation and Singly Implicit Runge–Kutta Methods

The DIRK methods require a special zero-structure for the coefficient matrix A, which implies the restrictions discussed above. A clever alternative is to seek implicit Runge–Kutta methods where A can be transformed by a similarity transformation T into a particularly simple form,

$$T^{-1}AT = S.$$

It can be shown that, upon transforming the variables

$$\hat{\boldsymbol{\delta}} = (T^{-1} \otimes I)\boldsymbol{\delta}$$

and multiplying equations (4.27) by $T^{-1} \otimes I$, the matrix problem to be solved has the form

$$(I - hS \otimes J)\hat{\boldsymbol{\delta}} = -\hat{\mathbf{r}}, \tag{4.30}$$

where $\hat{\mathbf{r}} = (T^{-1} \otimes I)\mathbf{r}$. Thus, any lower triangular matrix S yields the DIRK structure in (4.30) for the transformed variables $\hat{\boldsymbol{\delta}}$. An efficient implementation of Radau collocation can be obtained in this way (or using another simple form of S). We can go further and require that S be a scalar multiple of the identity matrix; i.e., we can look for methods in which A has a single s-fold real eigenvalue, a. This yields the SIRK (singly implicit Runge–Kutta) methods. Here, at most one $m \times m$ matrix needs to be formed and decomposed per step. A good s-stage method of this sort has order $s + 1$, and unlike DIRK this order does not reduce in the very stiff limit.

4.8 Software, Notes, and References

4.8.1 Notes

Runge [81] and Kutta [61] did not collaborate to invent the methods bearing their names: rather, Runge was first, and Kutta gave the general form. But together they were responsible for much of the development of the early explicit Runge–Kutta methods. Chapter II of Hairer, Norsett, and Wanner [50] gives a full exposition which we do not repeat here.

As noted earlier, the order of an efficient Runge–Kutta method can be a challenge to verify if it is higher than 3 or 4, and our exposition in Section 4.3 aims to give a taste of the issues involved, rather than to cover the most general case. Excellent and elaborate expositions of the general theory for order conditions can be found in Butcher [26] or in [50]. These references also discuss order barriers, i.e., limitations on the attainable order for explicit Runge–Kutta as a function of the number of stages.

Error estimation and control are discussed in all modern texts, e.g., [62, 43, 50, 52, 85]. Shampine [85] has many examples and elaborates on points of practical concern.

For a comprehensive treatment of implicit Runge–Kutta methods and related collocation methods, we refer to Hairer and Wanner [52]. It contains the theorems, proofs, and references which we have alluded to in Section 4.7. The concepts of stiff accuracy and stiff decay were introduced in [73].

A large number of topics have been omitted in our presentation, despite their importance. Below we briefly mention some of these topics. Others have made their way into the exercises.

The Runge–Kutta family is not the only practical choice for obtaining high-order one-step methods. Another family of methods is based on *extrapolation.* Early efforts in the 1960s are due to W.B. Gragg, J. Stoer and R. Bulirsch, and H. Stetter [89]. See [50] for a full description. These methods have performed well, but overall they do not appear to outperform the families of methods discussed in this book. The extrapolation idea is discussed in Chapter 8.

A great deal of theory relating to stability and convergence for stiff problems has been developed in the past few decades. We have described some of the more accessible work. Many different stability definitions have been proposed over the years; perhaps not all of them have stood the test of time. But we mention in particular the theories of *order stars* (and rational approximations to the exponential) and *B-convergence*, for their elegance and ability to explain the behavior of numerical methods [52].

A topic of practical importance is *dense output*, or *continuous extension*; see, e.g., [50] or [85]. Normally, a discretization method yields approximate solution values only at discrete mesh points. If every point at which a solution is required is taken to be a mesh point, an inefficient procedure may result. A better, obvious idea is to use cubic Hermite interpolation (which is a cubic polynomial interpolant using two function values and two derivative values) because at mesh points we have both $\mathbf{y}_n$ and $\mathbf{f}(t_n, \mathbf{y}_n)$, which approximates $\mathbf{y}'(t_n)$. Dense output can often be accomplished more efficiently, and at higher order, using Runge–Kutta methods which have been specifically designed for this purpose.

One important application, in addition to plotting solutions, for which dense output is needed, is for *event location*. Recall from Section 3.7 that if $\mathbf{f}$ has discontinuities, determined according to a change of sign of some switching function, then it is desirable to place mesh points at or very near such switching points (which are points of discontinuity) in t. Hence the "event" of the discontinuity occurrence requires detection. This is done by solving a nonlinear equation iteratively, and the function values needed are supplied by the dense output mechanism.

Another important instance where dense output may be required is in solving *delay differential equations*. A simple prototype is

$$\mathbf{y}'(t) = \mathbf{f}(t, \mathbf{y}(t), \mathbf{y}(t-\tau)),$$

where τ is the delay. Delay equations can be very complicated; see, e.g., [50]. We do not pursue this except to say that when discretizing in a straightforward way at $t = t_n$, say, the value of $\mathbf{y}$ at $t_n - \tau$ is required as well. If that does not fall on a past mesh point then the dense output extension must be called upon. An IVP implementation is described in [50]. For possibilities of converting delay differential equations to standard ODEs, see Exercise 7.6.

We have commented in Section 4.4 on the dissipativity of explicit Runge–Kutta methods of order > 2 and the use of these methods as a smoother in PDE solvers. A reference is Jameson [57].

A lot of attention has been devoted in recent years to *symplectic methods*. Recall from Section 2.5 that a Hamiltonian system provides a symplectic map. As a corollary, considering a set of initial values, each spawning a trajectory of a Hamiltonian system, the volume of this set at a later time remains constant under the flow. Next, we may ask if the property of the symplectic map is retained by a given numerical method. A numerical discretization that preserves symplecticity for a constant step size is called a symplectic method. Such methods are particularly desirable for applications involving *long time integration*, i.e., $Nh = b \gg 1$, where h is the step size and N is the number of steps taken. Examples appear in molecular dynamics and celestial mechanics simulations. For much more on this topic, see Sanz-Serna and Calvo [82] and [52]. See also Exercises 4.10, 4.11, and 4.19. As it turns out, there are difficulties in constructing general, efficient methods of this sort, and varying the step size is also a challenge, yet there are instances where symplectic methods impressively outperform standard methods.

Viewing the discretization of a nonlinear ODE as a continuous dynamical system yields a *discrete dynamical system.* One may wonder if the dynamical properties of the two systems are qualitatively similar, especially if the discretization step size h is not very small or the number of steps taken, N, is very large. We have already indicated above that this is not necessarily so, e.g., when a nonsymplectic method is applied to a Hamiltonian system. Another instance is where the discrete dynamical system has more solutions than its continuous counterpart. Viewed as a function of the step size h there are principal solutions which tend towards the corresponding genuine solutions, and in addition there may be *spurious solutions.* The latter would tend to 0 as $h \to 0$, but they may be confusingly present for a finite h. For more on this, see, e.g., Stuart and Humphries [93].

Many efforts have been invested since the 1980s in the development of Runge–Kutta methods suitable for the solution of large ODE systems on *parallel* computer architectures. The book by Burrage [23] covers such methods well. Basic ideas include the design of Runge–Kutta methods where different stages are sufficiently independent from one another that they can be evaluated in parallel. This is parallelism in time t. Another direction exploits parallelism also in the large system being integrated, and often leads to more specialized methods. We mention the closely related *multirate methods* and *waveform relaxation methods.* In the former, different components of the system that vary more slowly are discretized over larger elements (time steps) than more rapidly varying components, allowing for parallel evaluations of different parts of the system at a given stage. (The system being integrated must be such that a block decoupling of this sort is possible; this is often the case in very large scale integration circuit simulation, for example.) In the latter, a global iterative method in time, such as

$$(\mathbf{y}^{\nu+1})' - M\mathbf{y}^{\nu+1} = \mathbf{f}(t, \mathbf{y}^{\nu}) - M\mathbf{y}^{\nu},$$

is considered, where $\mathbf{y}^{\nu}(t)$ is known at the start of iteration ν, $\nu = 0, 1, \ldots$. This is really considered not for all time but over a time window (such as the large step size in a multirate method), and the matrix M is now chosen to allow parallelism in the evaluation of the iteration.

4.8.2 Software

Here we briefly mention some (certainly not all) general-purpose codes and the methods on which they are based. All of these codes provide error control and step-size selection. Some of them are available through NETLIB.

For **nonstiff problems**:

- Many Runge–Kutta codes have been based on the Fehlberg 4(5) embedded pair [40]. An early influential code was RKF45 by Shampine and Watts [87].

- The code DOPRI5 presented in the monumental book [50] is based on the Dormand–Prince 4(5) formulae [39], which uses local extrapolation. The code ODE45 used in MATLAB 5 is also based on these formulae, a switch from earlier MATLAB versions where it was based on the Fehlberg pair. This reflects the current accumulated experience, which suggests that the Dormand–Prince pair performs better in practice.

- Other codes based on embedded Runge–Kutta pairs are DVERK by Hull, Enright, and Jackson [56], which uses a pair of formulae by Verner [94], and RKSUITE by Brankin, Gladwell, and Shampine [18], which implements three Runge–Kutta pairs, 2(3), 4(5), and 7(8), and uses local extrapolation. The latter has an option for global error estimation and it also automatically checks for stiffness.

For **stiff problems**:

- The code RADAU5 [52] uses the Radau three-stage formula of order 5 with an implementation of the linear algebra along the lines described in Section 4.7.4.

- The code STRIDE by Burrage, Butcher, and Chipman [25] uses a family of SIRK methods.

The codes DOPRI5, ODE45, RKSUITE, RADAU5, and STRIDE all have a dense output option. The code ODE45 also has a built-in event location module.

4.9 Exercises

4.1. Show that a general Runge–Kutta method (4.8) can be written in the form (4.10). What is the relationship between $\mathbf{K}_i$ of (4.10a) and $\mathbf{Y}_i$ of (4.8a)?

4.2. Show that the explicit Runge–Kutta methods described in Section 4.1 can all be written in the form (4.11), with ψ Lipschitz continuous in $\mathbf{y}$ if $\mathbf{f}$ is.

4.3. Prove that the one-step method (4.11) is 0-stable if ψ satisfies a Lipschitz condition in $\mathbf{y}$.

4.4. Write a computer program that will find an upper bound on the order of a given Runge–Kutta method by checking the order conditions (4.13), *or alternatively*, only (4.14) and (4.15) (and (4.9)). For a given number of stages s, your program should check these conditions for $k = 1, 2, \ldots$ until one is violated (this will happen before k reaches $2s+1$). Note: Do not use any matrix-matrix multiplications!

(a) Apply your program to all the embedded methods given in Section 4.5 (both methods of each pair).

(b) Apply your program to the Lobatto methods given in Section 4.7.

(c) What does your program give for the counterexample of Example 4.4 (Section 4.3)? What is the actual order of that method?

4.5. For a four-stage explicit Runge–Kutta method of order 4:

(a) Show

$$\sum_{i=1}^{s} b_i a_{ij} = b_j(1 - c_j), \qquad j = 1, \ldots, s.$$

(This is a useful additional design requirement in general.) [The proof is a bit tricky and is given in [50].]

(b) Using this result, show that we must have $c_4 = 1$ in this case.

4.6. It has been argued that displaying absolute stability regions as in Figure 4.4 is misleading: since a step of an s-stage explicit method costs essentially s times a forward Euler step, its stability region should be compared with what forward Euler can do in s steps. Thus, the *scaled stability region* of an s-stage explicit method is the stability region shrunk by a factor s.

For all Runge–Kutta methods with $p = s$, $s = 1, 2, 3, 4$, plot the scaled stability regions. Observe that forward Euler looks mighty good: no other method's scaled stability region fully contains the forward Euler circle [52].

4.7. A Runge–Kutta method with a nonsingular A satisfying

$$a_{sj} = b_j, \qquad j = 1, \ldots, s, \tag{4.31}$$

is called *stiffly accurate* [73].

(a) Show that a stiffly accurate method has stiff decay.

(b) Show that a collocation method is stiffly accurate iff $c_1 = 1$ and $c_0 > 0$.

(c) Not all Runge–Kutta methods which have stiff decay satisfy (4.31). Show that stiff decay is obtained also if A is nonsingular and its first column is $b_1\mathbf{1}$.

4.8. For a given ODE

$$\mathbf{y}' = \mathbf{f}(\mathbf{y}),$$

consider the θ-method

$$\mathbf{y}_n = \mathbf{y}_{n-1} + h_n[\theta \mathbf{f}(\mathbf{y}_n) + (1-\theta)\mathbf{f}(\mathbf{y}_{n-1})]$$

for some value θ, $0 \le \theta \le 1$.

(a) Which methods are obtained for the values (i) $\theta = 0$, (ii) $\theta = 1$, and (iii) $\theta = 1/2$?

(b) Find a range of θ-values, i.e., an interval $[\alpha, \beta]$, such that the method is A-stable for any $\alpha \le \theta \le \beta$.

(c) For what values of θ does the method have stiff decay?

(d) For a given δ, $0 \le \delta < 1$, let us call a method δ-damping if

$$|y_n| \le \delta |y_{n-1}|$$

for the test equation $y' = \lambda y$ as $\lambda h \to -\infty$. (Thus, if $y_0 = 1$ then for any tolerance TOL > 0, $|y_n| \le$ TOL after n steps when n exceeds $\frac{\log TOL}{\log \delta}$.)

Find the range of θ-values such that the θ-method is δ-damping.

(e) Write the θ-method as a general Runge–Kutta method; i.e., specify A, $\mathbf{b}$, and $\mathbf{c}$ in the tableau

$$\begin{array}{c|c} \mathbf{c} & A \\ \hline & \mathbf{b}^T \end{array}$$

(f) What is the order of the θ-method?

[If you managed to answer the previous question, then try to answer this one *without* any Taylor expansions.]

4.9. The solution of the problem $\mathbf{y}' = \mathbf{f}(\mathbf{y})$, $\mathbf{y}(0) = \mathbf{c}$, where

$$\mathbf{f}(\mathbf{y}) = (-y_2, y_1)^T, \qquad \mathbf{c} = (1, 0)^T,$$

satisfies

$$y_1^2 + y_2^2 = 1;$$

i.e., it is a circle of radius 1 centered at the origin. The numerical solution, though, does not necessarily satisfy this invariant, and the obtained curve in the $y_1 \times y_2$ plane does not necessarily close.

Show that when using collocation based on Gaussian points, the approximate solution does satisfy the invariant; i.e., the obtained approximate solution stays on the circle. [See also Exercise 4.15, where a hint is provided.]

4.10. In molecular dynamics simulations using classical mechanics modeling, one is often faced with a large nonlinear ODE system of the form

$$M\mathbf{q}'' = \mathbf{f}(\mathbf{q}), \quad \text{where } \mathbf{f}(\mathbf{q}) = -\nabla U(\mathbf{q}). \tag{4.32}$$

Here $\mathbf{q}$ are generalized positions of atoms, M is a constant, diagonal, positive mass matrix, and $U(\mathbf{q})$ is a scalar potential function. Also, $\nabla U(\mathbf{q}) = (\frac{\partial U}{\partial q_1}, \ldots, \frac{\partial U}{\partial q_m})^T$. A small (and somewhat nasty) instance of this is given by the Morse potential [83] where $\mathbf{q} = q(t)$ is scalar, $U(q) = D(1-e^{-S(q-q_0)})^2$, and we use the constants $D = 90.5 * .4814\text{e}-3$, $S = 1.814$, $q_0 = 1.41$, and $M = 0.9953$.

(a) Defining the momenta $\mathbf{p} = M\mathbf{q}'$, the corresponding first-order ODE system for $\mathbf{q}$ and $\mathbf{p}$ is given by

$$M\mathbf{q}' = \mathbf{p}, \tag{4.33a}$$
$$\mathbf{p}' = \mathbf{f}(\mathbf{q}). \tag{4.33b}$$

Show that the Hamiltonian

$$H(\mathbf{q}, \mathbf{p}) = \mathbf{p}^T M^{-1}\mathbf{p}/2 + U(\mathbf{q})$$

is constant for all $t > 0$.

(b) Use a library nonstiff Runge–Kutta code based on a 4(5) embedded pair to integrate this problem for the Morse potential on the interval $0 \le t \le 2000$, starting from $q(0) = 1.4155$, $p(0) = \frac{1.545}{48.888}M$. Using a tolerance TOL $= 1.\text{e}{-4}$, the code should require a little more than 1000 time steps. Plot the obtained values for $H(q(t), p(t)) - H(q(0), p(0))$. Describe your observations.

4.11. The system (4.33) is in *partitioned form.* It is also a Hamiltonian system with a separable Hamiltonian; i.e., the ODE for $\mathbf{q}$ depends only on $\mathbf{p}$ and the ODE for $\mathbf{p}$ depends only on $\mathbf{q}$. This can be used to design special discretizations. Consider a constant step size h.

(a) The *symplectic Euler* method applies backward Euler to (4.33a) and forward Euler to (4.33b). Show that the resulting method is explicit and first-order accurate.

(b) The *leapfrog,* or *Verlet,* method can be viewed as a staggered midpoint discretization:

$$M(\mathbf{q}_{n+1/2} - \mathbf{q}_{n-1/2}) = h\,\mathbf{p}_n,$$
$$\mathbf{p}_n - \mathbf{p}_{n-1} = h\,\mathbf{f}(\mathbf{q}_{n-1/2});$$

i.e., the mesh on which the $\mathbf{q}$-approximations "live" is staggered by half a step compared to the $\mathbf{p}$-mesh. The method can be kick-started by

$$\mathbf{q}_{1/2} = \mathbf{q}_0 + h/2M^{-1}\mathbf{p}_0.$$

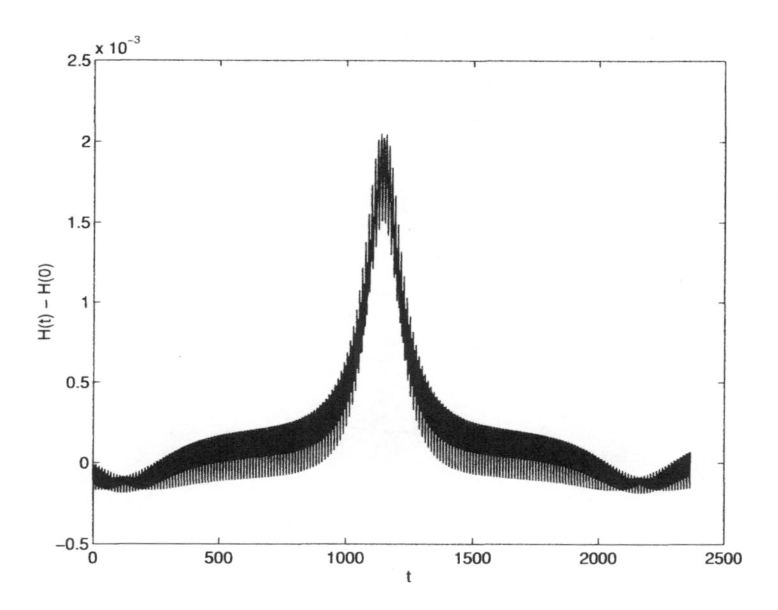

Figure 4.7: *Energy error for the Morse potential using leapfrog with* $h = 2.3684$.

To evaluate $\mathbf{q}_n$ at any mesh point, the expression

$$\mathbf{q}_n = \frac{1}{2}(\mathbf{q}_{n-1/2} + \mathbf{q}_{n+1/2})$$

can be used.

Show that this method is explicit and second-order accurate.

(c) Integrate the Morse problem defined in the previous exercise using 1000 uniform steps h. Apply three methods: forward Euler, symplectic Euler, and leapfrog. Try the values $h = 2$, $h = 2.3684$, and $h = 2.3685$ and plot in each case the discrepancy in the Hamiltonian (which equals 0 for the exact solution). The plot for $h = 2.3684$ is given in Figure 4.7.

What are your observations? [The surprising increase in leapfrog accuracy from $h = 2.3684$ to $h = 2.3685$ relates to a phenomenon called *resonance instability*.]

[Both the symplectic Euler and the leapfrog method are *symplectic*—like the exact ODE they conserve certain volume projections for Hamiltonian systems (Section 2.5). We refer to [82, 50, 93] for much more on symplectic methods.]

4.12. The following classical example from astronomy gives a strong motivation to integrate initial value ODEs with error control.

Consider two bodies of masses $\mu = 0.012277471$ and $\hat{\mu} = 1 - \mu$ (earth and sun) in a planar motion, and a third body of negligible mass

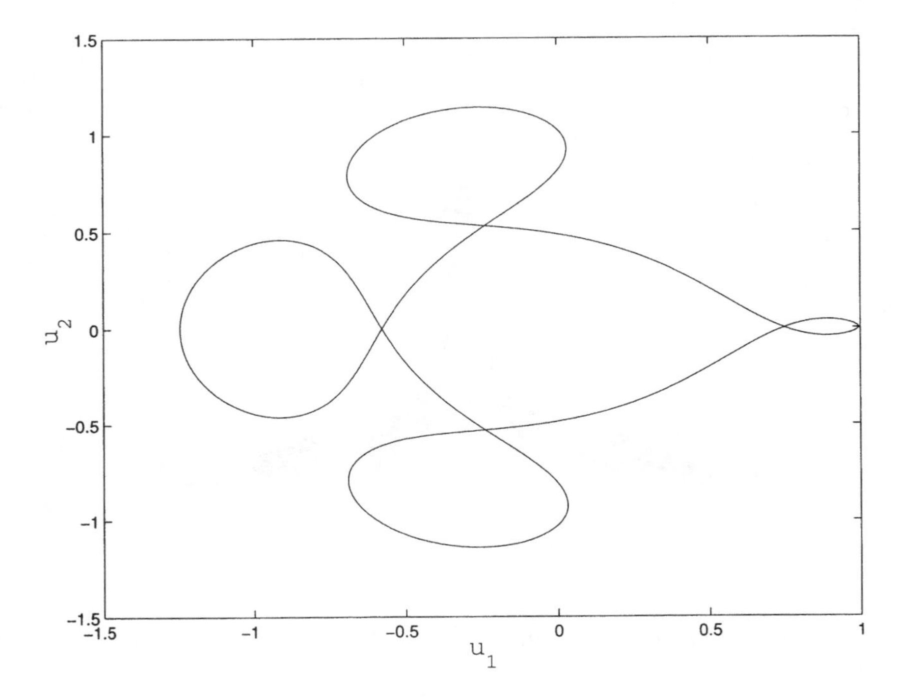

Figure 4.8: *Astronomical orbit using a Runge–Kutta* 4(5) *embedded pair method.*

(moon) moving in the same plane. The motion is governed by the equations

$$\begin{aligned}
u_1'' &= u_1 + 2u_2' - \hat{\mu}\frac{u_1 + \mu}{D_1} - \mu\frac{u_1 - \hat{\mu}}{D_2}, \\
u_2'' &= u_2 - 2u_1' - \hat{\mu}\frac{u_2}{D_1} - \mu\frac{u_2}{D_2}, \\
D_1 &= ((u_1 + \mu)^2 + u_2^2)^{3/2}, \\
D_2 &= ((u_1 - \hat{\mu})^2 + u_2^2)^{3/2}.
\end{aligned}$$

Starting with the initial conditions

$$u_1(0) = 0.994,\ u_2(0) = 0,\ u_1'(0) = 0,$$
$$u_2'(0) = -2.00158510637908252240537862224,$$

the solution is periodic with period < 17.1. Note that $D_1 = 0$ at $(-\mu, 0)$ and $D_2 = 0$ at $(\hat{\mu}, 0)$, so we need to be careful when the orbit passes near these singularity points.

The orbit is depicted in Figure 4.8. It was obtained using a 4(5) embedded pair with a local error tolerance $1.e - 6$. This necessitated 204 time steps.

Using the classical Runge–Kutta method of order 4, integrate this problem on $[0, 17.1]$ with a *uniform* step size, using 100, 1000, 10,000, and 20,000 steps. Plot the orbit for each case. How many uniform steps are needed before the orbit appears to be *qualitatively* correct?

4.13. For an s-stage Runge–Kutta method (4.8), define the $s \times s$ matrix M by

$$m_{ij} = b_i a_{ij} + b_j a_{ji} - b_i b_j.$$

The method is called *algebraically stable* [24] if $\mathbf{b} \geq \mathbf{0}$ (componentwise) and M is nonnegative definite. Show that

(a) Radau collocation is algebraically stable.

(b) Gauss collocation is algebraically stable. In fact, $M = 0$ in this case.

(c) The trapezoidal method, hence Lobatto collocation, is not algebraically stable.

(d) Algebraic stability is equivalent to AN-stability; i.e., for the nonautonomous test equation

$$y' = \lambda(t)y,$$

one gets $|y_n| \leq |y_{n-1}|$ whenever $\mathcal{R}e\lambda < 0$, all t.

[This exercise is difficult. The basic idea is to write the expression for $|y_n|_2^2$ and substitute y_{n-1} in terms of Y_i in it.]

4.14. A Runge–Kutta method (4.8) is *symmetric* if it remains invariant under a change of direction of integration. Thus, letting $\mathbf{z}_n \leftarrow \mathbf{y}_{n-1}$, $\mathbf{z}_{n-1} \leftarrow \mathbf{y}_n$, $\mathbf{Z}_j \leftarrow \mathbf{Y}_{s+1-j}$, and $h \leftarrow -h$, the same method (4.8) is obtained for $\mathbf{z}_n$.

(a) Let

$$E = \begin{pmatrix} 0 & & & 1 \\ & & 1 & \\ & \cdot & & \\ 1 & & & 0 \end{pmatrix}.$$

Show that the Runge–Kutta method (4.8) is symmetric if

$$\mathbf{c} + E\mathbf{c} = \mathbf{1}, \quad \mathbf{b} = E\mathbf{b},$$
$$EAE + A = \mathbf{1}\mathbf{b}^T.$$

(These conditions are essentially necessary for symmetry as well.)

(b) Show that a symmetric Runge–Kutta method is algebraically stable iff

$$M = 0.$$

4.15. The problem considered in Exercise 4.9 is a simple instance of a system with an *invariant* [2]. More generally, an ODE system $\mathbf{y}' = \mathbf{f}(\mathbf{y})$ may have an invariant defined by algebraic equations

$$\mathbf{h}(\mathbf{y}) = \mathbf{0} \tag{4.34}$$

meaning that for the exact solution $\mathbf{y}(t)$ of the ODE we have $\mathbf{h}(\mathbf{y}(t)) = \mathbf{0}$, provided the initial values satisfy $\mathbf{h}(\mathbf{y}(0)) = \mathbf{0}$. The question is, which numerical discretization of the ODE (if any) satisfies the invariant precisely, i.e.,

$$\mathbf{h}(\mathbf{y}_n) = \mathbf{h}(\mathbf{y}_{n-1}), \qquad n = 1, 2, \ldots, N.$$

Denote the Jacobian $H = \mathbf{h_y}$ and assume it has a full row rank for all relevant $\mathbf{y}$. We say that we have an *integral invariant* if

$$H\mathbf{f} = \mathbf{0} \quad \forall \mathbf{y}.$$

(See, e.g., [93].)

(a) Show that any Runge–Kutta method preserves linear invariants.

(b) Show that collocation at Gaussian points, and only at Gaussian points, preserves quadratic integral invariants. [Hint: Write $\mathbf{h}(\mathbf{y}_n) = \mathbf{h}(\mathbf{y}_{n-1}) + \int_{t_{n-1}}^{t_n} \mathbf{h}'$ and use your knowledge of quadrature.]

(More generally, for Runge–Kutta methods the needed condition is $M = 0$.)

(c) The nondimensionalized equations in Cartesian coordinates for the simple pendulum can be written as

$$\dot{q}_1 = v_1, \quad \dot{v}_1 = -q_1\lambda, \quad \dot{q}_2 = v_2, \quad \dot{v}_2 = -q_2\lambda - 1,$$
$$q_1^2 + q_2^2 = 1.$$

Differentiating the constraint twice and eliminating λ yields the ODE

$$\dot{q}_1 = v_1, \quad \dot{v}_1 = \frac{-q_1}{q_1^2 + q_2^2}(v_1^2 + v_2^2 - q_2),$$

$$\dot{q}_2 = v_2, \quad \dot{v}_2 = \frac{-q_2}{q_1^2 + q_2^2}(v_1^2 + v_2^2 - q_2) - 1$$

with the invariants

$$q_1^2 + q_2^2 = 1,$$
$$q_1 v_1 + q_2 v_2 = 0.$$

Show that the midpoint method preserves the second of these invariants but not the first. [You may show this by a numerical demonstration.]

4.16. This exercise builds on the previous one.

(a) Consider the matrix differential system

$$\begin{aligned} \dot{U} &= A(t,U)U, \qquad 0 < t < b, \\ U(0) &= I, \end{aligned} \tag{4.35}$$

where A and U are $m \times m$ and A is skew-symmetric for all U, t:

$$A^T = -A\,.$$

It can be shown that the solution $U(t)$ is then an orthogonal matrix for each t.

Show that collocation at Gaussian points (including the midpoint method) preserves this orthogonality. We note that collocation at Lobatto points (including the trapezoidal method) does not preserve orthogonality.

(b) A number of interesting applications lead to problems of *isospectral flow* [27], where one seeks a matrix function satisfying

$$\begin{aligned} \dot{L} &= AL - LA, \\ L(0) &= L_0 \end{aligned} \tag{4.36}$$

for a given initial value matrix L_0, where $A(L)$ is skew-symmetric. The eigenvalues of $L(t)$ are then independent of t.

Verify that

$$L = UL_0U^T,$$

where $U(t)$ is the orthogonal matrix function satisfying

$$\dot{U} = AU, \;\; U(0) = I,$$

and propose a discretization method that preserves the eigenvalues of L.

4.17. This exercise continues the previous two.

Collocation at Gaussian points is an implicit, expensive method. An alternative idea is to use an explicit Runge–Kutta method, orthogonalizing U at the end of each time step [37]. Consider a method of the form (4.11) for which the matrix $U(t)$ is written as an m^2-length vector of unknowns. Since the result of this step is not necessarily an orthogonal matrix, a step of this method starting with an orthogonal U_{n-1} approximating $U(t_{n-1})$ consists of two phases:

$$\begin{aligned} \hat{U}_n &= U_{n-1} + h\psi(t_{n-1}, U_{n-1}, h), \\ \hat{U}_n &= U_n R_n, \end{aligned}$$

where $U_n R_n$ is a QR decomposition of $\hat{U}_n$. The orthogonal matrix U_n is then the projection of the result of the Runge–Kutta step onto the invariant manifold, and it is taken as the end result of the step.

Write a program which carries out this algorithm using the classical fourth-order Runge–Kutta method. (A library routine from LINPACK or MATLAB can be used for the decomposition.) Try your program on the problem

$$U' = \omega \begin{pmatrix} 0 & 1 \\ -1 & 0 \end{pmatrix} U,$$

whose exact solution is the reflection matrix

$$U(t) = \begin{pmatrix} \cos \omega t & \sin \omega t \\ -\sin \omega t & \cos \omega t \end{pmatrix}$$

for various values of ω, h, and b. What are your conclusions?

[Note that the QR decomposition of a matrix is only determined up to the signs of the elements on the main diagonal of the upper triangular matrix R. You will have to ensure that U_n is that orthogonal matrix which is close to $\hat{U}_n$.]

4.18. If you are a MATLAB fan, like we are, then this exercise is for you.

MATLAB (version 5) offers the user a simple ODE integrator, called ODE45, which is based on the Dormand–Prince embedded pair. We used this facility to generate the plot of Figure 4.8 in less than one person-hour, in fact. In the interest of keeping things simple, the designers of MATLAB kept the interface for this routine on an elementary level, and the user simply obtains "the solution."

(a) Use MATLAB to solve the problem of Example 4.6. Plot the obtained solution. Does it look like the exact one, $y(t) = \sin t$? Explain your observations.

(b) It can be argued that the solution that MATLAB produces for this example does not look plausible (or "physical"); i.e., we could guess it's wrong even without knowing the exact one. Can you construct an example that will make MATLAB produce a plausible-looking solution which nonetheless is 100% in error? [This question is somewhat more difficult.]

4.19. The modified Kepler problem [82, 51] is a Hamiltonian system, i.e.,

$$\mathbf{q}' = H_{\mathbf{p}}, \quad \mathbf{p}' = -H_{\mathbf{q}},$$

with the Hamiltonian

$$H(\mathbf{q},\mathbf{p}) = \frac{p_1^2 + p_2^2}{2} - \frac{1}{r} - \frac{\alpha}{2r^3},$$

where $r = \sqrt{q_1^2 + q_2^2}$, and we take $\alpha = 0.01$. Clearly, $H' = H_{\mathbf{q}}\mathbf{q}' + H_{\mathbf{p}}\mathbf{p}' = 0$, so $H(\mathbf{q}(t), \mathbf{p}(t)) = H(\mathbf{q}(0), \mathbf{p}(0))\ \forall t$. We consider simulating this system over a long time interval with a relatively coarse, uniform step size h, i.e., $bh \gg 1$. The mere accumulation of local errors may then become a problem. For instance, using the explicit midpoint method with $h = 0.1$ and $b = 500$, the approximate solution for r becomes larger than the exact one by two orders of magnitude.

But some methods perform better than would normally be expected. In Figure 4.9 we plot q_1 vs. q_2 ("phase plane portrait") for (a) the implicit midpoint method using $h = 0.1$, (b) the classical explicit Runge–Kutta method of order 4 using $h = 0.1$, and (c) the exact solution (or rather, a sufficiently close approximation to it). The initial conditions are

$$q_1(0) = 1 - \beta,\ q_2(0) = 0,\ p_1(0) = 0,\ p_2(0) = \sqrt{(1+\beta)/(1-\beta)}$$

with $\beta = 0.6$. Clearly, the midpoint solution with this coarse step size outperforms not only the explicit midpoint method but also the fourth-order method. Even though the pointwise error reaches close to 100% when t is close to b, the midpoint solution lies on a torus, like the exact solution, whereas the RK4 (classical fourth-order) picture is noisy. Thus, we see yet again that truncation error is not everything, even in some nonstiff situations, and the theory in this case must include other aspects.

Integrate these equations using the two methods of Figure 4.9 with a constant step size $h = 0.1$ and $h = 0.01$ (four runs in total), monitoring the maximum deviation $|H(\mathbf{q}(t), \mathbf{p}(t)) - H(\mathbf{q}(0), \mathbf{p}(0))|$. (This is a simple error indicator which typically *underestimates* the error in the solution components, and is of interest in its own right.) What are your conclusions?

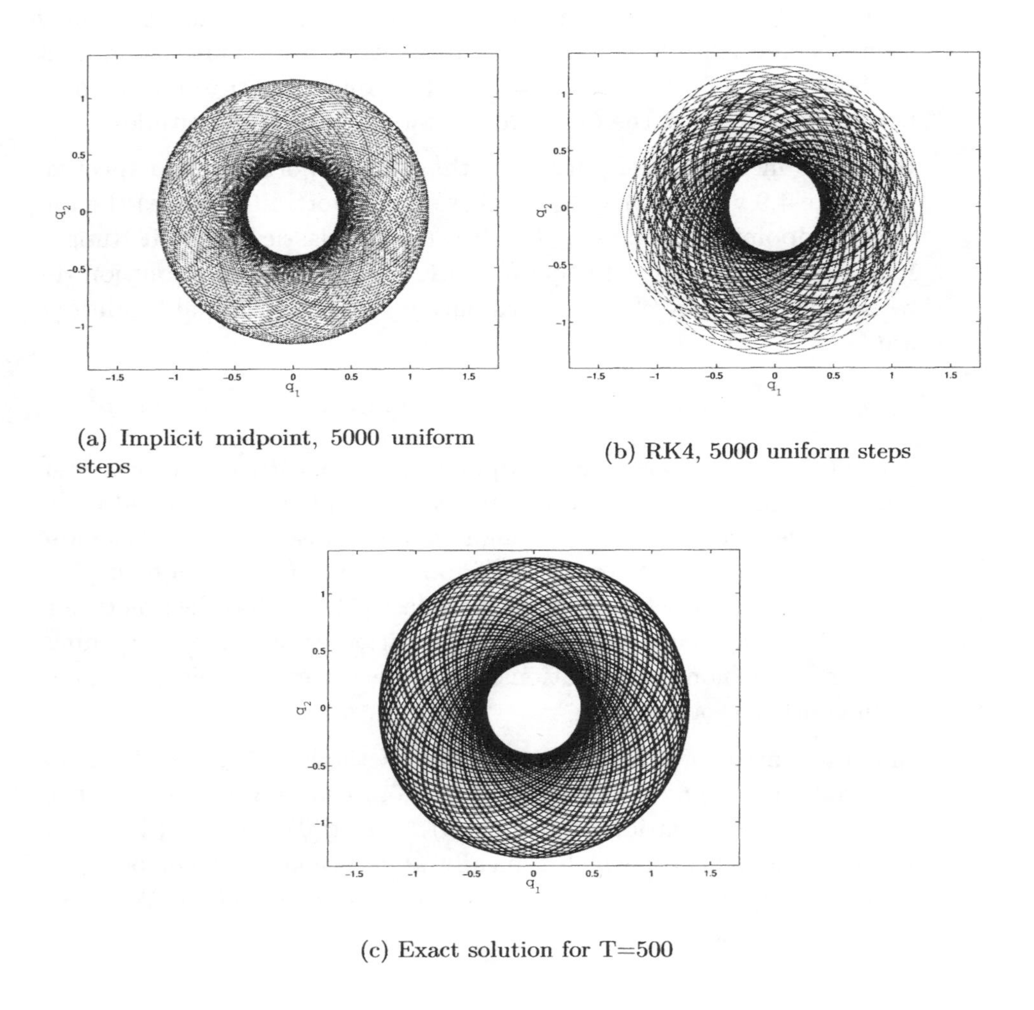

(a) Implicit midpoint, 5000 uniform steps

(b) RK4, 5000 uniform steps

(c) Exact solution for T=500

Figure 4.9: *Modified Kepler problem: Approximate and exact solutions.*

Chapter 5

Linear Multistep Methods

In this chapter we consider another group of methods extending the basic methods of Chapter 3 to higher order; see Figure 4.1 The methods considered here use information from previous integration steps to construct higher-order approximations in a simple fashion. Compared to the Runge–Kutta methods of the previous chapter the methods here typically require fewer function evaluations per step, and they allow a simpler, more streamlined method design, at least from the point of view of order and error estimate. On the other hand, the associated overhead is higher as well, e.g., when wanting to change the step size, and some of the flexibility of one-step methods is lost.

For our prototype ordinary differential equation (ODE) system

$$\mathbf{y}' = \mathbf{f}(t, \mathbf{y}), \qquad t \geq 0,$$

it is customary to denote

$$\mathbf{f}_l = \mathbf{f}(t_l, \mathbf{y}_l),$$

where $\mathbf{y}_l$ is the approximate solution at $t = t_l$. The general form of a k-step linear multistep method is given by

$$\sum_{j=0}^{k} \alpha_j \mathbf{y}_{n-j} = h \sum_{j=0}^{k} \beta_j \mathbf{f}_{n-j},$$

where α_j, β_j are the method's coefficients. We will assume that $\alpha_0 \neq 0$, and $|\alpha_k| + |\beta_k| \neq 0$. To eliminate arbitrary scaling, we set $\alpha_0 = 1$. The linear multistep method is *explicit* if $\beta_0 = 0$ and *implicit* otherwise. Note that the past k integration steps are assumed to be *equally spaced.*

Throughout most of this chapter we again consider a scalar ODE

$$y' = f(t, y)$$

to simplify the notation. The extension to ODE systems is straightforward unless otherwise noted. We also assume, as before, that f has as many bounded derivatives as needed. The general form of the method is rewritten for the scalar ODE for later reference,

$$\sum_{j=0}^{k} \alpha_j y_{n-j} = h \sum_{j=0}^{k} \beta_j f_{n-j}. \tag{5.1}$$

The method is called *linear* because, unlike general Runge–Kutta, the expression in (5.1) is linear in f. Make sure you understand that this *does not mean* that f is a linear function of y or t; i.e., it is the method which is linear, not the ODE problem to be solved. A consequence of this linearity is that the local truncation error, to be defined later, always has the simple expression

$$d_n = C_{p+1} h^p y^{(p+1)}(t_n) + O(h^{p+1}), \tag{5.2}$$

where p is the method's order and C_{p+1} is a computable constant. We will show this in Section 5.2.

The most popular linear multistep methods are based on polynomial interpolation, and even methods which are not based on interpolation use interpolation for such purposes as changing the step size. So be sure that you're up on polynomial interpolation in Newton's form.

5.1 The Most Popular Methods

Linear multistep methods typically come in families. The most popular for nonstiff problems is the *Adams family* and the most popular for stiff problems is the *backward differentiation formula* (BDF) *family.* In this section we derive these methods via the interpolating polynomial. In the next section we give an alternative derivation which is applicable for general multistep methods. We note that although the derived formulae in this section are for a constant step size h, the derivations themselves also suggest how to obtain formulae for a variable step size.

5.1.1 Adams Methods

Starting with the differential equation

$$y' = f(t, y),$$

we can integrate both sides to obtain

$$y(t_n) = y(t_{n-1}) + \int_{t_{n-1}}^{t_n} f(t, y(t))dt.$$

For Adams methods, the integrand $f(t, y)$ is approximated by an interpolating polynomial through previously computed values of $f(t_l, y_l)$. In the

Review: The interpolating polynomial and divided differences. Let $f(t)$ be a function to be interpolated at the k distinct points $t_1, t_2, \ldots, t_k$ by the unique polynomial $\phi(t)$ of degree $< k$ which satisfies the relations

$$\phi(t_l) = f(t_l), \quad l = 1, 2, \ldots, k.$$

The polynomial can be written down explicitly in Lagrangian form as we did in Chapter 4. Here, though, it is more convenient to write $\phi(t)$ in Newton's form:

$$\begin{aligned}\phi(t) &= f[t_1] + f[t_1, t_2](t - t_1) + \cdots \\ &\quad + f[t_1, t_2, \ldots, t_k](t - t_1)(t - t_2) \cdots (t - t_{k-1}),\end{aligned}$$

where the *divided differences* are defined recursively by $f[t_l] = f(t_l)$,

$$f[t_l, \ldots, t_{l+i}] = \frac{f[t_{l+1}, \ldots, t_{l+i}] - f[t_l, \ldots, t_{l+i-1}]}{t_{l+i} - t_l}.$$

The interpolation error at any point t is then

$$f(t) - \phi(t) = f[t_1, \ldots, t_k, t]\Pi_{i=1}^{k}(t - t_i).$$

If the points t_i and t are all in an interval of size $O(h)$ and f has k bounded derivatives, then the interpolation error is $O(h^k)$. If h is small then $k!f[t_1, \ldots, t_k, t] \approx f^{(k)}(t)$. Finally, for the case where the points t_l are equally spaced the expression for divided differences obviously simplifies. We define for future reference the backward differences

$$\begin{aligned}\nabla^0 f_l &= f_l, \\ \nabla^i f_l &= \nabla^{i-1} f_l - \nabla^{i-1} f_{l-1}.\end{aligned} \tag{5.3}$$

An important property of the backward differences of f is that they approximate the derivatives, $\nabla^k f \approx h^k f^{(k)}$.

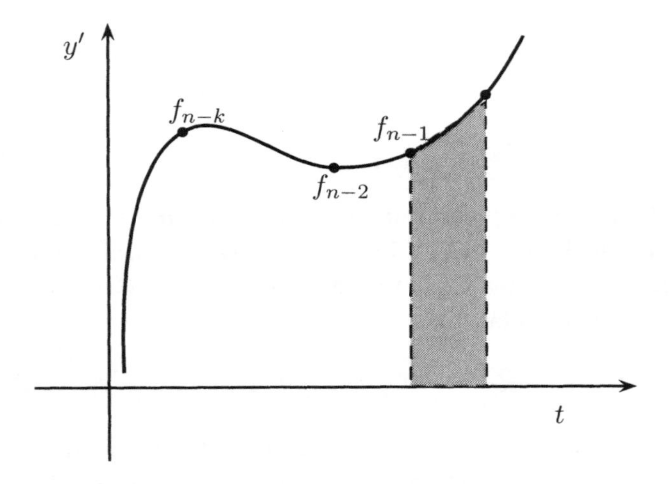

Figure 5.1: *Adams–Bashforth methods.*

general form (5.1) we therefore set, *for all Adams methods*, $\alpha_0 = 1$, $\alpha_1 = -1$, and $\alpha_j = 0$, $j > 1$.

The k-step **explicit Adams method** is obtained by interpolating f through the previous points $t = t_{n-1}, t_{n-2}, \ldots, t_{n-k}$; see Figure 5.1.

The explicit Adams methods[16] are the most popular among explicit multistep methods. A simple exercise in polynomial interpolation yields the formula

$$y_n = y_{n-1} + h \sum_{j=1}^{k} \beta_j f_{n-j},$$

where[17]

$$\begin{aligned} \beta_j &= (-1)^{j-1} \sum_{i=j-1}^{k-1} \binom{i}{j-1} \gamma_i, \\ \gamma_i &= (-1)^i \int_0^1 \binom{-s}{i} ds\,. \end{aligned}$$

This formula is a k-step method because it uses information at the k points $t_{n-1}, t_{n-2}, \ldots, t_{n-k}$. It is sometimes also called a $(k+1)$-value method, because the total information per step, which determines storage requirements, also involves y_{n-1}.

---

[16] Called Adams–Bashforth after J. C. Adams, who invented them to solve a problem of capillary action in collaboration with F. Bashforth. They were published in 1883.

[17] Recall

$$\binom{s}{i} = \frac{s(s-1)\cdots(s-i+1)}{i!}, \qquad \binom{s}{0} = 1.$$

The local truncation error turns out to be $C_{p+1}h^p y^{(p+1)}(t_n) + O(h^{p+1})$, where $p = k$. Note that there is only one function evaluation per step.

Example 5.1
The first-order Adams–Bashforth method is the forward Euler method. The second-order Adams–Bashforth method is given by the above formula with $k = 2$ and $\gamma_0 = 1$, $\gamma_1 = 1/2$. This yields

$$y_n = y_{n-1} + h\left(f_{n-1} + \frac{1}{2}\nabla f_{n-1}\right)$$

or, equivalently,

$$y_n = y_{n-1} + h\left(\frac{3}{2}f_{n-1} - \frac{1}{2}f_{n-2}\right).$$

♦

Table 5.1 gives the coefficients of the Adams–Bashforth methods for k up to 6.

p	k	$j \rightarrow$	1	2	3	4	5	6	C_{p+1}
1	1	β_j	1						$\frac{1}{2}$
2	2	$2\beta_j$	3	-1					$\frac{5}{12}$
3	3	$12\beta_j$	23	-16	5				$\frac{3}{8}$
4	4	$24\beta_j$	55	-59	37	-9			$\frac{251}{720}$
5	5	$720\beta_j$	1901	-2774	2616	-1274	251		$\frac{95}{288}$
6	6	$1440\beta_j$	4277	-7923	9982	-7298	2877	-475	$\frac{19087}{60480}$

Table 5.1: *Coefficients of Adams–Bashforth methods up to order* 6.

The Adams–Bashforth methods are explicit methods with very small regions of absolute stability. This has inspired the implicit versions of Adams methods, also called Adams–Moulton.

The k-step **implicit Adams method** is derived similarly to the explicit method. The difference is that for this method, the interpolating polynomial is of degree $\leq k$, and it interpolates f at the unknown value t_n as well; see Figure 5.2.

This yields an implicit multistep method

$$y_n = y_{n-1} + h\sum_{j=0}^{k} \beta_j f_{n-j}.$$

The order of the k-step Adams–Moulton method is $p = k+1$ (that $p \geq k+1$ follows immediately from the fact that $k+1$ points are used in the underlying

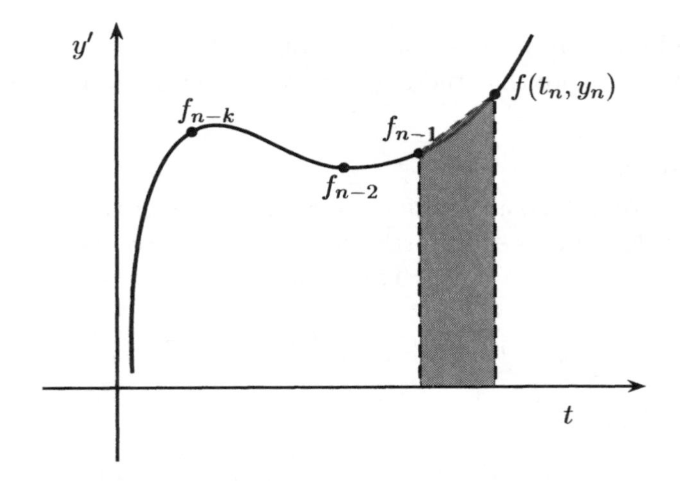

Figure 5.2: *Adams–Moulton methods.*

polynomial interpolation). An exception is the case for $k = 1$, where f_{n-1} is not used, yielding $p = k = 1$. A straightforward use of interpolation yields the coefficients summarized in Table 5.2.

p	k	$j \to$	0	1	2	3	4	5	C_{p+1}
1	1	β_j	1						$-\frac{1}{2}$
2	1	$2\beta_j$	1	1					$-\frac{1}{12}$
3	2	$12\beta_j$	5	8	-1				$-\frac{1}{24}$
4	3	$24\beta_j$	9	19	-5	1			$-\frac{19}{720}$
5	4	$720\beta_j$	251	646	-264	106	-19		$-\frac{3}{160}$
6	5	$1440\beta_j$	475	1427	-798	482	-173	27	$-\frac{863}{60480}$

Table 5.2: *Coefficients of Adams–Moulton methods up to order* 6.

Example 5.2
Here are some examples of Adams–Moulton methods:

- $k = 1$ with $\beta_1 = 0$ gives the backward Euler method;
- $k = 1$ with $\beta_1 \neq 0$ gives the implicit trapezoidal method;
- $k = 2$ gives $y_n = y_{n-1} + \frac{h}{12}[5f_n + 8f_{n-1} - f_{n-2}]$. ♦

The Adams–Moulton methods have smaller error constants than the Adams–Bashforth methods of the same order, and use one fewer step for the same order. They have much larger stability regions than the Adams–Bashforth methods. But they are implicit. Adams–Moulton methods are often used together with Adams–Bashforth methods for the solution of non-stiff ODEs. This type of implementation is called *predictor-corrector* and will be described later, in Section 5.4.2.

5.1.2 BDF

The most popular multistep methods for stiff problems are the BDF. Their distinguishing feature is that $f(t, y)$ is evaluated only at the right end of the current step, (t_n, y_n). A motivation behind this is to obtain formulae with the stiff decay property (recall Section 3.5). Applying the general linear multistep method (5.1) to the ODE $y' = \lambda(y - g(t))$ and considering the limit $h\mathcal{R}e(\lambda) \to -\infty$, we have $\sum_{j=0}^{k} \beta_j(y_{n-j} - g(t_{n-j})) \to 0$. To obtain $y_n - g(t_n) \to 0$ for an arbitrary function $g(t)$, we *must* therefore set $\beta_0 \neq 0$ and $\beta_j = 0, j > 0$. This leaves treating y' in the differential equation $y'(t) = f(t, y(t))$. In contrast to the Adams methods, which were derived by *integrating* the polynomial which interpolates past values of f, the BDF methods are derived by *differentiating* the polynomial which interpolates past values of y, and setting the derivative at t_n to $f(t_n, y_n)$. This yields the k-step BDF, which has order $p = k$,

$$\sum_{i=1}^{k} \frac{1}{i} \nabla^i y_n = hf(t_n, y_n).$$

This can be written in scaled form where $\alpha_0 = 1$,

$$\sum_{i=0}^{k} \alpha_i y_{n-i} = h\beta_0 f(t_n, y_n).$$

The BDF methods are implicit and are usually implemented together with a modified Newton method to solve the nonlinear system at each time step. The first six members of this family are listed in Table 5.3. The first, one-step method is again backward Euler.

5.1.3 Initial Values for Multistep Methods

For one-step methods we set $y_0 = y(0)$, the given initial value. Nothing else is needed to start up the iteration in time for $n = 1, 2, \ldots$.

With a k-step method, in contrast, the method is applied for $n = k, k + 1, \ldots$. Thus, k initial values $y_0, y_1, \ldots, y_{k-1}$ are needed to start it up. The additional initial values $y_1, \ldots, y_{k-1}$ must be $O(h^p)$ accurate for a method of order p, if the full convergence order is to be realized (Section 5.2.3). If error control is used, these additional starting values must be accurate to a given error tolerance.

p	k	β_0	α_0	α_1	α_2	α_3	α_4	α_5	α_6
1	1	1	1	-1					
2	2	$\frac{2}{3}$	1	$-\frac{4}{3}$	$\frac{1}{3}$				
3	3	$\frac{6}{11}$	1	$-\frac{18}{11}$	$\frac{9}{11}$	$-\frac{2}{11}$			
4	4	$\frac{12}{25}$	1	$-\frac{48}{25}$	$\frac{36}{25}$	$-\frac{16}{25}$	$\frac{3}{25}$		
5	5	$\frac{60}{137}$	1	$-\frac{300}{137}$	$\frac{300}{137}$	$-\frac{200}{137}$	$\frac{75}{137}$	$-\frac{12}{137}$	
6	6	$\frac{60}{147}$	1	$-\frac{360}{147}$	$\frac{450}{147}$	$-\frac{400}{147}$	$\frac{225}{147}$	$-\frac{72}{147}$	$\frac{10}{147}$

Table 5.3: *Coefficients of BDF methods up to order* 6.

To obtain these additional initial values, an appropriate Runge–Kutta method can be used. Another approach, utilized in all modern multistep packages, is to recursively use a $(k-1)$-step method. As we have seen, linear multistep methods tend to come in families, so a general-purpose code can be written which implements the first methods of such a family, for $k = 1, 2, \ldots, p$, say. Then the code can at a starting (or a restarting) point t gradually and adaptively increase the method's number of steps (and correspondingly, its order).

Example 5.3
We compute the solution of the simple Example 3.1:

$$y' = -5ty^2 + \frac{5}{t} - \frac{1}{t^2}, \qquad y(1) = 1.$$

The exact solution is $y(t) = 1/t$. We record results parallel to Example 4.1, i.e., for the same constant step sizes, measuring absolute errors and convergence rates at $t = 25$. Results for some Adams–Bashforth, Adams–Moulton, and BDF methods are displayed in Tables 5.4, 5.5, and 5.6, respectively. The initial values for the k-step method are obtained using the exact solution. The symbol $*$ denotes an "infinite" error, which occurs when the absolute stability restriction is strongly violated.

In Table 5.4 we can observe the high accuracy that the higher-order methods achieve for this very smooth problem. However, these small errors are wiped out by an explosion of the roundoff error if the step size is so large that $-10hy$ is not in the absolute stability region of the method. The region of absolute stability is seen to be shrinking as the order of the method increases, in contrast to the Runge–Kutta results of Table 4.1.

In the first column $k = 1$ of Table 5.5 we see the results for the backward Euler method. For h small they are very close to those of the forward Euler method (Table 4.1), but for the larger values of h they are much better. Newton's method was used to obtain convergence of the nonlinear

Step h	$(1,1)$ error	Rate	$(2,2)$ error	Rate	$(4,4)$ error	Rate
0.2	.40e-2		*		*	
0.1	.65e-6	12.59	.70e-3		*	
0.05	.32e-6	1.00	.16e-8	18.7	.16e-1	
0.02	.13e-6	1.00	.26e-9	2.00	.35e-14	31.8
0.01	.65e-7	1.00	.65e-10	2.00	.16e-14	1.17
0.005	.32e-7	1.00	.16e-10	2.00	.11e-14	0.54
0.002	.13e-7	1.00	.26e-11	2.00	.47e-14	-1.61

Table 5.4: *Example* 5.3*: Errors and calculated convergence rates for Adams–Bashforth methods;* (k,p) *denotes the* k*-step method of order* p*.*

Step h	$(1,1)$ error	Rate	$(1,2)$ error	Rate	$(3,4)$ error	Rate
0.2	.13e-5		.52e-8		.22e-11	
0.1	.65e-6	1.01	.13e-8	2.00	.14e-12	4.03
0.05	.32e-6	1.00	.33e-9	2.00	.87e-14	3.96
0.02	.13e-6	1.00	.52e-10	2.00	.50e-15	3.12
0.01	.65e-7	1.00	.13e-10	2.00	.17e-14	-1.82
0.005	.32e-7	1.00	.33e-11	2.00	.11e-14	0.73
0.002	.13e-7	1.00	.52e-12	2.00	.47e-14	-1.62

Table 5.5: *Example* 5.3*: Errors and calculated convergence rates for Adams–Moulton methods;* (k,p) *denotes the* k*-step method of order* p*.*

iteration for $h > .02$, and functional iteration was used for the smaller step sizes. The column of $p = 2$ describes the performance of the second-order trapezoidal method. For the fourth-order method the error reaches roundoff level already for $h = .02$. The BDF methods perform similarly to the Adams–Moulton methods for this nonstiff problem. The order of the methods, before the onset of roundoff error, is clearly reflected in the results. The absolute value of the errors is unusually small. ♦

5.2 Order, 0-Stability, and Convergence

As in the two previous chapters, the basic convergence theory requires that a method have a certain (positive) order of accuracy (i.e., consistency) and that it be 0-stable. The emphasis, though, is somewhat different here from

Step h	$(1,1)$ error	Rate	$(2,2)$ error	Rate	$(4,4)$ error	Rate
0.2	.13e-5		.21e-7		.17e-10	
0.1	.65e-6	1.01	.53e-8	2.02	.10e-11	4.06
0.05	.32e-6	1.00	.13e-8	2.01	.65e-13	4.02
0.02	.13e-6	1.00	.21e-9	2.00	.94e-15	4.62
0.01	.65e-7	1.00	.52e-10	2.00	.18e-14	-0.96
0.005	.32e-7	1.00	.13e-10	2.00	.10e-14	0.86
0.002	.13e-7	1.00	.21e-11	2.00	.46e-14	-1.66

Table 5.6: *Example 5.3: Errors and calculated convergence rates for BDF methods; (k,p) denotes the k-step method of order p.*

what we had for Runge–Kutta methods: whereas there 0-stability was trivial and attaining useful methods with a high order of accuracy was tricky, here 0-stability is not automatic (although it is not difficult to check), whereas attaining high order is straightforward, provided only that we are prepared to use sufficiently many past values and provide sufficiently accurate initial values. Note also that the restriction to a constant step size, which is not needed in Section 4.3, simplifies life considerably in this section.

5.2.1 Order

The simple derivation below is incredibly general: it will give us a tool not only for checking a method's order but also for finding its leading local truncation error term and even for designing linear multistep methods, given some desired criteria.

Define the linear operator $\mathcal{L}_h y(t)$ by

$$\mathcal{L}_h y(t) = \sum_{j=0}^{k} [\alpha_j y(t - jh) - h\beta_j y'(t - jh)], \tag{5.4}$$

where $y(t)$ is an arbitrary continuously differentiable function on $[0, b]$. The *local truncation error* is naturally defined as the defect obtained when plugging the exact solution into the difference equation (which here is (5.1) divided by h; see Section 3.2). This can be written as

$$d_n = h^{-1}\mathcal{L}_h y(t_n), \tag{5.5}$$

where $y(t)$ is the exact solution. In particular, the exact solution satisfies $y' = f(t, y(t))$, so

$$\mathcal{L}_h y(t) = \sum_{j=0}^{k} [\alpha_j y(t - jh) - h\beta_j f(t - jh, y(t - jh))].$$

If we now expand $y(t-jh)$ and $y'(t-jh)$ in Taylor series about t and collect terms, we have

$$\mathcal{L}_h y(t) = C_0 y(t) + C_1 h y'(t) + \cdots + C_q h^q y^{(q)}(t) + \cdots ,$$

where the C_q are computable constants. Recall that the order of the method is p if $d_n = O(h^p)$. Thus:

- the order of the linear multistep method is p iff

$$C_0 = C_1 = \cdots = C_p = 0, \quad C_{p+1} \neq 0;$$

- the local truncation error is given, as advertised in (5.2), by

$$d_n = C_{p+1} h^p y^{(p+1)}(t_n) + O(h^{p+1}).$$

From the Taylor series expansions, it can be easily seen that the coefficients are given by:

$$\begin{aligned} C_0 &= \sum_{j=0}^{k} \alpha_j, \\ C_i &= (-1)^i \left[\frac{1}{i!} \sum_{j=1}^{k} j^i \alpha_j + \frac{1}{(i-1)!} \sum_{j=0}^{k} j^{i-1} \beta_j \right], \quad i = 1, 2, \ldots. \end{aligned} \tag{5.6}$$

To obtain a method of order p, therefore, the first p of these expressions must be set to 0. The first few of these conditions read

$$\begin{aligned} 0 &= \alpha_0 + \alpha_1 + \alpha_2 + \cdots + \alpha_k, \\ 0 &= (\alpha_1 + 2\alpha_2 + \cdots + k\alpha_k) + (\beta_0 + \beta_1 + \beta_2 + \cdots + \beta_k), \\ 0 &= \frac{1}{2}(\alpha_1 + 4\alpha_2 + \cdots + k^2\alpha_k) + (\beta_1 + 2\beta_2 + \cdots + k\beta_k), \end{aligned}$$

etc. When the order is p, C_{p+1} is called the *error constant* of the method.

Example 5.4
For the forward Euler method, $\alpha_1 = -1$, $\beta_1 = 1$. So,

$$C_0 = 1 - 1 = 0, \quad C_1 = 1 - 1 = 0, \quad C_2 = -\frac{1}{2} + 1 = \frac{1}{2}.$$

For the two-step Adams–Bashforth method, $\alpha_1 = -1$, $\beta_1 = \frac{3}{2}$, $\beta_2 = -\frac{1}{2}$. So,

$$C_0 = 1 - 1 = 0, \quad C_1 = 1 - \frac{3}{2} + \frac{1}{2} = 0, \quad C_2 = -\frac{1}{2} + \frac{3}{2} - 1 = 0,$$

$$C_3 = -\left(-\frac{1}{6} + \frac{1}{2}\left(\frac{3}{2} - 2\right)\right) = \frac{5}{12}.$$ ♦

Example 5.5
The coefficients of the methods of the previous section can be obtained by applying their family design criteria to select some method coefficients and then using the order conditions to choose the remaining coefficients such that the order is maximized.

For instance, consider a two-step BDF, $\beta_0 \neq 0$, $\beta_1 = \beta_2 = 0$. The method is

$$y_n + \alpha_1 y_{n-1} + \alpha_2 y_{n-2} = h\beta_0 f_n.$$

The order conditions give the linear equations

$$\begin{aligned} 1 + \alpha_1 + \alpha_2 &= 0, \\ \alpha_1 + 2\alpha_2 + \beta_0 &= 0, \\ \alpha_1 + 4\alpha_2 &= 0. \end{aligned}$$

This system can be easily solved to yield $\beta_0 = \frac{2}{3}$, $\alpha_1 = -\frac{4}{3}$, $\alpha_2 = \frac{1}{3}$, as per Table 5.3. The coefficient of the leading term of the local truncation error is $C_3 = -\frac{1}{6}\left(-\frac{4}{3} + \frac{8}{3}\right) = -\frac{2}{9}$. ♦

Given some of the α's and β's we can obviously use these relations to find the remaining α's and β's for the method of maximal order (see Exercise 5.3; note, though, that this method may not be optimal, or even usable, due to stability considerations).

A linear multistep method is *consistent* if it has order $p \geq 1$. Thus, the method is consistent iff

$$\sum_{j=0}^{k} \alpha_j = 0, \quad \sum_{j=1}^{k} j\alpha_j + \sum_{j=0}^{k} \beta_j = 0.$$

Sometimes it is more convenient to express the linear multistep method in terms of the *characteristic polynomials*

$$\rho(\xi) = \sum_{j=0}^{k} \alpha_j \xi^{k-j}, \tag{5.7a}$$

$$\sigma(\xi) = \sum_{j=0}^{k} \beta_j \xi^{k-j}. \tag{5.7b}$$

In terms of these polynomials, the linear multistep method is consistent iff $\rho(1) = 0$, $\rho'(1) = \sigma(1)$.

Note: The material that follows below is important for the fundamental understanding of linear multistep methods. We derive simple conditions on the roots of the characteristic polynomial $\rho(\xi)$ which guarantee that a method is 0-stable. This, together with consistency, then gives convergence. Recall that $\rho(1) = 0$ by consistency, so this determines one root. The bottom line of the following discussion is that a usable method must have all other roots of the polynomial $\rho(\xi)$ strictly inside the unit circle. A reader who is interested mainly in practical aspects may therefore skip the next few pages until after Example 5.7, at least on first reading.

5.2.2 Stability: Difference Equations and the Root Condition

One way of looking at a linear multistep method is that it is a difference equation which approximates the differential equation. The stability of the linear multistep method and the essential theoretical difference between multistep and one-step methods are given by the stability of the difference equation.

Before discussing stability for linear multistep methods, we review some basic facts about linear difference equations with constant coefficients. Given such a scalar difference equation

$$a_k y_{n-k} + a_{k-1} y_{n-k+1} + \cdots + a_0 y_n = q_n, \qquad n = k, k+1, \ldots ,$$

if $\{v_n\}$ is a particular solution for this equation then the general solution is $y_n = x_n + v_n$, where x_n is the general solution to the homogeneous difference equation,[18]

$$a_k x_{n-k} + a_{k-1} x_{n-k+1} + \cdots + a_0 x_n = 0, \qquad n = k, k+1, \ldots .$$

There are k linearly independent solutions to the homogeneous equation. To find them, we try the educated guess (ansatz) $x_n = \xi^n$. Substituting into the homogeneous difference equation we have

$$\phi(\xi) = \sum_{j=0}^{k} a_j \xi^{k-j} = 0, \tag{5.8}$$

thus ξ must be a zero of the polynomial $\phi(\xi)$. If all k roots are distinct, then the general solution is given by

$$y_n = \sum_{i=1}^{k} c_i \xi_i^n + v_n,$$

[18] For example, $\{v_n\}$ can be the solution of the difference equation with zero initial conditions, $v_0 = v_1 = \cdots = v_{k-1} = 0$.

where the c_i, $i = 1, \ldots, k$, are arbitrary constants which are determined by the k initial conditions required for the difference equation. If the roots are not distinct, say $\xi_1 = \xi_2$ is a double root, then the solution is given by

$$y_n = c_1\xi_1^n + c_2 n\xi_2^n + \sum_{i=3}^{k} c_i\xi_i^n + v_n.$$

For a triple root, we have $\xi^n, n\xi^n, n(n-1)\xi^n$ as solution modes, etc. Thus the solutions to the difference equation are intimately related to the roots of the characteristic polynomial which is associated with it.

We can define *stability* for this difference equation similarly to stability for a differential equation (see Sections 2.1–2.2, particularly Example 2.2). Clearly, for a perturbation in the c_i not to grow unboundedly with n, we need to bound the roots ξ_i. We give the following definitions, in complete analogy to the constant-coefficient ODE case.

- The difference equation is *stable* if all k roots of $\phi(\xi)$ satisfy $|\xi_i| \le 1$, and if $|\xi_i| = 1$, then ξ_i is a simple root.
- The difference equation is *asymptotically stable* if all roots satisfy $|\xi_i| < 1$.

This completes our review of difference equations.

For multistep methods applied to the test equation $y' = \lambda y$, the difference equation is given by

$$\sum_{j=0}^{k} (\alpha_j - h\lambda\beta_j) y_{n-j} = 0. \tag{5.9}$$

This is a homogeneous, constant-coefficient difference equation, like what we have just treated, with $a_j = \alpha_j - h\lambda\beta_j$. A solution to this difference equation is $\{\xi_i^n\}$ if ξ_i is a root of the polynomial $\phi(\xi) = \rho(\xi) - h\lambda\sigma(\xi) = 0$. Since the solution to the ODE (with $y(0) = 1$) is $y = e^{\lambda t} = (e^{h\lambda})^n$, we expect one root to approximate $e^{h\lambda}$ so that y_n can approximate $y(t_n)$ (i.e., this is a consistency requirement). That root is called the *principal root.* The other roots are called *extraneous roots.*

What should strike you in the above review is how closely the solution procedure for the difference equation is related to that of a scalar differential equation *of order k.* The source of these extraneous roots (also referred to at times as *parasitic roots*) is the discrepancy between the ODE of order 1, which should be approximated, and the ODE of order k, which is approximated instead by the multistep method. A good multistep method must therefore ensure that these extraneous roots, which cannot do any good, do not cause any harm either. This is what 0-stability (and strong stability, to be defined below) is about.

5.2.3 0-Stability and Convergence

Recall that in the previous two chapters convergence followed from accuracy using a perturbation bound, i.e., 0-stability. Consider an ODE system, $\mathbf{y}' = \mathbf{f}(t, \mathbf{y})$, on the interval $[0, b]$. The definition (3.9) of 0-stability for one-step methods must be updated here to read that the linear multistep method is 0-stable if there are positive constants h_0 and K such that for any mesh functions $\mathbf{x}_h$ and $\mathbf{z}_h$ with $h \le h_0$,

$$|\mathbf{x}_l - \mathbf{z}_l| \le K \left\{ \sum_{i=0}^{k-1} |\mathbf{x}_i - \mathbf{z}_i| + \max_{k \le n \le N} \left| h^{-1} \sum_{j=0}^{k} \alpha_j (\mathbf{x}_{n-j} - \mathbf{z}_{n-j}) \right.\right.$$
$$\left.\left. - \sum_{j=0}^{k} \beta_j (\mathbf{f}(t_{n-j}, \mathbf{x}_{n-j}) - \mathbf{f}(t_{n-j}, \mathbf{z}_{n-j})) \right| \right\}, \quad 1 \le l \le N. \qquad (5.10)$$

If we have this bound then convergence follows immediately. In fact, by plugging $\mathbf{x}_n \leftarrow \mathbf{y}_n$ and $\mathbf{z}_n \leftarrow \mathbf{y}(t_n)$ into the stability bound (5.10) we obtain that if the k initial values are accurate to order p and the method has order p, then the global error is $O(h^p)$.

The 0-stability bound is cumbersome to check for a given linear multistep method. Fortunately, it turns out that it is equivalent to a simple condition on the roots of the characteristic polynomial $\rho(\xi)$ of (5.7a). The complete proof is technical and appears in classical texts. Instead, we bring its essence.

As the name implies, 0-stability is concerned with what happens in the limit $h \to 0$. In this limit, it is sufficient to consider the ODE $\mathbf{y}' = \mathbf{0}$ (corresponding to the fact that $\mathbf{y}'$ is the dominant part of the differential operator $\mathbf{y}' - \mathbf{f}(t, \mathbf{y})$). Now, the ODE $\mathbf{y}' = \mathbf{0}$ is decoupled, so we can consider a scalar component $y' = 0$. For the latter ODE, the method reads

$$\alpha_k y_{n-k} + \alpha_{k-1} y_{n-k+1} + \cdots + \alpha_0 y_n = 0.$$

This is a difference equation of the type considered in the previous subsection. It must be stable for the multistep method to be 0-stable. Identifying $\phi(\xi)$ of (5.8) with $\rho(\xi)$ of (5.7a), we obtain the following theorem.

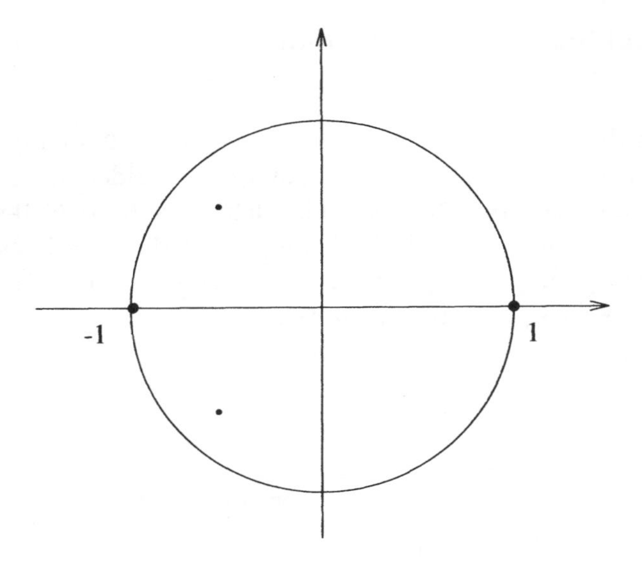

Figure 5.3: *Zeros of $\rho(\xi)$ for a 0-stable method.*

Theorem 5.1

- *The linear multistep method is 0-stable iff all roots ξ_i of the characteristic polynomial $\rho(\xi)$ satisfy*

$$|\xi_i| \leq 1,$$

and if $|\xi_i| = 1$, then ξ_i is a simple root, $1 \leq i \leq k$.

- *If the root condition is satisfied, the method is accurate to order p, and the initial values are accurate to order p, then the method is convergent to order p.*

Note that the root condition guaranteeing 0-stability relates to the characteristic polynomial $\rho(\xi)$ alone; see Figure 5.3. Also, for any consistent method the polynomial ρ has the root 1. One-step methods have no other roots, which again highlights the fact that they are automatically 0-stable.

Example 5.6
Instability is a disaster. Here is an example of an unstable method,

$$y_n = -4y_{n-1} + 5y_{n-2} + 4hf_{n-1} + 2hf_{n-2}\,.$$

In terms of the local truncation error, this is the most accurate explicit two-step method. However, $\rho(\xi) = \xi^2 + 4\xi - 5 = (\xi - 1)(\xi + 5)$. The extraneous root is $\xi_2 = -5$ and the root condition is violated.

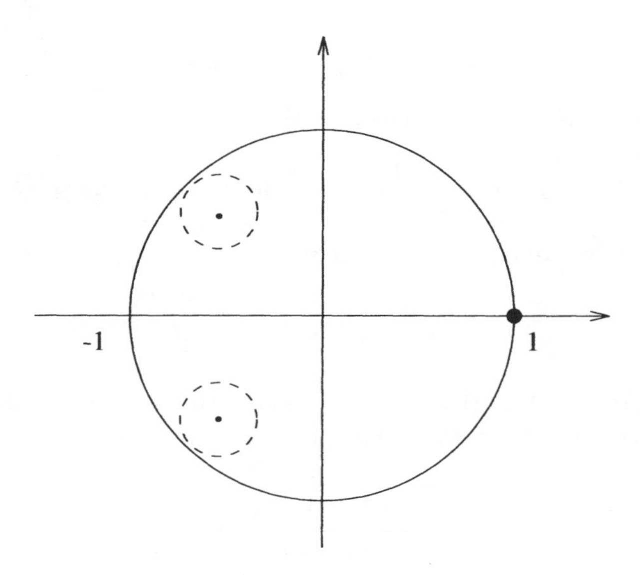

Figure 5.4: *Zeros of $\rho(\xi)$ for a strongly stable method. It is possible to draw a circle contained in the unit circle about each extraneous root.*

Consider solving $y' = 0$ with initial values $y_0 = 0$, $y_1 = \epsilon$. Then

$$\begin{aligned} y_2 &= -4y_1 = -4\epsilon, \\ y_3 &= -4y_2 + 5y_1 = 21\epsilon, \\ y_4 &= -4y_3 + 5y_2 = -104\epsilon, \\ &\vdots \end{aligned}$$

There is no hope for convergence here. ♦

Consider again the test equation, and its discretization (5.9). If $\mathcal{R}e(\lambda) < 0$ then the exact solution decays and we must prevent any growth in the approximate solution. This is not possible for all such λ if there are extraneous roots of the polynomial ρ with magnitude 1. For $h > 0$ sufficiently small the difference equation (5.9) must be asymptotically stable in this case; see Figure 5.4. We define a linear multistep method to be

- *strongly stable* if all roots of $\rho(\xi) = 0$ are inside the unit circle except for the root $\xi = 1$;
- *weakly stable* if it is 0-stable but not strongly stable.

Example 5.7
Weak stability can be a disaster for some problems, too. Consider Milne's method,

$$y_n = y_{n-2} + \frac{1}{3}h(f_n + 4f_{n-1} + f_{n-2})$$

for $y' = \lambda y$. The error satisfies the equation

$$e_n = e_{n-2} + \frac{1}{3}h\lambda(e_n + 4e_{n-1} + e_{n-2}).$$

Substituting as before $e_n = \xi^n$, we have

$$\left(1 - \frac{1}{3}h\lambda\right)\xi^2 - \frac{4}{3}h\lambda\xi - \left(1 + \frac{1}{3}h\lambda\right) = 0$$

$(\rho(\xi) - h\lambda\sigma(\xi) = 0)$. Clearly, $\rho = \xi^2 - 1$ has a root at $+1$ and a root at -1. The roots of the full polynomial equation are given by

$$\xi = \frac{\frac{2}{3}h\lambda \pm \sqrt{1 + \frac{1}{3}(h\lambda)^2}}{(1 - \frac{1}{3}h\lambda)}.$$

By expanding ξ into a power series in $h\lambda$, we find that

$$\begin{aligned} \xi_1 &= e^{h\lambda} + O(h\lambda)^5, \\ \xi_2 &= -e^{-(\frac{h\lambda}{3})} + O(h^3). \end{aligned}$$

For $\lambda < 0$, the extraneous root dominates, so the solution is unstable. ♦

A practically minded reader must conclude that any useful linear multistep method must be strongly stable. We shall not be interested henceforth in any other methods. But this restricts the attainable order of accuracy. G. Dahlquist showed that strongly stable k-step methods can have at most order $k + 1$.

Example 5.8
The Adams methods, both explicit and implicit, have the characteristic polynomial

$$\rho(\xi) = \xi^k - \xi^{k-1} = (\xi - 1)\xi^{k-1},$$

so the extraneous roots are all 0 for any k. These methods are all strongly stable. The implicit methods have the highest order attainable.

This explains in part the popularity of Adams methods. ♦

Example 5.9
The BDF methods were motivated in Section 5.1 by the desire to achieve stiff decay. This, however, does not automatically mean that they are strongly stable. Exercise 5.4 shows that BDF methods are 0-stable for $1 \le k \le 6$ and unstable for $k > 6$. Thus, only the first six members of this family are usable. ♦

5.3 Absolute Stability

Recall that the general linear multistep method

$$\sum_{j=0}^{k} \alpha_j y_{n-j} = h \sum_{j=0}^{k} \beta_j f_{n-j}$$

applied to the test equation $y' = \lambda y$ gives (5.9), i.e.,

$$\sum_{j=0}^{k} \alpha_j y_{n-j} = h\lambda \sum_{j=0}^{k} \beta_j y_{n-j}.$$

If we let $y_n = \xi^n$, then ξ must satisfy

$$\sum_{j=0}^{k} \alpha_j \xi^{k-j} = h\lambda \sum_{j=0}^{k} \beta_j \xi^{k-j}, \tag{5.11}$$

or $\rho(\xi) = h\lambda\sigma(\xi)$.

Now, the method is absolutely stable for those values of $z = h\lambda$ such that $|y_n|$ does not grow with n. This corresponds to values for which all roots of (5.11) satisfy $|\xi| \leq 1$.

For differential equations with positive eigenvalues, it is sometimes convenient to define an alternate concept: the *region of relative stability*. This is a region where the extraneous roots may be growing, but they are growing more slowly than the principal root, so that the principal root still dominates. We will not pursue this further.

Finding the region of absolute stability is simple for linear multistep methods. Just look for the boundary

$$z = \frac{\rho(e^{\imath\theta})}{\sigma(e^{\imath\theta})}$$

and plot (the complex scalar) z for θ ranging from 0 to 2π.

In Figure 5.5 we plot absolute stability regions for the Adams methods. The first two Adams–Moulton methods are missing because they are A-stable. Notice how much larger the stability regions for the Adams–Moulton methods are compared to the Adams–Bashforth methods for the same order (or for the same number of steps): interpolation is a more stable process than extrapolation.

Recall the definition of A-stability: a numerical method is *A-stable* if its region of absolute stability contains the left half-plane $h\mathcal{R}e(\lambda) < 0$. Unfortunately, A-stability is very difficult to attain for multistep methods. It can be shown that:

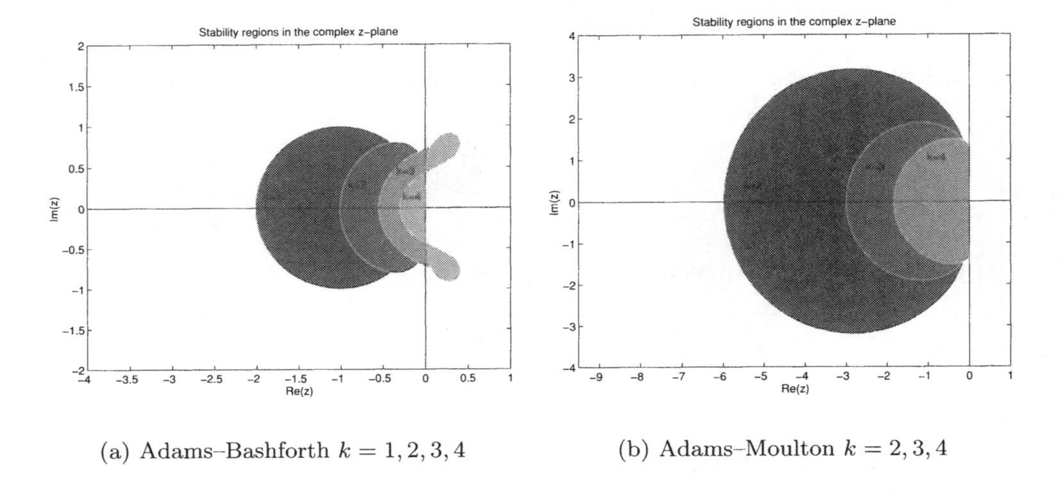

(a) Adams–Bashforth $k = 1, 2, 3, 4$ (b) Adams–Moulton $k = 2, 3, 4$

Figure 5.5: *Absolute stability regions of Adams methods.*

- an explicit linear multistep method cannot be A-stable;
- the order of an A-stable linear multistep method cannot exceed 2;
- the second-order A-stable implicit linear multistep method with smallest error constant ($C_3 = \frac{1}{12}$) is the trapezoidal method.

The utility of the trapezoidal method has already been discussed in Section 3.6.[19]

If we want to use linear multistep methods for stiff problems, the A-stability requirement must be relaxed. Moreover, the discussion in Chapter 3 already reveals that in the very stiff limit $h\mathcal{R}e(\lambda) \to -\infty$, the A-stability bound may not be sufficient and the concept of stiff decay is more useful. The BDF methods introduced in Section 5.1 trade in chunks of absolute stability regions near the imaginary axis for stiff decay. The size of these chunks increases with the number of steps k, until the methods become unstable for $k > 6$; see Figure 5.6.

[19]These results were given by Dahlquist in the 1960s. They had a major impact on research in this area in the 1960s and 1970s. Today it is still easy to appreciate their mathematical beauty and the sophistication that went into the proofs, even though a glance at the methods used in successful implementations makes it clear that A-stability is *not* the property that separates the winners from the also-rans.

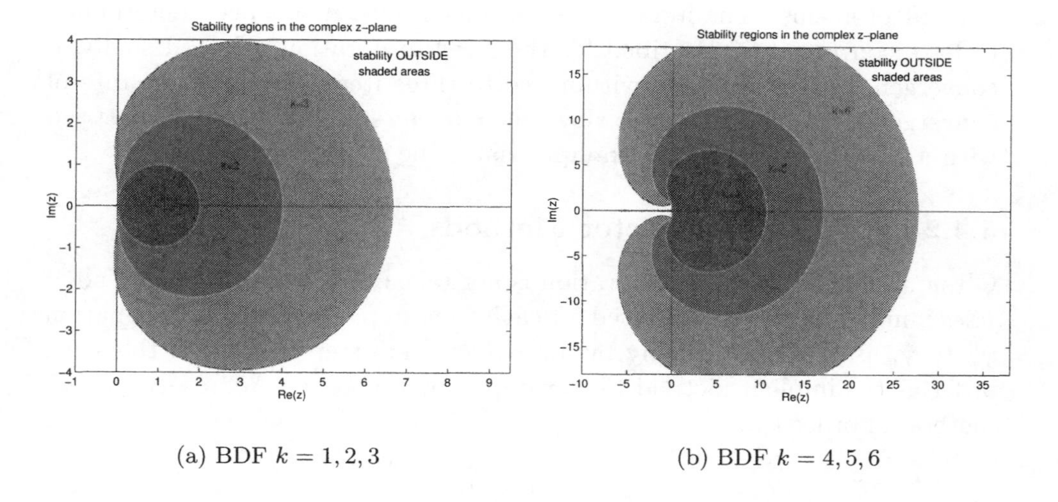

(a) BDF $k = 1, 2, 3$ (b) BDF $k = 4, 5, 6$

Figure 5.6: *BDF absolute stability regions. The stability regions are outside the shaded area for each method.*

5.4 Implementation of Implicit Linear Multistep Methods

When using an implicit, k-step linear multistep method, i.e., $\beta_0 \neq 0$ in the formula

$$\sum_{j=0}^{k} \alpha_j \mathbf{y}_{n-j} = h \sum_{j=0}^{k} \beta_j \mathbf{f}_{n-j},$$

a system of m nonlinear equations for $\mathbf{y}_n$ has to be solved at each step (recall Section 3.4). We can solve this system by some variant of functional iteration (for nonstiff systems), or by a modified Newton iteration (for stiff systems). For any of these iterative methods we must "guess," or *predict* a starting iterate $\mathbf{y}_n^0$, usually by evaluating an interpolant passing through past values of $\mathbf{y}$ and/or $\mathbf{f}$ at t_n, or via an explicit multistep method.

5.4.1 Functional Iteration

The simplest way to solve the nonlinear algebraic system for $\mathbf{y}_n$ is via functional iteration. The iteration is given by[20]

$$\mathbf{y}_n^{\nu+1} = h\beta_0 \mathbf{f}(t_n, \mathbf{y}_n^{\nu}) - \sum_{j=1}^{k} \alpha_j \mathbf{y}_{n-j} + h \sum_{j=1}^{k} \beta_j \mathbf{f}_{n-j}, \quad \nu = 0, 1, \ldots. \quad (5.12)$$

This is a fixed point iteration. It converges to the fixed point $\mathbf{y}_n$ if it is a contraction, i.e., if $\|h\beta_0 \frac{\partial \mathbf{f}}{\partial \mathbf{y}}\| \leq r < 1$. Hence it is appropriate only for

[20]Do not confuse the notation $\mathbf{y}_n^{\nu+1}$ for the $(\nu+1)$st iterate of $\mathbf{y}_n$ with the notation for the $(\nu+1)$st power. Which of these is the correct interpretation should be clear from the context. We have reserved the superscript ν for an iteration counter in this chapter.

nonstiff problems. The iteration is continued until it has been determined to have converged as described for the Newton iteration below. Usually, if convergence is not attained within two to three iterations, or if the rate of convergence is found to be too slow, the current step is rejected and retried with a smaller step size (for example, halve the step size).

5.4.2 Predictor-Corrector Methods

Often in nonstiff codes, the iteration is not taken to convergence. Instead, a *fixed* number of iterations is used for each time step. First, an approximation $\mathbf{y}_n^0$ to $\mathbf{y}_n$ is *predicted*, usually by an explicit multistep method of the same order as the implicit method (for example, by the k-step Adams–Bashforth method of order k),

$$P : \mathbf{y}_n^0 + \hat{\alpha}_1 \mathbf{y}_{n-1} + \cdots + \hat{\alpha}_k \mathbf{y}_{n-k} = h(\hat{\beta}_1 \mathbf{f}_{n-1} + \cdots + \hat{\beta}_k \mathbf{f}_{n-k}).$$

Then the function is evaluated at $\mathbf{y}_n^0$:

$$E : \mathbf{f}_n^0 = \mathbf{f}(t_n, \mathbf{y}_n^0),$$

and inserted into the *corrector* formula (for example, Adams–Moulton of order k or $k+1$) to obtain a new approximation to $\mathbf{y}_n$. Setting $\nu = 0$,

$$C : \mathbf{y}_n^{\nu+1} + \alpha_1 \mathbf{y}_{n-1} + \cdots + \alpha_k \mathbf{y}_{n-k} = h(\beta_0 \mathbf{f}_n^\nu + \beta_1 \mathbf{f}_{n-1} + \cdots + \beta_k \mathbf{f}_{n-k}).$$

The procedure can be stopped here (this is called a PEC method), or the function can be evaluated at $\mathbf{y}_n^1$ to give

$$E : \mathbf{f}_n^1 = \mathbf{f}(t_n, \mathbf{y}_n^1)$$

(this is called a PECE method), or the steps E and C can be iterated ν times to form a P(EC)$^\nu$ method or a P(EC)$^\nu$E method. The final function evaluation in a P(EC)$^\nu$E method yields a better value for $\mathbf{f}$ to be used in the next time step (i.e., $n \leftarrow n+1$) as the new $\mathbf{f}_{n-1}$. Although it appears that the method might be expensive, the final function evaluation is usually advantageous because it yields a significant increase in the region of absolute stability over the corresponding P(EC)$^\nu$ method.

It should be noted that because the corrector formula is not iterated to convergence, the order, error, and stability properties of the P(EC)$^\nu$E or P(EC)$^\nu$ methods are not necessarily the same as for the corrector formula alone. The methods of this subsection are different, in principle, from the methods of the previous subsection 5.4.1 for the same implicit formula. Predictor-corrector methods are explicit methods[21] which are members of

[21]Unlike an implicit method, an explicit method evaluates the next $\mathbf{y}_n$ at each step precisely (in the absence of roundoff error) in a finite number of elementary operations. This relates to the fact that no such predictor-corrector formula has an unbounded absolute stability region, even if the implicit corrector formula has one. These predictor-corrector methods are suitable only for nonstiff problems.

a class of methods called *general linear methods.* This class also contains linear multistep methods.

Example 5.10
Combining the two-step Adams–Bashforth method (i.e., $\hat{\alpha}_1 = -1$, $\hat{\alpha}_2 = 0$, $\hat{\beta}_1 = 3/2$, $\hat{\beta}_2 = -1/2$) with the second-order one-step Adams–Moulton method (i.e., the trapezoidal method, $\alpha_1 = -1, \beta_0 = \beta_1 = 1/2$), we obtain the following method.

Given $\mathbf{y}_{n-1}, \mathbf{f}_{n-1}, \mathbf{f}_{n-2}$,

1. $\mathbf{y}_n^0 = \mathbf{y}_{n-1} + \frac{h}{2}(3\mathbf{f}_{n-1} - \mathbf{f}_{n-2})$,
2. $\mathbf{f}_n^0 = \mathbf{f}(t_n, \mathbf{y}_n^0)$,
3. $\mathbf{y}_n = \mathbf{y}_{n-1} + \frac{h}{2}(\mathbf{f}_{n-1} + \mathbf{f}_n^0)$,
4. $\mathbf{f}_n = \mathbf{f}(t_n, \mathbf{y}_n)$.

This is an explicit, second-order method which has the local truncation error

$$\mathbf{d}_n = -\frac{1}{12}h^2\mathbf{y}'''(t_n) + O(h^3). \qquad \blacklozenge$$

The most-used variant of predictor-corrector methods is PECE. In the common situation where the order of the predictor formula is equal to the order of the corrector formula, the principal term of the local truncation error for the PECE method is the same as that of the corrector:

$$\mathbf{d}_n = C_{p+1}h^p\mathbf{y}^{(p+1)}(t_n) + O(h^{p+1}).$$

The local error is given by a similar expression (see (3.14)). Roughly speaking, the principal terms of the error are the same for the corrector as for the PECE method because $\mathbf{y}_n^0$, which is already accurate to the order of the corrector, enters into the corrector formula multiplied by h, hence the error which is contributed by this term is $O(h^{p+1})$.

5.4.3 Modified Newton Iteration

For stiff systems, a variant of Newton's method is used to solve the nonlinear algebraic equations at each time step. For the general linear multistep method we write

$$\mathbf{y}_n - h\beta_0\mathbf{f}(t_n, \mathbf{y}_n) = -\sum_{j=1}^{k}\alpha_j\mathbf{y}_{n-j} + h\sum_{j=1}^{k}\beta_j\mathbf{f}_{n-j},$$

where the right-hand side is known. Newton's iteration yields

$$\mathbf{y}_n^{\nu+1} = \mathbf{y}_n^{\nu} - \left(I - h\beta_0\frac{\partial\mathbf{f}}{\partial\mathbf{y}}\right)^{-1}\left[\sum_{j=0}^{k}\alpha_j\mathbf{y}_{n-j} - h\sum_{j=0}^{k}\beta_j\mathbf{f}_{n-j}\right],$$

where $\mathbf{y}_n$, $\mathbf{f}_n$, and $\partial\mathbf{f}/\partial\mathbf{y}$ are all evaluated at $\mathbf{y}_n^\nu$. The initial guess $\mathbf{y}_n^0$ is usually obtained by evaluating an interpolant passing through past values of $\mathbf{y}$ at t_n. For a simple implementation, this method does the job. However, it is often not the cheapest possible.

A modified Newton method is usually employed in stiff ODE packages, where the Jacobian matrix $\partial\mathbf{f}/\partial\mathbf{y}$ and its LU decomposition are evaluated (updated) only when deemed necessary. The matrix may be evaluated whenever

1. the iteration fails to converge, or
2. the step size has changed by a significant amount or the order has changed, or
3. a certain number of steps have passed.

Since forming and LU-decomposing the matrix in Newton's iteration are often the major computational expense in carrying out the next step's approximation, relatively large savings are realized by the modified Newton method.

The iteration is considered to have converged, e.g., when

$$\frac{\rho}{1-\rho}|\mathbf{y}_n^{\nu+1} - \mathbf{y}_n^\nu| < \mathrm{NTOL},$$

where the Newton iteration tolerance NTOL is usually taken to be a fraction of ETOL, the user error tolerance, say NTOL $= .33\,$ETOL, and ρ is an indication of the rate of convergence of the iteration, which can be estimated by

$$\rho = \left(\frac{|\mathbf{y}_n^{\nu+1} - \mathbf{y}_n^\nu|}{|\mathbf{y}_n^1 - \mathbf{y}_n^0|}\right)^{\frac{1}{\nu}}.$$

Note: The next section deals with some of the nuts and bolts of writing general-purpose software based on multistep methods. Depending on your orientation, you may wish to read it with special care, or to skip it.

5.5 Designing Multistep General-Purpose Software

The design of an effective general-purpose code for solving initial value problems using multistep methods is a challenging task. It involves decisions regarding error estimation and control, varying the step size, varying the method's order, and solving the nonlinear algebraic equations. The latter has been considered already. Here we outline some of the important options for the resolution of the remaining issues.

5.5.1 Variable Step-Size Formulae

We have seen in the previous chapters that in some applications, varying the step size is crucial for the effective performance of a discretization method. The general k-step linear methods that we have seen so far,

$$\sum_{j=0}^{k} \alpha_j \mathbf{y}_{n-j} = h \sum_{j=0}^{k} \beta_j \mathbf{f}_{n-j},$$

assume that we know the past values $(\mathbf{y}_{n-j}, \mathbf{f}_{n-j})$, $j = 1, \ldots, k$, at a sequence of equally spaced mesh points defined by the step length h. Now, if at $t = t_{n-1}$ we want to take a step of size h_n which is different from the step size h_{n-1} used before, then we need solution values at past times $t_{n-1} - jh_n$, $1 \leq j \leq k-1$, whereas what we have from previous steps are values at $t_{n-1} - jh_{n-1}$, $1 \leq j \leq k-1$. To obtain approximations for the missing values, there are three main options. We will illustrate them in terms of the second-order BDF method. Note that for Adams methods the interpolations will be of past values of $\mathbf{f}$ instead of past values of $\mathbf{y}$, and errors are estimated via differences of $\mathbf{f}$ and not $\mathbf{y}$.

Fixed-Coefficient Strategy

The constant step-size, second-order BDF (for step size h_n) is given by

$$\frac{3}{2}\left(\mathbf{y}_n - \frac{4}{3}\mathbf{y}_{n-1} + \frac{1}{3}\mathbf{y}_{n-2}\right) = h_n \mathbf{f}(t_n, \mathbf{y}_n),$$

where $t_n = t_{n-1} + h_n$. The BDF requires values of $\mathbf{y}$ at t_{n-1} and $t_{n-1} - h_n$. The fixed-coefficient method computes these values from the values at t_{n-1}, $t_{n-1} - h_{n-1}$, and $t_{n-1} - h_{n-1} - h_{n-2}$ by quadratic (more generally, polynomial) interpolation. The interpolated values of $\mathbf{y}$ at $t_{n-1} - h_n$ become the "new" past values $\mathbf{y}_{n-2}$, and are used in the fixed-coefficient BDF to advance the step.

Fixed-coefficient formulae have the advantage of simplicity. However, there is an error due to the interpolation, and they are less stable than variable-coefficient formulae. Stability of the variable step-size formulae is an important consideration for problems where the step size must be changed frequently or drastically (i.e., $h_n \ll h_{n-1}$ or $h_n \gg h_{n-1}$).

Variable-Coefficient Strategy

Better stability properties are obtained by directly deriving the formulae which are based on unequally spaced data. Recall that the BDF were derived by first approximating $\mathbf{y}$ by an interpolating polynomial, and then differentiating the interpolating polynomial and requiring it to satisfy the ODE at t_n. The variable-coefficient BDF are derived in exactly the same way, using an interpolating polynomial which is based on unequally spaced data. The Adams methods can also be directly extended, using polynomial interpolation of unequally spaced **f**-values.

For example, to derive the variable-coefficient form of the second-order BDF method, we can first construct the interpolating quadratic polynomial $\phi(t)$ based on unequally spaced data (here it is written in Newton form):

$$\phi(t) = \mathbf{y}_n + (t - t_n)[\mathbf{y}_n, \mathbf{y}_{n-1}] + (t - t_n)(t - t_{n-1})[\mathbf{y}_n, \mathbf{y}_{n-1}, \mathbf{y}_{n-2}].$$

Next we differentiate the interpolating polynomial to obtain[22]

$$\phi'(t_n) = [\mathbf{y}_n, \mathbf{y}_{n-1}] + (t_n - t_{n-1})[\mathbf{y}_n, \mathbf{y}_{n-1}, \mathbf{y}_{n-2}].$$

Then the variable-coefficient form of the second-order BDF is given by

$$\mathbf{f}(t_n, \mathbf{y}_n) = [\mathbf{y}_n, \mathbf{y}_{n-1}] + h_n[\mathbf{y}_n, \mathbf{y}_{n-1}, \mathbf{y}_{n-2}].$$

Note that on an equally spaced mesh, this formula reduces to the fixed step-size BDF method. The coefficients in this method depend on h_n and h_{n-1}.

The variable-coefficient method has the advantage for problems which require frequent or drastic changes of step size. However, in the case of implicit methods, it can be less efficient than the alternatives. To see this, rewrite the formula in terms of past steps:

$$h_n\mathbf{f}(t_n, \mathbf{y}_n) = \mathbf{y}_n - \mathbf{y}_{n-1} + \frac{h_n^2}{h_n + h_{n-1}}\left(\frac{\mathbf{y}_n - \mathbf{y}_{n-1}}{h_n} - \frac{\mathbf{y}_{n-1} - \mathbf{y}_{n-2}}{h_{n-1}}\right).$$

Then the iteration matrix for Newton's method is given by

$$\left(\left(1 + \frac{h_n}{h_n + h_{n-1}}\right) I - h_n \frac{\partial \mathbf{f}}{\partial \mathbf{y}}\right).$$

So the coefficients of the iteration matrix depend not only on the current step size, but also on the previous one, and more generally on the sequence of $k - 1$ past step sizes. For economy, it is advantageous to try to save and reuse the iteration matrix and/or its factorization from one step to the next. However, if the coefficients in the matrix change frequently, then this is not possible.

This changing Jacobian is a serious shortcoming of the variable-coefficient strategy, in the case of implicit methods for stiff problems. In the design of codes for nonstiff problems, for example, in Adams codes, the Jacobian matrix does not arise and there is no need to consider the next alternative.

[22] Recall that the divided differences are defined by

$$\begin{aligned} [\mathbf{y}_n] &= \mathbf{y}_n, \\ [\mathbf{y}_n, \mathbf{y}_{n-1}, \ldots, \mathbf{y}_{n-i}] &= \frac{[\mathbf{y}_n, \mathbf{y}_{n-1}, \ldots, \mathbf{y}_{n-i+1}] - [\mathbf{y}_{n-1}, \mathbf{y}_{n-2}, \ldots, \mathbf{y}_{n-i}]}{t_n - t_{n-i}}. \end{aligned}$$

Fixed Leading-Coefficient Strategy

This is a compromise which incorporates the best features of both previous methods. We describe it for the k-step BDF. First a polynomial $\phi(t)$ of degree $\leq k$, which is sometimes called a *predictor polynomial*, is constructed such that it interpolates $\mathbf{y}_{n-i}$ at the last $k+1$ values on the unequally spaced mesh

$$\phi(t_{n-i}) = \mathbf{y}_{n-i}, \quad i = 1, \ldots, k+1.$$

Then the fixed leading-coefficient form of the k-step BDF is given by requiring that a second polynomial $\psi(t)$ of degree $\leq k$, which interpolates the predictor polynomial on a fixed mesh $t_{n-1}, t_{n-1} - h_n, \ldots, t_{n-1} - kh_n$, satisfies the ODE at t_n,

$$\begin{aligned} \psi(t_n - ih_n) &= \phi(t_n - ih_n), \qquad 1 \leq i \leq k, \\ \psi'(t_n) &= \mathbf{f}(t_n, \psi(t_n)), \end{aligned}$$

and setting

$$\mathbf{y}_n = \psi(t_n).$$

The fixed leading-coefficient form has stability properties which are intermediate between the other two forms, but is as efficient as the fixed-coefficient form.

Whichever method is chosen to vary the step size, it is clear that the effort is more significant than what is required for Runge–Kutta methods. On the other hand, estimating the local truncation error is easier with linear multistep methods, as we will see next.

5.5.2 Estimating and Controlling the Local Error

As was the case for Runge–Kutta methods, the errors made at each step are much easier to estimate than the global error. Thus, even though the global error is more meaningful, the local truncation error is the one that general-purpose multistep codes usually estimate in order to control the step size and to decide on the order of the method to be used. We recall from (3.14) that the local truncation error is related to the local error by

$$h_n(|\mathbf{d}_n| + O(h^{p+1})) = |\mathbf{l}_n|(1 + O(h_n)).$$

Thus, to control the local error, multistep codes attempt to estimate and control $h_n \mathbf{d}_n$.

In developing estimates below using the local truncation error, we will pretend that there is no error in previous steps. This is of course not true in general, but it turns out that the errors in previous time steps are often correlated so as to create a higher-order contribution, so the expressions derived by ignoring these past errors do yield the leading term of the current local error. There are more difficulties with the theory when the order is varied.

Estimating the Local Truncation Error

In the case of predictor-corrector methods (Section 5.4.2), the error estimate can be expressed in terms of the difference between the predictor and the corrector. Let the local truncation error of the predictor formula be given by

$$\hat{\mathbf{d}}_n = \hat{C}_{p+1} h^p \mathbf{y}^{(p+1)}(t_n) + O(h^{p+1}).$$

Subtracting the predicted from the corrected values, we obtain

$$\mathbf{y}_n - \mathbf{y}_n^0 = (C_{p+1} - \hat{C}_{p+1}) h^p \mathbf{y}^{(p+1)}(t_n) + O(h^{p+1}).$$

Hence an estimate for the local truncation error of the corrector formula or of the PECE formula is given in terms of the predictor-corrector difference by

$$C_{p+1} h^p \mathbf{y}^{(p+1)}(t_n) + O(h^{p+1}) = \frac{C_{p+1}}{C_{p+1} - \hat{C}_{p+1}} (\mathbf{y}_n - \mathbf{y}_n^0).$$

This is called *Milne's estimate.*[23] In an Adams predictor-corrector pair a k-step Adams–Bashforth predictor is used together with a $(k-1)$-step Adams–Moulton corrector to obtain a PECE method of order $p = k$ with a local error estimate, at the cost of two function evaluations per step. See Example 5.10.

Alternatively, it is also possible to use a predictor of order $p - 1$. This is an instance of local extrapolation, as defined in the previous chapter.

The local truncation error for more general multistep methods can be estimated directly by approximating $\mathbf{y}^{(p+1)}$ using divided differences.[24] For example, for second-order BDF, if $\phi(t)$ is the quadratic interpolating $\mathbf{y}_n$, $\mathbf{y}_{n-1}$, and $\mathbf{y}_{n-2}$, then

$$\mathbf{f}(t_n, \mathbf{y}_n) = \phi'(t_n) = [\mathbf{y}_n, \mathbf{y}_{n-1}] + h_n[\mathbf{y}_n, \mathbf{y}_{n-1}, \mathbf{y}_{n-2}] + \mathbf{r}_n,$$

where

$$\mathbf{r}_n = h_n(h_n + h_{n-1})[\mathbf{y}_n, \mathbf{y}_{n-1}, \mathbf{y}_{n-2}, \mathbf{y}_{n-3}].$$

The principal term of the local truncation error is then given by $\beta_0 \mathbf{r}_n$.

The error estimate is used to decide whether to accept the results of the current step or to redo the step with a smaller step size. The step is accepted based on a test

$$\text{EST} \leq \text{ETOL},$$

where EST is h_n times the estimated local truncation error.

---

[23] Note that in contrast to the methods used in the Runge–Kutta context to evaluate two approximations to $\mathbf{y}(t_n)$, here the predictor and the corrector methods have the same order.

[24] In the case of Adams methods, $\mathbf{y}^{(p+1)}$ is approximated via the divided difference of $\mathbf{f}$, using $\mathbf{y}^{(p+1)} = \mathbf{f}^{(p)}$.

Choosing the Step Size and Order for the Next Step

Once the current step has been accepted, the next task is to choose the step size and order for the next step. We begin by forming estimates of the error which we expect would be incurred on the next step, if it were taken with a method of order $\hat{p}$, for several possible orders, for example, $p-2$, $p-1$, p, and $p+1$, where p is the current order.

There are several philosophies for choosing the next order.

1. Choose the next order so that the step size at that order is the largest possible. We will show how to compute these step sizes.

2. Raise or lower the order depending on whether

$$|h^{p-1}\mathbf{y}^{(p-1)}|,\ |h^{p}\mathbf{y}^{(p)}|,\ |h^{p+1}\mathbf{y}^{(p+1)}|,\ |h^{p+2}\mathbf{y}^{(p+2)}|$$

form an increasing or decreasing sequence, where h is the current step size. The philosophy behind this type of order selection strategy is that the Taylor series expansion is behaving as expected for higher orders only if the magnitudes of successive higher-order terms form a decreasing sequence. If the terms fail to form a monotone decreasing sequence, the order is lowered. The effect is to bias the formulae towards lower orders, especially in situations where the higher-order formulae are unstable (thus causing the higher-order differences to grow).

Given the order $\hat{p}$, the step size for the next step is computed as follows. Because the error for the next step is a highly nonlinear function of the step size to be chosen, a simplifying assumption is made. The step size expected for a step of order $\hat{p}$ is computed as if the last $\hat{p}+1$ steps were taken at the current step size, and the step size is chosen so that the error estimate satisfies the tolerance. More precisely, the new step size $h_{n+1} = rh_n$ is chosen conservatively so that the estimated error is a fraction of the desired integration error tolerance ETOL,

$$|r^{\hat{p}+1}h_n^{\hat{p}+1}C_{\hat{p}+1}\mathbf{y}^{(\hat{p}+1)}| = \mathrm{frac\,ETOL},$$

with frac $= 0.9$, say. If $\hat{\mathrm{EST}} = |h_n^{\hat{p}+1}C_{\hat{p}+1}\mathbf{y}^{(\hat{p}+1)}|$ is the error estimate, then

$$r^{\hat{p}+1}\hat{\mathrm{EST}} = \mathrm{frac\,ETOL}.$$

Thus

$$r = \left(\frac{\mathrm{frac\,ETOL}}{\hat{\mathrm{EST}}}\right)^{\frac{1}{\hat{p}+1}}.$$

5.5.3 Approximating the Solution at Off-Step Points

In many applications, the approximate solution is needed at intermediate times which may not coincide with the mesh points chosen by the code. Generally, it is easy and cheap to construct polynomial interpolants based on solution values at mesh points. Then, we just evaluate the interpolant at the off-step points. However, we note that the natural interpolant for BDF is continuous but not differentiable, and the natural interpolant for Adams methods is not even continuous (although its derivative is)! Although the natural interpolants yield the requested accuracy, for applications where more smoothness is required of the numerical solution, interpolants which match the solution with greater continuity have been derived.

5.6 Software, Notes, and References

5.6.1 Notes

The Adams–Bashforth methods date back to 1883; J. C. Adams also designed the implicit formulae known as Adams–Moulton. Both F. R. Moulton and W. E. Milne used these formulae in 1926 in predictor-corrector combinations. The BDF methods were introduced in the 1950s, if not earlier, but they came to prominence only much later, due to the work of C. W. Gear [43]. See [50] for more background and early references.

The material in Sections 5.2–5.4 is standard, although different methods have been used to prove the basic Stability Theorem 5.1. It is covered (along with other topics) in a number of other texts, e.g., [50, 52, 62, 43, 85]. Early works of G. Dahlquist and others, reported in [54], laid the foundations of this material.

For our presentation we chose a different order than the other texts by combining the nonstiff and the stiff cases. This reflects our belief that stiff equations should not be considered advanced material to be taught only towards the end of a course. Also, as in Chapter 4, we have omitted many stability concepts that have been proposed in the literature in the past 30 years, and have instead concentrated only on the properties of stiff decay and A-stability.

Writing a general-purpose code based on multistep methods is a more complicated endeavor than for one-step methods, as Section 5.5 may already suggest. The books by Shampine and Gordon [86] and Brenan, Campbell, and Petzold [19] describe such implementations in detail.

While most recent developments in the numerical ODE area seem to have related more to Runge–Kutta methods, you should not conclude that linear multistep methods may be forgotten. In fact, there are still some serious holes in the theory behind the practical implementation issues in Section 5.5 on one hand, and on the other hand these methods are winners for certain (but not all) applications, both stiff and nonstiff. Software exists as well. Note also that a number of the features and additional topics described in the previous chapter, including, for example, global error estimation, dense output, and waveform relaxation, are equally relevant here.

5.6.2 Software

A variety of excellent and widely used software based on linear multistep methods is readily available. A few of the codes are described here.

- ODE, written by Shampine and described in detail in [86], is based on variable-coefficient Adams PECE methods. It is useful for nonstiff problems, and has a feature to diagnose stiffness.
- VODE, written by Hindmarsh, Brown, and Byrne [21], offers fixed leading-coefficient Adams and BDF methods. The implicit formulae are solved via functional iteration or modified Newton, depending on the option selected. Thus, this code has options for dealing with both stiff and nonstiff problems.
- DIFSUB, written by Gear [43], solves stiff problems and was a very influential code popularizing the BDF methods.
- VODPK is an extension of VODE for large-scale stiff systems. In addition to the direct methods for solving linear systems used in VODE, VODPK offers the option of preconditioned Krylov iterative methods (see, e.g., [48, 14]; the user must write a routine which gives the preconditioner, and this in some applications is a major task).
- DASSL and DASPK [19] are based on fixed leading-coefficient BDF and can accommodate differential-algebraic equations as well as stiff ODEs (see Chapter 10).

5.7 Exercises

5.1. (a) Construct a consistent, unstable multistep method of order 2 (other than the one in Example 5.6).

(b) Is it possible to construct a consistent, unstable one-step method of order 2? Why?

5.2. For the numerical solution of the problem

$$y' = \lambda(y - \sin t) + \cos t, \quad y(0) = 1, \quad 0 \le t \le 1,$$

whose exact solution is $y(t) = e^{\lambda t} + \sin t$, consider using the following four two-step methods, with $y_0 = 1$ and $y_1 = y(h)$ (i.e., using the exact solution so as not to worry here about y_1).

(a) Your unstable method from the previous question.

(b) The midpoint two-step method

$$y_n = y_{n-2} + 2hf_{n-1}.$$

(c) Adams–Bashforth

$$y_n = y_{n-1} + \frac{h}{2}(3f_{n-1} - f_{n-2}).$$

(d) BDF

$$y_n = \frac{(4y_{n-1} - y_{n-2})}{3} + \frac{2h}{3}f_n.$$

Consider using $h = .01$ for $\lambda = 10$, $\lambda = -10$, and $\lambda = -500$. Discuss the expected quality of the obtained solutions in these twelve calculations. Try to do this without calculating any of these solutions. Then confirm your predictions by doing the calculations.

5.3. Write a program which, given k and the values of *some* of the coefficients $\alpha_1, \alpha_2, \ldots, \alpha_k$, $\beta_0, \beta_1, \ldots, \beta_k$ of a linear k-step method, will

- find the rest of the coefficients, i.e., determine the method, such that the order of the method is maximized,
- find the error coefficient C_{p+1} of the leading local truncation error term.

Test your program to verify the second and the last rows in each of the Tables 5.1 and 5.2.

Now use your program to find C_{p+1} for each of the six BDF methods in Table 5.3.

5.4. Write a program which, given a linear multistep method, will test whether the method is

- 0-stable,
- strongly stable.

[Hint: This is a very easy task using MATLAB, for example.]

Use your program to show that the first six BDF methods are strongly stable, but the seven-step and eight-step BDF methods are unstable. (For this you may want to combine your program with the one from the previous exercise.)

5.5. The famous Lorenz equations provide a simple example of a chaotic system (see, e.g., [92, 93]). They are given by

$$\mathbf{y}' = \mathbf{f}(\mathbf{y}) = \begin{pmatrix} \sigma(y_2 - y_1) \\ ry_1 - y_2 - y_1y_3 \\ y_1y_2 - by_3 \end{pmatrix},$$

where σ, r, b are positive parameters. Following Lorenz we set $\sigma = 10$, $b = 8/3$, $r = 28$ and integrate starting from $\mathbf{y}(0) = (0, 1, 0)^T$. Plotting y_3 vs. y_1 we obtain the famous "butterfly" depicted in Figure 5.7.

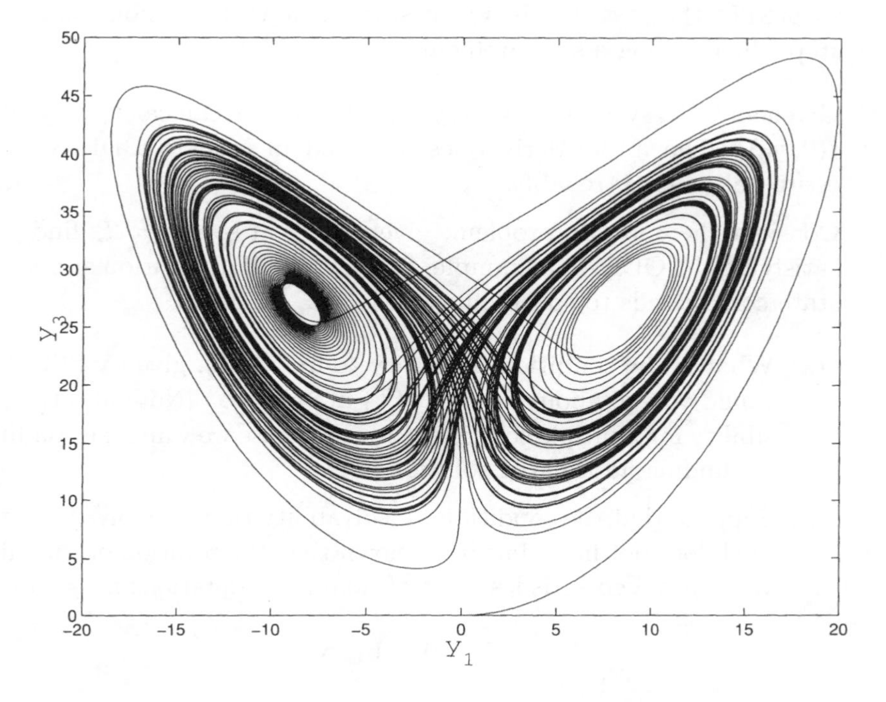

Figure 5.7: *Lorenz "butterfly" in the $y_1 \times y_3$ plane.*

(a) Using a software package of your choice, integrate these equations for $0 \leq t \leq 100$ with an error tolerance 1.e − 6, and plot y_3 vs. y_1, as well as y_2 as a function of t. What do you observe?

(b) Plot the resulting trajectory in the three-dimensional phase space (i.e., the three y-coordinates; if in MATLAB, type `help plot3`). Observe the *strange attractor* that the trajectory appears to settle into.

(c) Chaotic solutions are famous for their highly sensitive dependence on initial data. This leads to *unpredictability* of the solution (and the physical phenomena it represents). When solving numerically we also expect large errors to result from the numerical discretization. Recompute your trajectory with the same initial data using the same package, changing only the error tolerance to 1.e − 7. Compare the values of $\mathbf{y}(100)$ for the two computations, as well as the plots in phase plane. Discuss.

5.6. (a) Show that the only k-step method of order k which has the stiff decay property is the k-step BDF method.

(b) Is it possible to design a strongly stable linear multistep method of order 7 which has stiff decay?

5.7. Explain why it is not a good idea to use an Adams–Bashforth method to predict the first iterate $\mathbf{y}_n^0$ to start a Newton method for a BDF step when solving a stiff problem.

5.8. Given an ODE system $\mathbf{y}' = \mathbf{f}(t, \mathbf{y})$, $\mathbf{y}(0) = \mathbf{y}_0$, we can calculate $\mathbf{y}'(0) = \mathbf{f}(0, \mathbf{y}_0)$. The initial derivatives are used in modern BDF codes to estimate the error reliably.

Consider the opposite problem: given $\mathbf{y}'(T)$ at some $t = T$, find $\mathbf{y}(T)$ satisfying the ODE. For example, finding the ODE solution at a steady state corresponds to specifying $\mathbf{y}'(T) = \mathbf{0}$.

(a) What is the condition necessary to find $\mathbf{y}(T)$, given $\mathbf{y}'(T)$? How would you go about finding $\mathbf{y}(T)$ in practice? [Note also the possibility for multiple solutions, in which case we want the condition for finding an isolated solution.]

(b) Suppose that the condition for solvability that you have just specified does not hold, but it is known that the solution of the initial value problem satisfies a set of nonlinear equations at each t,

$$\mathbf{0} = \mathbf{h}(t, \mathbf{y}).$$

How would you modify the solvability condition? How would you implement it? [Hint: Exercise 5.9 provides an example.]

5.9. The following ODE system due to H. Robertson models a chemical reaction system and has been used extensively as a test problem for stiff solvers [85, 52]:

$$\begin{aligned} y_1' &= -\alpha y_1 + \beta y_2 y_3, \\ y_2' &= \alpha y_1 - \beta y_2 y_3 - \gamma y_2^2, \\ y_3' &= \gamma y_2^2 . \end{aligned}$$

Here $\alpha = 0.04$, $\beta = 1.e+4$, and $\gamma = 3.e+7$ are slow, fast, and very fast reaction rates. The starting point is $\mathbf{y}(0) = (1, 0, 0)^T$.

(a) It is known that this system reaches a steady state, i.e., where $\mathbf{y}' = \mathbf{0}$. Show that $\sum_{i=1}^{3} y_i(t) = 1$, $0 \le t \le b$; then find the steady state.

(b) Integrate the system using a nonstiff code with a permissive error tolerance (say $1.e-2$) for the interval length $b = 3$, just to see how inadequate a nonstiff solver can be. How far is $\mathbf{y}(b)$ from the steady state?

(c) The steady state is reached very slowly for this problem. Use a stiff solver to integrate the problem for $b = 1.e+6$ and plot the solution on a semilog scale in t. How far is $\mathbf{y}(b)$ from the steady state?

5.10. Consider the following two-step method [11],

$$y_n - y_{n-1} = \frac{h}{16}(9f_n + 6f_{n-1} + f_{n-2}). \tag{5.13}$$

Investigate the properties of this method in comparison with the one-step and the two-step Adams–Moulton formulae. Does this method have any advantage?

[Hint: When $h\mathcal{R}e(\lambda) \to -\infty$ one must consider the roots of the characteristic polynomial $\sigma(\xi)$ of (5.7b).]

5.11. The border of the absolute stability region is the curve in the λh-plane where $|y_n| = |\xi^n| = |\xi^{n-1}|$, for ξ satisfying $\rho(\xi) = h\lambda\sigma(\xi)$. Occasionally it is interesting to plot the region where the approximate solution for $y' = \lambda y$ is actually *dampened* by a factor $\delta < 1$, i.e., $|\xi| = \delta$ (recall Exercise 4.8).

(a) Show that the boundary of this δ-region is given by

$$h\lambda = \frac{\rho(\delta e^{\imath\theta})}{\sigma(\delta e^{\imath\theta})}.$$

(b) Plot the δ-curves with $\delta = .9, .5, .1$ for the backward Euler, trapezoidal, and (5.13) methods. Discuss your observations.

5.12. Often in practice one has to solve an ODE system of the form

$$\mathbf{y}' = \mathbf{f}(t, \mathbf{y}) + \mathbf{g}(t, \mathbf{y}), \qquad t \geq 0, \tag{5.14}$$

where $\mathbf{f}$ and $\mathbf{g}$ have significantly different characteristics. For instance, $\mathbf{f}$ may be nonlinear and the ODE $\mathbf{z}' = \mathbf{f}(t, \mathbf{z})$ is nonstiff, while $\mathbf{g}$ is linear but the ODE $\mathbf{z}' = \mathbf{g}(t, \mathbf{z})$ is stiff. This suggests mixing an explicit method for $\mathbf{f}$ with an implicit method, suitable for stiff problems, for $\mathbf{g}$. An *implicit-explicit* (IMEX) [11] k-step method has the form

$$\sum_{j=0}^{k} \alpha_j \mathbf{y}_{n-j} = h\sum_{j=1}^{k} \beta_j \mathbf{f}_{n-j} + h\sum_{j=0}^{k} \gamma_j \mathbf{g}_{n-j}. \tag{5.15}$$

The combination of two-step Adams–Bashforth for $\mathbf{f}$ and the trapezoidal rule for $\mathbf{g}$ is common in the partial differential equation (PDE) literature (especially in combination with spectral methods).

Show that:

(a) The method (5.15) has order p if

$$\begin{aligned}\sum_{j=0}^{k}\alpha_j &= 0, \qquad (5.16)\\ \frac{1}{i!}\sum_{j=1}^{k} j^i\alpha_j &= -\frac{1}{(i-1)!}\sum_{j=1}^{k} j^{i-1}\beta_j = -\frac{1}{(i-1)!}\sum_{j=0}^{k} j^{i-1}\gamma_j\end{aligned}$$

for $i = 1, 2, \ldots, p$, and such a condition does not hold for $i = p+1$.

(b) The $2p + 1$ constraints (5.16) are linearly independent, provided that $p \le k$; thus, there exist k-step IMEX methods of order k.

(c) A k-step IMEX method cannot have order greater than k.

(d) The family of k-step IMEX methods of order k has k parameters.

5.13. A *convection-diffusion* PDE in one space variable has the form (recall Examples 1.3 and 1.7)

$$\frac{\partial u}{\partial t} = u\frac{\partial u}{\partial x} + \frac{\partial}{\partial x}\left(p(x)\frac{\partial u}{\partial x}\right), \quad 0 \le x \le 1, \quad t \ge 0,$$

where $p = p(x) > 0$ is a given function which may be small in magnitude (in which case the equation is said to be convection-dominated).

We now apply the method of lines (cf. Sections 1.1, 1.3). Discretizing in x on a mesh $0 = x_0 < x_1 < \cdots < x_J = 1$, $\Delta x_i = x_i - x_{i-1}$, $\Delta x = \max_i \Delta x_i$, let $y_i(t)$ be the approximation along the line of $u(x_i, t)$, and obtain the ODE system of the form (5.14),

$$\begin{aligned}y_i' &= y_i\left(\frac{y_{i+1} - y_{i-1}}{\Delta x_i + \Delta x_{i+1}}\right)\\ &\quad + \frac{2}{\Delta x_i + \Delta x_{i+1}}\left(\frac{p_{i+1/2}}{\Delta x_{i+1}}(y_{i+1} - y_i) - \frac{p_{i-1/2}}{\Delta x_i}(y_i - y_{i-1})\right)\\ &= f_i(\mathbf{y}) + g_i(\mathbf{y}), \; i = 1, \ldots, J-1. \qquad (5.17)\end{aligned}$$

Here, $p_{i-1/2} = \frac{1}{2}(p(x_i) + p(x_{i-1}))$ or, if p is a rough function, we choose the *harmonic average*

$$p_{i-1/2} \approx \Delta x_i \left[\int_{x_{i-1}}^{x_i} p^{-1}(x)dx\right]^{-1}.$$

If p is small then the centered discretization leading to $f_i(\mathbf{y})$ is questionable, but we do not pursue this further here.

It is natural to apply an IMEX method to (5.17), since the nonlinear convection term typically yields an absolute stability requirement of the form $h \le \text{const}\Delta x$, which is not difficult to live with, whereas the linear diffusion term is stiff (unless p is very small). Moreover, due to the hyperbolic nature of the convection term and the parabolic nature of the diffusion term, an appropriate test equation to investigate the stability properties of the IMEX method (5.15) is

$$y' = (a + \imath b)y \tag{5.18}$$

with a, b real constants, $a < 0$ ($\imath = \sqrt{-1}$), and where we identify $f(y) = \imath by$ and $g(y) = ay$ in (5.14), (5.15).

(a) What is the domain of absolute stability for an IMEX method with respect to this test equation? What corresponds to a δ-region as in Exercise 5.11?

(b) Plot δ-curves with $\delta = 1, .9, .5, .1$ for the following two-step IMEX methods.

- Adams–Bashforth with trapezoidal method

$$\mathbf{y}_n = \mathbf{y}_{n-1} + \frac{h}{2}(3\mathbf{f}_{n-1} - \mathbf{f}_{n-2} + \mathbf{g}_n + \mathbf{g}_{n-1}).$$

- Adams–Bashforth with (5.13)

$$\mathbf{y}_n = \mathbf{y}_{n-1} + \frac{h}{16}(24\mathbf{f}_{n-1} - 8\mathbf{f}_{n-2} + 9\mathbf{g}_n + 6\mathbf{g}_{n-1} + \mathbf{g}_{n-2}).$$

- Semi-explicit BDF

$$\mathbf{y}_n = \frac{1}{3}(4\mathbf{y}_{n-1} - \mathbf{y}_{n-2}) + \frac{2h}{3}(2\mathbf{f}_{n-1} - \mathbf{f}_{n-2} + \mathbf{g}_n).$$

Discuss your observations.

Part III: Boundary Value Problems

Chapter 6

More Boundary Value Problem Theory and Applications

In this chapter and the next two we will consider an ordinary differential equation (ODE) system with m components,

$$\mathbf{y}' = \mathbf{f}(t, \mathbf{y}), \qquad 0 < t < b, \tag{6.1}$$

subject to m *two-point* boundary conditions

$$\mathbf{g}(\mathbf{y}(0), \mathbf{y}(b)) = \mathbf{0}. \tag{6.2}$$

We denote the Jacobian matrices of $\mathbf{g}(\mathbf{u}, \mathbf{v})$ with respect to its first and second argument vectors by

$$B_0 = \frac{\partial \mathbf{g}}{\partial \mathbf{u}}, \qquad B_b = \frac{\partial \mathbf{g}}{\partial \mathbf{v}}. \tag{6.3}$$

Often in applications, $\mathbf{g}$ is linear; i.e., the boundary conditions can be written as

$$B_0 \mathbf{y}(0) + B_b \mathbf{y}(b) = \mathbf{b} \tag{6.4}$$

for some given data $\mathbf{b}$,[25] and the $m \times m$ matrices B_0 and B_b are constant.

Also, often in applications the boundary conditions are *separated*; i.e., each of the components of $\mathbf{g}$ is given either at $t = 0$ or at $t = b$, but none involves both ends simultaneously.[26] In this case for each i, $1 \le i \le m$, either the ith row of B_0 or the ith row of B_b are identically zero.

[25] Note that the data vector $\mathbf{b}$ and the interval end b are not related. Alas, we seem to be running out of good notation.

[26] A notable exception is the case of periodic boundary conditions.

Example 6.1
Recall Example 1.4 (the vibrating spring),

$$-(p(t)u')' + q(t)u = r(t),$$

$$u(0) = 0, \qquad u'(b) = 0,$$

where $p(t) > 0$, $q(t) \geq 0$ for all $0 \leq t \leq b$. In more independent variables, a problem like this corresponds to an elliptic partial differential equation.

To convert this into a system we have two popular options.

- The standard option is to set $y_1 = u$, $y_2 = u'$, resulting in a system of the form (6.1) with $m = 2$.
- Often in practice the function $p(t)$ has discontinuities. In this case it is better to define the unknowns $y_1 = u$ and $y_2 = py_1'$ (this y_2 is sometimes referred to as the *flux*). This gives an ODE system in the form (6.1) with

$$\mathbf{f}(t, \mathbf{y}) = \begin{pmatrix} p^{-1}y_2 \\ qy_1 - r \end{pmatrix}.$$

The boundary conditions are separated and are given for both choices of unknowns by

$$B_0 = \begin{pmatrix} 1 & 0 \\ 0 & 0 \end{pmatrix}, \qquad B_b = \begin{pmatrix} 0 & 0 \\ 0 & 1 \end{pmatrix}, \qquad \mathbf{b} = \mathbf{0}.$$

♦

We have already seen in Chapter 1 that there is no chance for extending the general Existence and Uniqueness Theorem 1.1 for initial value problems (IVPs) to the boundary value problem (BVP) case. In particular, if $\mathbf{f}$ satisfies the conditions of that theorem, then for each initial value vector $\mathbf{c}$ the ODE (6.1) has a solution $\mathbf{y}(t) = \mathbf{y}(t; \mathbf{c})$ satisfying $\mathbf{y}(0; \mathbf{c}) = \mathbf{c}$. Substituting into (6.2) we have

$$\mathbf{g}(\mathbf{c}, \mathbf{y}(b; \mathbf{c})) = \mathbf{0}.$$

This gives a set of m nonlinear algebraic equations for the m unknowns $\mathbf{c}$ (unknown, because we are asking what initial conditions would yield a solution that satisfies the boundary conditions). It is well known that in general such a system may have many solutions, one, or none at all.

Example 6.2
The problem

$$u'' + e^{u+1} = 0,$$

$$u(0) = u(1) = 0$$

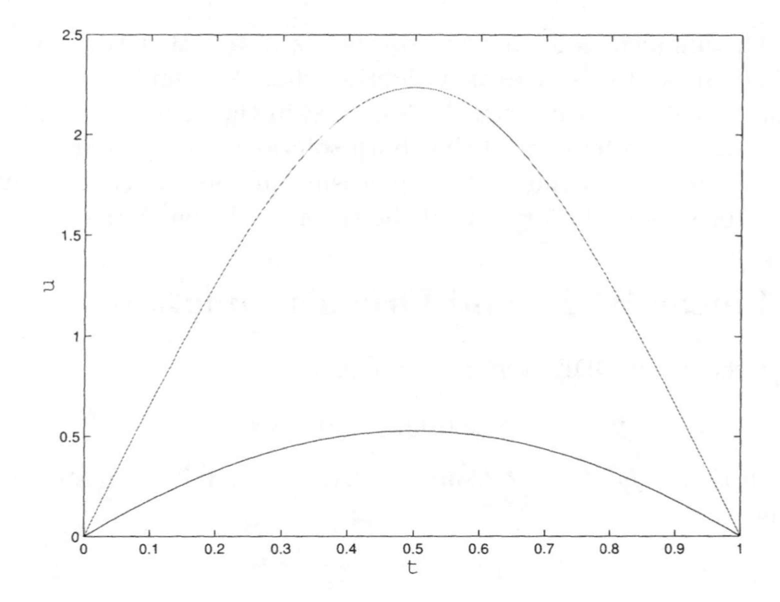

Figure 6.1: *Two solutions $u(t)$ for the BVP of Example* 6.2.

has two solutions of the form

$$u(t) = -2\ln\left\{\frac{\cosh[(t-1/2)\theta/2]}{\cosh(\theta/4)}\right\},$$

where θ satisfies

$$\theta = \sqrt{2e}\cosh(\theta/4)\,.$$

This nonlinear algebraic equation has two solutions for θ (Exercise 6.1). The corresponding two solutions of the BVP are plotted in Figure 6.1. ♦

The possibility of having more than one solution does not in itself prevent us from expecting to be able to find them. The question of existence of unique solutions for a nonlinear BVP must be considered in a local sense. The important question is whether a BVP solution is *isolated*, i.e., if there is a neighborhood about it in which it is the only solution. For this purpose we look at the *variational problem* for the BVP (6.1)–(6.2): assuming for simplicity of notation that $\mathbf{g}$ is linear, i.e., that the boundary conditions are in the form (6.4), the variational problem corresponding to linearizing the problem about an exact solution $\mathbf{y}(t)$ is

$$\begin{aligned} \mathbf{z}' &= A(t,\mathbf{y}(t))\mathbf{z}, \\ B_0\mathbf{z}(0) + B_b\mathbf{z}(b) &= \mathbf{0}, \end{aligned} \tag{6.5}$$

where $A = \frac{\partial \mathbf{f}}{\partial \mathbf{y}}$ is the Jacobian matrix. Now, if the variational problem has the unique solution $\mathbf{z} = \mathbf{0}$ then the solution $\mathbf{y}(t)$ of the given nonlinear problem is isolated, or locally unique. We will show this claim following (6.14) below. The uniqueness of the zero solution $\mathbf{z}$ means that the linearization is nonsingular, and this gives us a fighting chance at finding isolated solutions using the numerical methods described in the next two chapters. For Example 6.2, it can be verified that both solutions are isolated.

In order to understand the issues arising in the numerical solution of BVPs, we must get a better idea of the theory of linear BVPs.

6.1 Linear BVPs and Green's Function

Consider the linear ODE system of m equations,

$$\mathbf{y}' = A(t)\mathbf{y} + \mathbf{q}(t), \qquad 0 < t < b, \tag{6.6}$$

and recall that a *fundamental solution* $Y(t)$ is the $m \times m$ matrix function satisfying

$$Y' = A(t)Y, \qquad 0 < t < b,$$

and $Y(0) = I$. Using this fundamental solution, the general solution of the ODE (6.6) is

$$\mathbf{y}(t) = Y(t)\left[\mathbf{c} + \int_0^t Y^{-1}(s)\mathbf{q}(s)ds\right]. \tag{6.7}$$

The parameter vector $\mathbf{c}$ in (6.7) is determined by the linear boundary conditions (6.4). Substituting, we get

$$[B_0Y(0) + B_bY(b)]\mathbf{c} = \mathbf{b} - B_bY(b)\int_0^b Y^{-1}(s)\mathbf{q}(s)ds.$$

The right-hand side in the above expression depends on the given data. Thus we have obtained a basic existence and uniqueness theorem for linear BVPs.

Theorem 6.1 *Let $A(t)$ and $\mathbf{q}(t)$ be continuous and define the matrix*

$$Q = B_0 + B_bY(b) \tag{6.8}$$

(remember, $Y(0) = I$). Then

- *the linear BVP (6.6), (6.4) has a unique solution iff Q is nonsingular;*
- *if Q is nonsingular then the solution is given by (6.7) with*

$$\mathbf{c} = Q^{-1}\left[\mathbf{b} - B_bY(b)\int_0^b Y^{-1}(s)\mathbf{q}(s)ds\right].$$

Example 6.3
Returning to the first example in Chapter 1,

$$u'' = -u, \qquad u(0) = b_1, \quad u(b) = b_2,$$

we write this in first-order form with

$$A = \begin{pmatrix} 0 & 1 \\ -1 & 0 \end{pmatrix}, \; B_0 = \begin{pmatrix} 1 & 0 \\ 0 & 0 \end{pmatrix}, \; B_b = \begin{pmatrix} 0 & 0 \\ 1 & 0 \end{pmatrix}, \; \mathbf{b} = \begin{pmatrix} b_1 \\ b_2 \end{pmatrix}.$$

It is easy to verify that

$$Y(t) = \begin{pmatrix} \cos t & \sin t \\ -\sin t & \cos t \end{pmatrix}$$

so

$$Q = B_0 + B_b \begin{pmatrix} \cos b & \sin b \\ -\sin b & \cos b \end{pmatrix} = \begin{pmatrix} 1 & 0 \\ \cos b & \sin b \end{pmatrix}.$$

This matrix is singular iff $b = j\pi$ for some integer j. Theorem 6.1 now implies that a unique solution exists if $b \neq j\pi$ for any integer j; see Figure 1.1. ♦

The fundamental solution $Y(t)$ satisfies $Y(0) = I$; i.e., it is scaled for an IVP. A better scaled fundamental solution for the BVP at hand is

$$\Phi(t) = Y(t)Q^{-1}. \tag{6.9}$$

Note that Φ satisfies the homogeneous ODE; i.e. it is indeed a fundamental solution. We have

$$\begin{aligned} \Phi' &= A\Phi, \qquad 0 < t < b, \\ B_0\Phi(0) &+ B_b\Phi(b) = I. \end{aligned} \tag{6.10}$$

So $\Phi(t)$ plays the same role for the BVP as $Y(t)$ plays for the IVP.

We often refer to the columns of the scaled fundamental solution $\Phi(t)$ (or $Y(t)$ in the IVP case) as *solution modes*, or just *modes* for short. They indicate the solution sensitivity to perturbation in the initial data (recall Chapter 2).

If we carry out the suggestion in Theorem 6.1 and substitute the expression for $\mathbf{c}$ into (6.7) then we get an expression for the solution $\mathbf{y}(t)$ in terms of the data $\mathbf{b}$ and $\mathbf{q}(t)$. Rearranging, this gives

$$\mathbf{y}(t) = \Phi(t)\mathbf{b} + \int_0^b G(t,s)\mathbf{q}(s)ds, \tag{6.11}$$

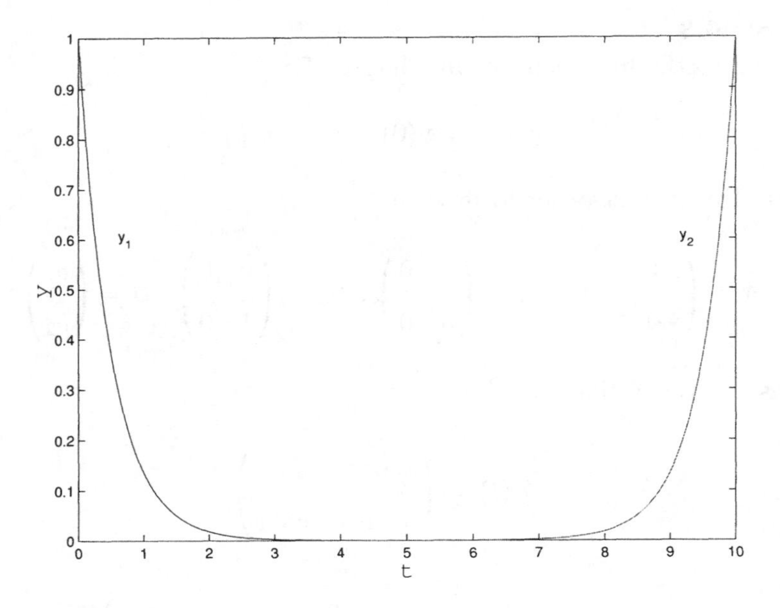

Figure 6.2: *The function* $y_1(t)$ *and its mirror image* $y_2(t) = y_1(b-t)$, *for* $\lambda = -2$, $b = 10$.

where $G(t,s)$ is the *Green's function*

$$G(t,s) = \begin{cases} \Phi(t)B_0\Phi(0)\Phi^{-1}(s), & s \le t\,, \\ -\Phi(t)B_b\Phi(b)\Phi^{-1}(s), & s > t. \end{cases} \tag{6.12}$$

Green's function may be loosely viewed as the inverse of the differential operator (or as the solution operator).

6.2 Stability of BVPs

To understand the fundamental issues in stability for BVPs, the reader must be familiar with the rudiments of stability of IVPs. Therefore, please make sure that you are familiar with the contents of Chapter 2.

Consider the test equation $y_1' = \lambda y_1$ for $0 < t < b$ and regard b as very large. The IVP (e.g., $y_1(0) = 1$) is stable if $\mathcal{R}\mathrm{e}(\lambda) \le 0$. Now apply a variable transformation $\tau = b - t$. The same problem in τ then reads $\frac{dy_2}{d\tau} = -\lambda y_2$, with $y_2(b) = 1$; i.e., this is a terminal value problem and we are integrating from b to 0 (see Figure 6.2). Of course, reversing the direction of time does not affect the stability, which has to do with the effect of small changes in the data on the solution, so this terminal value problem is stable as well (for $\mathcal{R}\mathrm{e}(-\lambda) \ge 0$). Putting the two together, we obtain that the following BVP is stable:

$$\mathbf{y}' = A\mathbf{y}, \quad A = \begin{pmatrix} \lambda & 0 \\ 0 & -\lambda \end{pmatrix},$$

$$y_1(0) = 1, \quad y_2(b) = 1,$$

although the IVP for the same ODE is unstable when $\mathcal{R}e(\lambda) \neq 0$. Thus, the stability of solutions for a given ODE depends on how (and where) the boundary conditions are specified.

For a general linear BVP (6.6), (6.4), the sensitivity of the solution to perturbations in the data is immediately given by introducing bounds in (6.11), because this formula gives the solution in terms of the data.

Define the *stability constant* of the BVP by

$$\kappa = \max(\|\Phi\|_\infty, \|G\|_\infty). \tag{6.13}$$

Then from (6.11),

$$\|\mathbf{y}\| = \max_{0 \le t \le b} |\mathbf{y}(t)| \le \kappa \left[|\mathbf{b}| + \int_0^b |\mathbf{q}(s)| ds \right]. \tag{6.14}$$

Rather than considering families of problems with b becoming unbounded, we shall say qualitatively that the linear BVP is *stable* if the stability constant κ is *of moderate size*. Roughly, "moderate size" means not much larger than the magnitude of the problem's coefficients, $\|A(t)\|b$.

Why is (6.14) a stability bound? Consider a perturbed problem, $\hat{\mathbf{y}}' = A(t)\hat{\mathbf{y}} + \hat{\mathbf{q}}(t)$, $B_0\hat{\mathbf{y}}(0) + B_b\hat{\mathbf{y}}(b) = \hat{\mathbf{b}}$. Thus, the inhomogeneities are perturbed by $\boldsymbol{\delta}(t) = \hat{\mathbf{q}}(t) - \mathbf{q}(t)$ and $\boldsymbol{\beta} = \hat{\mathbf{b}} - \mathbf{b}$. Then the perturbation in the solution, $\mathbf{x}(t) = \hat{\mathbf{y}}(t) - \mathbf{y}(t)$, satisfies the same linear BVP (6.6), (6.4) for the perturbation in the data,

$$\mathbf{x}' = A(t)\mathbf{x} + \boldsymbol{\delta}(t), \quad 0 < t < b,$$
$$B_0\mathbf{x}(0) + B_b\mathbf{x}(b) = \boldsymbol{\beta}.$$

So (6.14) bounds $\mathbf{x}$ in terms of the perturbations in the data,

$$\|\mathbf{x}\| \le \kappa \left[|\boldsymbol{\beta}| + \int_0^b |\boldsymbol{\delta}(s)| ds \right]. \tag{6.15}$$

Now we can further explain the concept of an isolated solution for the nonlinear problem (6.1), (6.4). Suppose that $\mathbf{y}(t)$ is a nonisolated solution; i.e., for any arbitrarily small $\epsilon > 0$ there is another solution $\hat{\mathbf{y}}$ which satisfies $\hat{\mathbf{y}}' = \mathbf{f}(t, \hat{\mathbf{y}})$, $B_0\hat{\mathbf{y}}(0) + B_b\hat{\mathbf{y}}(b) = \mathbf{b}$, $\|\hat{\mathbf{y}} - \mathbf{y}\| = \epsilon$. Then the difference $\mathbf{x}(t) = \hat{\mathbf{y}}(t) - \mathbf{y}(t)$ satisfies

$$\mathbf{x}' = \mathbf{f}(t, \hat{\mathbf{y}}) - \mathbf{f}(t, \mathbf{y}) = A(t, \mathbf{y}(t))\,\mathbf{x} + O(\epsilon^2), \quad 0 < t < b,$$
$$B_0\mathbf{x}(0) + B_b\mathbf{x}(b) = \mathbf{0}.$$

Note that the variational problem (6.5) has the unique zero solution iff the corresponding Q of (6.8) is nonsingular. But if Q is nonsingular then for some finite κ we get from (6.15) that

$$\epsilon = \|\mathbf{x}\| \leq \kappa O(\epsilon^2)\,.$$

This inequality cannot hold if ϵ is arbitrarily small and positive. Hence, the nonsingularity of the variational problem implies that $\mathbf{y}(t)$ is an isolated solution of the nonlinear problem.

Dichotomy

The stability of the problem essentially means that Green's function is nicely bounded. Consider next the case of separated boundary conditions; i.e., assume that the first k rows of B_b and the last $m-k$ rows of B_0 are all zeros. Then from (6.10), clearly

$$B_0\Phi(0) = P = \begin{pmatrix} I_k & 0 \\ 0 & 0 \end{pmatrix}, \qquad B_b\Phi(b) = I - P,$$

where I_k is the $k \times k$ identity, so P is an orthogonal projection matrix (meaning $P^2 = P$) of rank k. In this case we can write Green's function as

$$G(t,s) = \begin{cases} \Phi(t)P\Phi^{-1}(s), & s \leq t\,, \\ -\Phi(t)(I-P)\Phi^{-1}(s), & s > t. \end{cases}$$

The BVP is said to have *dichotomy* if there is a constant K of moderate size such that

$$\|\Phi(t)P\Phi^{-1}(s)\| \leq K, \quad s \leq t, \tag{6.16a}$$
$$\|\Phi(t)(I-P)\Phi^{-1}(s)\| \leq K, \quad s > t\,. \tag{6.16b}$$

The BVP has *exponential dichotomy* if there are positive constants α, β such that

$$\|\Phi(t)P\Phi^{-1}(s)\| \leq Ke^{\alpha(s-t)}, \quad s \leq t \tag{6.17a}$$
$$\|\Phi(t)(I-P)\Phi^{-1}(s)\| \leq Ke^{\beta(t-s)}, \quad s > t\,. \tag{6.17b}$$

Dichotomy and exponential dichotomy correspond to stability and asymptotic stability, respectively, in IVPs. (Compare (2.10) with (6.16a) for $k = m$.) Dichotomy implies that the first k columns of $\Phi(t)$ are nonincreasing (actually decreasing in the case of exponential dichotomy) as t grows, and that the last $m-k$ columns of $\Phi(t)$ are nondecreasing (actually increasing in the case of exponential dichotomy) as t grows. The k nonincreasing modes are controlled in size by the boundary conditions at 0, whereas the $m-k$ nondecreasing modes are controlled in size by the boundary conditions at b. Dichotomy is a necessary and sufficient condition for stability of the BVP.

The situation for nonseparated boundary conditions is much more complicated, although the conclusions remain essentially the same.

Example 6.4
For the problem

$$u'' = u, \qquad u(0) = b_1, \quad u(b) = b_2,$$

i.e., with the ODE different from Example 6.3 but the boundary conditions the same, we convert to first-order form with

$$A = \begin{pmatrix} 0 & 1 \\ 1 & 0 \end{pmatrix}.$$

The fundamental solution satisfying $Y(0) = I$ is

$$Y(t) = \begin{pmatrix} \cosh t & \sinh t \\ \sinh t & \cosh t \end{pmatrix}.$$

Clearly, $\|Y(t)\|$ grows exponentially with t, indicating that the IVP is unstable. For the BVP, however, we have

$$Q = B_0 Y(0) + B_b Y(b) = \begin{pmatrix} 1 & 0 \\ \cosh b & \sinh b \end{pmatrix},$$

so

$$\Phi(t) = Y(t)Q^{-1} = \frac{1}{\sinh b} \begin{pmatrix} \sinh(b-t) & \sinh t \\ -\cosh(b-t) & \cosh t \end{pmatrix}.$$

Thus, the first column of $\Phi(t)$ (here $k = 1$ and $m = 2$) is decreasing in t, and the second column of $\Phi(t)$ is increasing. Both of these columns are nicely scaled:

$$\|\Phi\| \approx 1,$$

even though Q becomes extremely ill conditioned as b grows. We leave it to Exercise 6.2 to show that this BVP is stable and has exponential dichotomy. ♦

6.3 BVP Stiffness

In Section 3.4 we introduce the notion of stiffness for IVPs. In the terminology of the previous section, a stiff (linear) problem is a stable problem which has very fast modes. For an IVP such modes can only be decreasing. But for a stable BVP we must entertain the possibility of both rapidly decreasing and rapidly increasing modes being present.

Corresponding to (3.23) in Section 3.4 we say that a stable BVP for the test equation

$$y' = \lambda y\,, \qquad 0 < t < b,$$

is *stiff* if

$$b|\mathcal{R}\mathrm{e}(\lambda)| \gg 1. \tag{6.18}$$

In contrast to the IVP case, here we no longer require $\mathcal{R}\mathrm{e}(\lambda) < 0$. Similarly to (3.24), this generalizes for a nonlinear system $\mathbf{y}' = \mathbf{f}(t, \mathbf{y})$ to

$$b|\mathcal{R}\mathrm{e}(\lambda_j)| \gg 1, \qquad j = 1, \ldots, m, \tag{6.19}$$

where λ_j are the eigenvalues of the local Jacobian matrix $\frac{\partial \mathbf{f}}{\partial \mathbf{y}}(t, \mathbf{y}(t))$.[27]

This extension of the IVP definition makes sense, in light of the discussion of dichotomy in the previous section. The practical understanding of the qualitative notion behind the inequalities in (6.18) and (6.19) is that we must look for numerical methods that work also when $h|\mathcal{R}\mathrm{e}(\lambda_j)| \gg 1$, where h is a typical discretization step size.

However, this is easier said than done. There are really no known discretization methods which have a similar robustness to that in the IVP case of backward Euler and its higher-order extensions (e.g., BDF methods and collocation at Radau points). The methods discussed in the next chapter, and other variants which are not discussed there, are not suitable for very stiff BVPs. Symmetric difference methods like midpoint, which are our methods of choice for BVPs and are discussed in Chapter 8, often perform well in practice for stiff BVPs, but their theoretical foundation is somewhat shaky in this case, as discussed further in Chapter 8. There are methods (e.g., Riccati) which attempt to *decouple* rapidly increasing and rapidly decreasing modes explicitly, and then integrate such modes only in their corresponding stable directions. But these methods appear more suitable for special applications than for general-purpose use for nonlinear problems. To explicitly decouple modes, especially for nonlinear problems, is no easy task.

[27] Of course, $\lambda_j = \lambda_j(t)$ may in general have a large real part in some parts of the interval and a small (in magnitude) real part in others, but let us assume here, for simplicity of the exposition, that this does not happen.

6.4 Some Reformulation Tricks

While general-purpose codes for BVPs usually assume a system of the form (6.1) subject to boundary conditions of the form (6.2) or, even more frequently, separated boundary conditions, the natural way in which BVPs arise in applications often does not conform to this standard form. An example that we have already seen is the conversion from a higher-order ODE system to a first-order system. There are other, less obvious situations, where a given BVP can be reformulated. Of course all this can be said of IVPs as well, but there is more diversity in the BVP case. There are a number of reformulation "tricks" that can be used to convert a given problem to standard form, of which we describe a few basic ones here.

In many applications, the ODE system depends on an unknown constant, a, and this gives rise to an additional boundary condition. One can then add the ODE

$$a' = 0$$

to the system. This means that the constant a is viewed as a function over the interval of integration which is independent of t.

Example 6.5
The flow in a channel can be modeled by the ODE

$$f''' - R[(f')^2 - ff''] + Ra = 0,$$

$$f(0) = f'(0) = 0,\ f(1) = 1,\ f'(1) = 0\,.$$

The constant R (Reynolds number) is known, but the constant a is undetermined. There are four boundary conditions on the potential function f which determine both it and a. To convert to standard form we write $y_1 = f$, $y_2 = f'$, $y_3 = f''$, $y_4 = a$, and obtain

$$\mathbf{y}' = \mathbf{f}(\mathbf{y}) = \begin{pmatrix} y_2 \\ y_3 \\ R[y_2^2 - y_1 y_3 - y_4] \\ 0 \end{pmatrix}.$$

The boundary conditions are obviously in separated, standard form as well. ♦

The unknown constant can be the size of the interval of integration. Assuming that the problem is given in the form (6.1) but with the integration range b unknown, we can apply the change of variable

$$\tau = t/b$$

to obtain the ODE system

$$
\begin{aligned}
\frac{d\mathbf{y}}{d\tau} &= b\mathbf{f}(b\tau, \mathbf{y}), \qquad 0 < \tau < 1, \\
\frac{db}{d\tau} &= 0\,.
\end{aligned}
$$

So, the new vector of unknowns is $\binom{\mathbf{y}(\tau)}{b}$ and its length is $m+1$. This is also the number of independent boundary conditions that should be given for this system.

The unknown constants trick can also be used to convert nonseparated boundary conditions to separated boundary conditions, at the expense of increasing the size of the ODE system. (In the end, representing a constant by an unknown function to be integrated throughout the interval in t is never very economical, so there has to be a good reason for doing this.) We have seen in the previous section that the theory is easier and simpler for the case of separated boundary conditions. This tends also to be reflected in simpler solution methods for the linear systems arising from a finite difference or multiple shooting discretization.

Given the boundary conditions $\mathbf{g}(\mathbf{y}(0), \mathbf{y}(b)) = \mathbf{0}$, let $\mathbf{a} = \mathbf{y}(0)$ be our unknown constants. Then we can rewrite the system in the form

$$
\begin{aligned}
\mathbf{y}' &= \mathbf{f}(t, \mathbf{y}), \\
\mathbf{a}' &= \mathbf{0},
\end{aligned}
$$

with

$$
\begin{aligned}
\mathbf{y}(0) &= \mathbf{a}(0), \\
\mathbf{g}(\mathbf{a}(b), \mathbf{y}(b)) &= \mathbf{0}.
\end{aligned}
$$

6.5 Notes and References

Chapter 3 of Ascher, Mattheij, and Russell [8] contains a much more detailed account of the material presented in Sections 6.1 and 6.2, which includes the various extensions and proofs mentioned here. Classical references on Green's function and on dichotomy are Stakgold [88] and Coppel [32], respectively. For periodic solutions, see Stuart and Humphries [93]. Stiffness and decoupling in the linear case are discussed at length in [8], where more reformulation examples and references can be found as well.

6.6 Exercises

6.1. Show that the equation

$$\theta = \sqrt{2e}\cosh(\theta/4)$$

has two solutions θ.

6.2. (a) Show that the problem in Example 6.4 is stable for all $b > 0$ and has exponential dichotomy. What are its Green's function and stability constant?

(b) Answer the same question for the periodic boundary conditions

$$u(0) = u(b), \quad u'(0) = u'(b).$$

6.3. Consider the problem

$$\begin{aligned} u''' &= 2u'' + u' - 2u, \qquad 0 < t < b, \\ u'(0) &= 1, \quad u(b) - u'(b) = 0, \\ u(*) &= 1 \end{aligned}$$

with $b = 100$.

(a) Convert the ODE to a first-order system and find its fundamental solution satisfying $Y(0) = I$.

[Hint: Another, well-scaled fundamental solution is

$$\Psi(t) = \begin{pmatrix} e^{-t} & e^{t-b} & e^{2(t-b)} \\ -e^{-t} & e^{t-b} & 2e^{2(t-b)} \\ e^{-t} & e^{t-b} & 4e^{2(t-b)} \end{pmatrix},$$

and recall that $Y(t) = \Psi(t)R$ for some constant matrix R.]

(b) It's not given whether the last boundary condition is prescribed at $* = 0$ or at $* = b$. But it is known that the BVP is stable (with stability constant $\kappa < 20$). Determine where this boundary condition is prescribed.

6.4. Consider an ODE system of size m,

$$\mathbf{y}' = \mathbf{f}(t, \mathbf{y}), \tag{6.20a}$$

where $\mathbf{f}$ has bounded first and second partial derivatives, subject to initial conditions

$$\mathbf{y}(0) = \mathbf{c} \tag{6.20b}$$

or boundary conditions

$$B_0\mathbf{y}(0) + B_b\mathbf{y}(b) = \mathbf{b}. \tag{6.20c}$$

It is often important to determine the *sensitivity* of the problem with respect to the data $\mathbf{c}$ or $\mathbf{b}$. For instance, if we change c_j to $c_j + \epsilon$ for some j, $1 \le j \le m$, where $|\epsilon| \ll 1$, and call the solution of the perturbed problem $\hat{\mathbf{y}}(t)$, what can be said about $|\hat{\mathbf{y}}(t) - \mathbf{y}(t)|$ for $t \ge 0$?

(a) Writing the solution of (6.20a), (6.20b) as $\mathbf{y}(t;\mathbf{c})$, define the $m \times m$ matrix function

$$Y(t) = \frac{\partial \mathbf{y}(t;\mathbf{c})}{\partial \mathbf{c}}.$$

Show that Y satisfies the IVP

$$\begin{aligned} Y' &= A(t)Y, \\ Y(0) &= I, \end{aligned}$$

where $A = \frac{\partial \mathbf{f}}{\partial \mathbf{y}}(t, \mathbf{y}(t;\mathbf{c}))$.

(b) Let $\hat{\mathbf{y}}(t)$ satisfy (6.20a) and

$$\hat{\mathbf{y}}(0) = \mathbf{c} + \epsilon \mathbf{d},$$

where $|\mathbf{d}| = 1$ and $|\epsilon| \ll 1$. Show that

$$\hat{\mathbf{y}}(t) = \mathbf{y}(t) + \epsilon Y(t)\mathbf{d} + O(\epsilon^2).$$

In particular, what can you say about the sensitivity of the problem with respect to the jth initial value?

(c) Answer questions analogous to (a) and (b) above regarding the sensitivity of the BVP (6.20a), (6.20c) with respect to the boundary values $\mathbf{b}$. How would a bound on $\|\hat{\mathbf{y}} - \mathbf{y}\|_\infty = \max_{0 \le t \le b} |\hat{\mathbf{y}}(t) - \mathbf{y}(t)|$ relate to the stability constant κ of (6.13)?

Chapter 7

Shooting

Shooting is a straightforward extension of the initial value techniques that we have seen so far in this book for solving boundary value problems (BVPs). Essentially, one "shoots" trajectories of the same ordinary differential equation (ODE) with different initial values until one "hits" the correct given boundary values at the other interval end. The advantages are conceptual simplicity and the ability to make use of the excellent, widely available, adaptive initial value ODE software. But there are fundamental disadvantages as well, mainly in that the algorithm inherits its stability properties from the stability of the initial value problems (IVPs) that it solves, not just the stability of the given BVP.

7.1 Shooting: A Simple Method and Its Limitations

For a system of ODEs of order m,

$$\mathbf{y}' = \mathbf{f}(t, \mathbf{y}), \qquad 0 < t < b, \tag{7.1}$$

subject to m two-point boundary conditions

$$\mathbf{g}(\mathbf{y}(0), \mathbf{y}(b)) = \mathbf{0}, \tag{7.2}$$

we denote by $\mathbf{y}(t) = \mathbf{y}(t; \mathbf{c})$ the solution of the ODE (7.1) satisfying the initial condition $\mathbf{y}(0; \mathbf{c}) = \mathbf{c}$. Substituting into (7.2) we have

$$\mathbf{h}(\mathbf{c}) \equiv \mathbf{g}(\mathbf{c}, \mathbf{y}(b; \mathbf{c})) = \mathbf{0}\,. \tag{7.3}$$

This gives a set of m nonlinear algebraic equations for the m unknowns $\mathbf{c}$.

The simple (or *single*) shooting method consists of a numerical implementation of these observations, which we have used in previous chapters for theoretical purposes. Thus, one couples a program module for solving nonlinear algebraic equations (such library routines are available) with a module that, for a given $\mathbf{c}$, solves the corresponding initial value ODE problem.

Example 7.1
Recall Example 6.2, which considers a very simple model of a chemical reaction

$$u'' + e^{u+1} = 0,$$

$$u(0) = u(1) = 0\,.$$

The two solutions are depicted in Figure 6.1 (only the lower one is a physically stable steady state). Converting to first-order form for $\mathbf{y} = (u, u')^T$, we know that $y_1(0) = 0 = c_1$, so only $y_2(0) = c_2$ is unknown. The IVP has a unique solution $\mathbf{y}(t; \mathbf{c})$ (or $u(t; c_2)$) for each value c_2, even though it is not guaranteed that this solution will reach $t = 1$ for any c_2. But, as it turns out, this problem is easy to solve using simple shooting. With a starting "angle" of shooting (for the nonlinear iteration) $c_2^0 = 0.5$, the lower curve of Figure 6.1 is obtained after a few Newton iterations to solve (7.3), and with a starting "angle" of shooting $c_2^0 = 10$, the high curve of Figure 6.1 is easily obtained as well (Exercise 7.1). ♦

Let us consider Newton's method for the solution of the nonlinear equations (7.3). The iteration is

$$\mathbf{c}^{\nu+1} = \mathbf{c}^{\nu} - \left(\frac{\partial \mathbf{h}}{\partial \mathbf{c}}\right)^{-1} \mathbf{h}(\mathbf{c}^{\nu}),$$

where $\mathbf{c}^0$ is a starting iterate[28] (different starting guesses can lead to different solutions, as in Example 7.1). To evaluate $\mathbf{h}(\mathbf{c}^{\nu})$ at a given iterate we have to solve an IVP for $\mathbf{y}(t; \mathbf{c})$ (see (7.3)). Moreover, to evaluate $\frac{\partial \mathbf{h}}{\partial \mathbf{c}}$ at $\mathbf{c} = \mathbf{c}^{\nu}$, we must differentiate the expression in (7.3) with respect to $\mathbf{c}$. Using the chain rule of differentiation and the notation of (6.3), this gives

$$\left(\frac{\partial \mathbf{h}}{\partial \mathbf{c}}\right) = B_0 + B_b Y(b) = Q,$$

where $Y(t)$ is the $m \times m$ fundamental solution matrix satisfying

$$Y' = A(t)Y, \qquad 0 < t < b,$$

$$Y(0) = I,$$

with $A(t, \mathbf{y}(t; \mathbf{c}^{\nu})) = \frac{\partial \mathbf{f}}{\partial \mathbf{y}}$ (see Chapter 6—this variational ODE should be familiar to you at this point).

We see therefore that using Newton's method, $m + 1$ IVPs are to be solved at each iteration (one for $\mathbf{h}$ and m linear ones for the columns of $Y(t)$). However, the m linear systems are simple and they share the same matrix $A(t)$ which can therefore be evaluated once for all m systems, so

[28] Note that the superscript ν is an iteration counter, not a power.

the solution of these IVPs typically costs much less than $m + 1$ times the solution of the IVP for $\mathbf{h}$.[29]

Once convergence has been obtained, i.e., the appropriate initial value vector $\mathbf{c}$ which solves $\mathbf{h}(\mathbf{c}) = \mathbf{0}$ has been (approximately) found, we integrate the corresponding IVP to evaluate the solution of the BVP at any given points.

To summarize, Algorithm 7.1 describes the combination of shooting with Newton's method for a nonlinear BVP (7.1)–(7.2). We note that to maximize the efficiency of a shooting code, methods other than Newton's (e.g., quasi-Newton) should be used. We do not pursue this further, though.

7.1.1 Difficulties

From the above description we also see the potential trouble that the simple shooting method may run into: the conditioning of each iteration depends on the IVP stability, *not only on the BVP stability.* The matrix that features in the iteration is $Q = B_0 + B_b Y(b)$, and this matrix can be extremely poorly conditioned (recall Example 6.4) even when the BVP is stable and not very stiff. Finding the solution, once the correct initial values are known, also involves integrating a potentially unstable IVP.

It is not difficult to see that if a method of order p is used for the initial value integrations (in the sense of the IVP methods studied in Chapters 3, 4, and 5) then a method of order p is obtained for the BVP. This follows directly if we assume that the nonlinear iteration for (7.3) converges, and in the absence of roundoff errors. The trouble in finding $\mathbf{c}$ (if there is any) does not arise because of truncation errors, because for a stable BVP, error growth along unstable modes gets cancelled (recall Section 6.2), and this effect is reproduced by a consistent, stable IVP discretization. Also, if the BVP is unstable then the shooting method is expected to have difficulties, but these will be shared by other standard methods; the case where the simple shooting method is particularly unsatisfactory is when other simple methods (discussed in the next chapter) work well, while this method does not. Such is the case in the following example.

Example 7.2
The following problem:

$$\mathbf{y}' = A(t)\mathbf{y} + \mathbf{q}(t),$$

$$A = \begin{pmatrix} 0 & 1 & 0 \\ 0 & 0 & 1 \\ -2\lambda^3 & \lambda^2 & 2\lambda \end{pmatrix},$$

[29]Solving the variational ODE is equivalent to computing the sensitivity of the solution to the original ODE (7.1) with respect to variations in the initial conditions; see Section 4.6.

Algorithm 7.1. Shooting with Newton

- Given
  1. $\mathbf{f}$, $\frac{\partial \mathbf{f}}{\partial \mathbf{y}}$ *for each* t *and* $\mathbf{y}$;
  2. $\mathbf{g}(\mathbf{u}, \mathbf{v})$, $\frac{\partial \mathbf{g}}{\partial \mathbf{u}}$, $\frac{\partial \mathbf{g}}{\partial \mathbf{v}}$ *for each* $\mathbf{u}$ *and* $\mathbf{v}$;
  3. *an initial value solver;*
  4. *an initial guess* $\mathbf{c}^0$; *and*
  5. *a convergence tolerance* TOL *for the nonlinear iteration.*
- *For* $s = 0, 1, \ldots,$ *until* $|\mathbf{c}^{s+1} - \mathbf{c}^s| < \text{TOL}$:
  1. Solve *the IVP* (7.1) *with* $\mathbf{y}(0) = \mathbf{c}^s$, *obtaining a mesh and solution values* $\mathbf{y}_n^s$, $n = 0, \ldots, N_s$.
  2. Construct $\mathbf{h}(\mathbf{c}^s) = \mathbf{g}(\mathbf{c}^s, \mathbf{y}_{N_s}^s)$.
  3. Integrate *the fundamental matrix* Y_n, $n = 0, \ldots, N_s$ $(Y_0 = I)$, *on the same mesh, using* $A(t_n) = \frac{\partial \mathbf{f}}{\partial \mathbf{y}}(t_n, \mathbf{y}_n^s)$.
  4. Form $Q = B_0 + B_b Y_{N_s}^s$ *using*

     $$B_0 = \frac{\partial \mathbf{g}}{\partial \mathbf{u}}(\mathbf{c}^s, \mathbf{y}_{N_s}^s), \qquad B_b = \frac{\partial \mathbf{g}}{\partial \mathbf{v}}(\mathbf{c}^s, \mathbf{y}_{N_s}^s),$$

     and solve *the linear system*

     $$Q\boldsymbol{\eta} = \mathbf{h}(\mathbf{c}^s)$$

     for the Newton correction vector $\boldsymbol{\eta}$.
  5. Set

     $$\mathbf{c}^{s+1} = \mathbf{c}^s + \boldsymbol{\eta}.$$
- Solve *the IVP for* (7.1) *for* $\mathbf{y}(0) = \mathbf{c}$, *with the values* $\mathbf{c}$ *obtained by the Newton iteration.*

$$y_1(0) = b_1,\ y_1(1) = b_2,\ y_2(1) = b_3,$$

has the exact solution

$$\mathbf{y}(t) = (u(t), u'(t), u''(t))^T, \quad u(t) = \frac{e^{\lambda(t-1)} + e^{2\lambda(t-1)} + e^{-\lambda t}}{2 + e^{-\lambda}} + \cos \pi t$$

(you can evaluate the expressions and values for $\mathbf{q}(t) = \mathbf{y}'(t) - A(t)\mathbf{y}(t)$ and the boundary values $\mathbf{b}$ from this exact solution). The problem is in the form

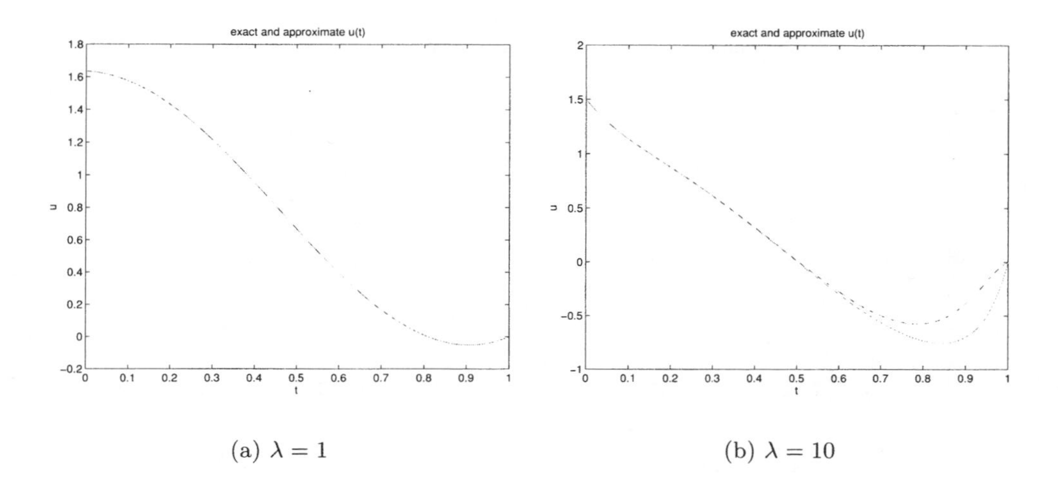

(a) $\lambda = 1$ (b) $\lambda = 10$

Figure 7.1: *Exact (solid line) and shooting (dashed line) solutions for Example* 7.2.

(7.7) with

$$B_0 = \begin{pmatrix} 1 & 0 & 0 \\ 0 & 0 & 0 \\ 0 & 0 & 0 \end{pmatrix}, \qquad B_b = \begin{pmatrix} 0 & 0 & 0 \\ 1 & 0 & 0 \\ 0 & 1 & 0 \end{pmatrix}.$$

For $\lambda = 20$, say, the BVP is stable but the IVP is not.

In Figures 7.1 and 7.2 we display the exact and approximate solutions (solid and dashed lines, respectively) for various values of λ ranging from a harmless 1 to a tough 50. We use the classical Runge–Kutta method of order 4 with a fixed step size $h = .004$ and a 14-hexadecimal-digit floating point arithmetic.

Note that the disastrous effect observed is due to the propagation of errors in the obtained initial values $\mathbf{c}$ by unstable modes (recall Example 2.1). The error in $\mathbf{c}$ is unavoidable and is due to *roundoff*, not truncation errors. We have chosen the discretization step size so small, in fact, that for the case $\lambda = 1$ the errors in the initial values vector are all below 10^{-9}, as is the maximum error in u in the ensuing integration for the approximate solution of the BVP. For $\lambda = 10$, already an $O(1)$ error is observed (the maximum error in u is .34). This case may be regarded as particularly worrisome, because the wrong solution obtained for a moderate value of λ may also look plausible. For $\lambda = 20$, this maximum error is 206.8 (although the error in the initial conditions is only less than 10^{-4}); for $\lambda = 50$, the overall error in u is $2.1\mathrm{e} + 32$, and in the initial conditions it is about 10^8.

The instability is already extreme for $\lambda = 20$, a value for which the BVP is not very stiff. ♦

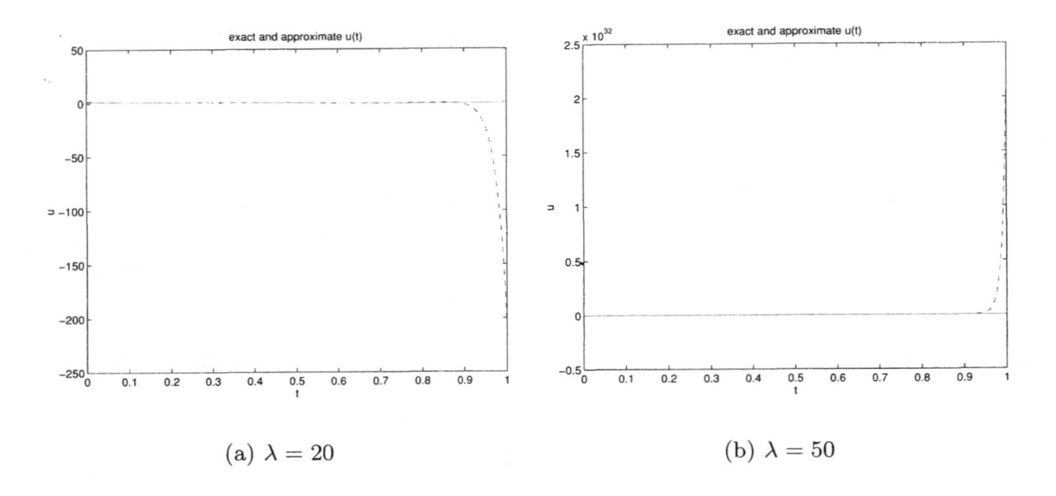

(a) $\lambda = 20$ (b) $\lambda = 50$

Figure 7.2: *Exact (solid line) and shooting (dashed line) solutions for Example* 7.2.

Another potential difficulty with the simple shooting method arises for nonlinear problems. The method assumes that the IVPs encountered will have solutions, even for inaccurate initial values, that reach all the way to $t = b$. For nonlinear problems, however, there is no guarantee that this would be the case. Initial value solutions with incorrect initial values are typically guaranteed to exist *locally* in t, but not necessarily globally. For another potential difficulty with the nonlinear iteration, see Exercise 7.7.

7.2 Multiple Shooting

Both disadvantages of the simple shooting method become worse for larger intervals of integration of the IVPs. In fact, a rough bound on the propagation error, which is approximately achieved in Example 7.2, is e^{Lb}, where $L = \max_t \|A(t)\|$. The basic idea of multiple shooting is then to restrict the size of intervals over which IVPs are integrated. Defining a mesh

$$0 = t_0 < t_1 < \cdots < t_{N-1} < t_N = b,$$

we consider approximating the solution of the ODE system $\mathbf{y}' = \mathbf{f}(t, \mathbf{y})$ by constructing an approximate solution on each subinterval $[t_{n-1}, t_n]$ and patching these approximate solutions together to form a global one (see Figure 7.3). Thus, let $\mathbf{y}_n(t; \mathbf{c}_{n-1})$ be the solution of the IVP

$$\mathbf{y}_n' = \mathbf{f}(t, \mathbf{y}_n), \qquad t_{n-1} < t < t_n, \tag{7.4a}$$

$$\mathbf{y}_n(t_{n-1}) = \mathbf{c}_{n-1} \tag{7.4b}$$

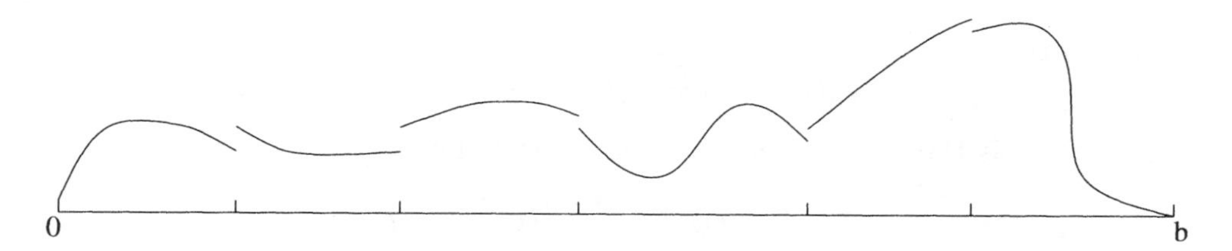

Figure 7.3: *Multiple shooting.*

for $1 \le n \le N$.[30] Assuming for the moment that the IVPs (7.4) are solved exactly, we then have that the exact solution of the problem (7.1)–(7.2) satisfies

$$\mathbf{y}(t) = \mathbf{y}_n(t; \mathbf{c}_{n-1}), \qquad t_{n-1} \le t \le t_n,\ 1 \le n \le N$$

if

$$\mathbf{y}_n(t_n; \mathbf{c}_{n-1}) = \mathbf{c}_n, \qquad 1 \le n \le N-1, \tag{7.5a}$$

$$\mathbf{g}(\mathbf{c}_0, \mathbf{y}_N(b; \mathbf{c}_{N-1})) = \mathbf{0}. \tag{7.5b}$$

The conditions (7.5a) are *patching* conditions which ensure that $\mathbf{y}(t)$ patched from the different pieces $\mathbf{y}_n(t; \mathbf{c}_{n-1})$ is continuous on the entire interval $[0, b]$, and (7.5b) is just the resulting expression for the boundary conditions (7.2).

The conditions (7.5) give Nm algebraic equations for the Nm coefficients

$$\mathbf{c} = (\mathbf{c}_0^T, \mathbf{c}_1^T, \ldots, \mathbf{c}_{N-1}^T)^T.$$

We write these equations, as before, as

$$\mathbf{h}(\mathbf{c}) = \mathbf{0}. \tag{7.6}$$

Applying Newton's method to solve the nonlinear equations (7.6) results at each iteration ν in a system of linear equations which can be viewed as arising from the same multiple shooting method for the linearized BVP

$$\mathbf{A}(\mathbf{c}^{\nu+1} - \mathbf{c}^{\nu}) = -\mathbf{h}(\mathbf{c}^{\nu}),$$

where $\mathbf{A} = \frac{\partial \mathbf{h}}{\partial \mathbf{c}}(\mathbf{c}^{\nu})$ has a sparse block structure, as in (7.10) below. An advantage of Newton's method here (not shared by quasi-Newton methods) is that this sparse block structure remains intact during the iteration process.

The Linear Multiple Shooting System

Since the system of linear equations is the same as the one obtained by applying the same multiple shooting method to the linearized problem, let us consider the latter further. For the linear problem

$$\begin{aligned} \mathbf{y}' &= A(t)\mathbf{y} + \mathbf{q}(t), \\ B_0\mathbf{y}(0) + B_b\mathbf{y}(b) &= \mathbf{b}, \end{aligned} \tag{7.7}$$

[30]It is important not to confuse this notation with what is used in the finite difference chapters 3, 4, 5, and 8 for a slightly different purpose. Here $\mathbf{y}_n$ is meant to be the *exact* solution on a subinterval $[t_{n-1}, t_n]$, provided we can find the right $\mathbf{c}_{n-1}$.

we can write

$$\mathbf{y}_n(t; \mathbf{c}_{n-1}) = Y_n(t)\mathbf{c}_{n-1} + \mathbf{v}_n(t),$$

where $Y_n(t)$ is the fundamental solution satisfying

$$Y_n' = A(t)Y_n, \qquad Y_n(t_{n-1}) = I$$

(in particular, $Y_1 \equiv Y$), and $\mathbf{v}_n(t)$ is a particular solution satisfying, e.g.,

$$\mathbf{v}_n' = A(t)\mathbf{v}_n + \mathbf{q}(t), \qquad \mathbf{v}_n(t_{n-1}) = \mathbf{0}\,.$$

The patching conditions and boundary conditions are then

$$\begin{aligned} I\mathbf{c}_n - Y_n(t_n)\mathbf{c}_{n-1} &= \mathbf{v}_n(t_n), \qquad 1 \le n \le N-1, &\text{(7.8a)}\\ B_0\mathbf{c}_0 + B_bY_N(b)\mathbf{c}_{N-1} &= \mathbf{b} - B_b\mathbf{v}_N(b)\,. &\text{(7.8b)}\end{aligned}$$

Writing these conditions as a linear system, we get

$$\mathbf{Ac} = \mathbf{r}, \tag{7.9}$$

where

$$\mathbf{A} = \begin{pmatrix} -Y_1(t_1) & I & & & \\ & -Y_2(t_2) & I & & \\ & & \ddots & \ddots & \\ & & & -Y_{N-1}(t_{N-1}) & I \\ B_0 & & & & B_bY_N(b) \end{pmatrix},$$

$$\mathbf{c} = \begin{pmatrix} \mathbf{c}_0 \\ \mathbf{c}_1 \\ \vdots \\ \mathbf{c}_{N-2} \\ \mathbf{c}_{N-1} \end{pmatrix}, \quad \mathbf{r} = \begin{pmatrix} \mathbf{v}_1(t_1) \\ \mathbf{v}_2(t_2) \\ \vdots \\ \mathbf{v}_{N-1}(t_{N-1}) \\ \mathbf{b} - B_b\mathbf{v}_N(b) \end{pmatrix}. \tag{7.10}$$

The matrix $\mathbf{A}$ is large and sparse when N is large, but there are well-known variants of Gaussian elimination which allow the solution of the linear system of equations (7.9) in $O(N)$ time. This will be discussed in Section 8.2. In fact, given N parallel processors in a computational model which ignores communication costs, the solution time for this linear system can be reduced to $O(\log N)$. Note that the blocks $Y_n(t_n)$ can be constructed in parallel too.[31] Initial value integration is applied for these constructions, as well as for the construction of the $\mathbf{v}_n$'s.

[31]For this reason the multiple shooting method is sometimes referred to as the *parallel shooting* method.

Stability

Turning to the question of whether the instability of the single shooting method has been improved upon, note that, assuming that the boundary matrices are scaled to $O(1)$,

$$\|\mathbf{A}\| = \text{const}\left(\max_{1\le n\le N}\{\|Y_n(t_n)\|\}+1\right).$$

It can also be verified directly that $\mathbf{A}$ has the inverse

$$\mathbf{A}^{-1} = \begin{pmatrix} G(t_0,t_1) & \cdots & G(t_0,t_{N-1}) & \Phi(t_0) \\ \vdots & & \vdots & \vdots \\ G(t_{N-1},t_1) & \cdots & G(t_{N-1},t_{N-1}) & \Phi(t_{N-1}) \end{pmatrix}, \tag{7.11}$$

where G and Φ are defined in (6.12) and (6.9), respectively. Therefore, with κ the stability constant of the given BVP (recall (6.13)),

$$\|\mathbf{A}^{-1}\| \le N\kappa,$$

so

$$\text{cond}(\mathbf{A}) = \|\mathbf{A}\|\|\mathbf{A}^{-1}\| \le \text{const}\,\kappa N\left(\max_{1\le n\le N}\{\|Y_n(t_n)\|\}+1\right) \tag{7.12}$$

for some moderate constant const.

The problem with simple shooting is that $\|Y(b)\|$ can be very large, even for stable BVPs, and it features prominently in the conditioning of the shooting algorithm (because $\|\mathbf{A}\|$ is very large). The bound on the condition number in (7.12) is often much more acceptable. For Example 7.2 with $\lambda = 20$, 10 equally spaced multiple shooting points produce an accurate solution (to 7 digits, using the same discretization for the IVPs), in contrast to the simple shooting results shown in Figure 7.2. The other disadvantage, resulting from finite escape time in nonlinear IVPs, is corrected to a large extent by multiple shooting as well.

However, with the significant improvement of the various deficiencies, the conceptual simplicity of simple shooting is also gone. Moreover, for very stiff BVPs the number of shooting points must grow unacceptably large with the stiffness parameter (e.g., it is proportional to λ, as $\lambda \to \infty$, in Example 7.2).

7.3 Software, Notes, and References

7.3.1 Notes

A detailed treatment of the techniques covered in this chapter can be found in Chapter 4 of Ascher, Mattheij, and Russell [8]. See also Mattheij and

Molenaar [67]. Earlier references include Keller [59]. Our presentation is deliberately short—we have chosen to concentrate more on finite difference methods in the next chapter.

The simple shooting method applied to a linear BVP (see (6.6)) can be viewed as a method of *superposition*, where the solution is composed of a linear combination of solution modes (columns of $Y(t)$) plus a particular solution of the nonhomogeneous problem (6.6) subject to (say) homogeneous initial conditions. There are more efficient, *reduced superposition* variants as well; see [8] and references therein.

There are other initial value techniques like stabilized march and Riccati methods which possess certain advantages (and disadvantages) over the multiple shooting method presented here. They can be viewed as achieving, for a linear(ized) problem, a *decoupling* of rapidly increasing modes (whose forward integration yields stability problems) from the other modes. The algorithm described in Exercise 7.8 can be made to be stable then. See Section 8.7 for more on decoupling. For reasons of space and bias, however, we do not explore these methods further. The interested reader can consult Chapter 4 of [8].

For use of a multiple shooting method for parameter identification, i.e., attempting to find unknown parameters which define the ODE given observations on its solution, see Bock [17].

7.3.2 Software

Many scientists and engineers seem to implement their own application-dependent shooting codes, making use of the excellent and abundant software which is available for IVPs. Shooting handles problems with nonseparated boundary conditions, and extends naturally to handle problems with parameters. Sparse linear algebra is avoided, at least when m is not large. The NAG library has a simple shooting code written by I. Gladwell. Another shooting code is being developed by L. Shampine, at the time of this writing, for MATLAB. However, we find the limited applicability of this method somewhat unsettling for general purposes.

A number of multiple shooting codes were developed in the 1970s and 1980s. We mention the code MUS by Mattheij and Staarink [68, 8] which is available from NETLIB. Earlier codes include SUPORT by Scott and Watts [84].

7.4 Exercises

7.1. Write a simple shooting code, using available software modules for initial value ODE integration, solution of nonlinear algebraic equations, and solution of linear algebraic equations as you find necessary. Apply your code to the following problems.

(a) Find both solutions of Example 6.2. What are the correct initial values for each of the two solutions?

(b) Use your program (only after verifying that it is correct) on some stable BVP of your choice where it is not supposed to work, and explain the observed results.

7.2. (a) Verify that the expression given in (7.11) is indeed the inverse of $\mathbf{A}$ given in (7.10).

(b) Estimate cond($\mathbf{A}$) for Example 7.2 with $\lambda = 20$, using 10 equally spaced multiple shooting points.

(c) How many multiple shooting points are needed to obtain a similar bound on cond($\mathbf{A}$) when $\lambda = 5000$?

7.3. Consider the car model of Example 4.7 with the same initial conditions as employed there. Given that $a = 100$, the task is to find a constant steering angle ψ so that the car will pass through the point $x(b) = 100$, $y(b) = 0$.

(a) Formulate this as a BVP (of order 6) in standard form.

(b) Solve this BVP numerically, using a package of your choice or your own home-grown program. Verify that the final speed is $v(b) = 137.63$. What is the required angle ψ? How long does it take the car to get to $x(b), y(b)$?

7.4. Consider the nonlinear problem

$$\begin{aligned} v'' + \frac{4}{t}v' + (tv - 1)v &= 0, \qquad 0 < t < \infty, \\ v'(0) &= 0, \qquad v(\infty) = 0. \end{aligned} \tag{7.13}$$

This is a well-behaved problem with a smooth, nontrivial solution. To solve it numerically, we replace $[0, \infty)$ by a finite, large interval $[0, L]$ and require

$$v(L) = 0\,.$$

For large t the solution is expected to decay exponentially, like $e^{-\alpha t}$, for some $\alpha > 0$.

(a) Find the asymptotic behavior of the solution for large t (i.e., find α). [You may assume that $v(t)$ is very (i.e., exponentially) small when t is large.]

(b) Show that the simple shooting method is unstable for this problem.

(c) Describe the application of the multiple shooting method for this problem. Estimate (roughly) the number and location of the needed shooting points.

(d) What would you do to obtain convergence of your scheme, avoiding convergence to the trivial solution?

7.5. The so-called SH equations, arising when calculating the ground displacements caused by a point moment seismic source in a layered medium, form a simple ODE system

$$\mathbf{y}' = A(t; \omega, k)\mathbf{y}, \qquad 0 < t < b,$$

where

$$A = \begin{pmatrix} 0 & \mu^{-1} \\ \mu k^2 - \rho\omega^2 & 0 \end{pmatrix}.$$

Here the angular frequency ω and the horizontal wave number k are parameters, $-\infty < \omega < \infty$, $0 \leq k < \infty$. The independent variable t corresponds to depth into the earth (which is the medium in this seismological application). See [60, 8] and references therein for more details, although you don't really need to understand the physics in order to solve this exercise. A hefty assumption is made that the earth in the area under consideration consists of horizontal layers. Thus, there is no horizontal variation in medium properties. Assume, moreover, that there is a partition

$$0 = t_0 < t_1 < \cdots < t_N = b$$

such that the S-wave velocity $\beta(t)$, the density $\rho(t)$, and thus also $\mu(t) = \rho\beta^2$ are constant in each layer:

$$\beta = \beta_n, \rho = \rho_n, \mu = \mu_n, \qquad t_{n-1} \leq t < t_n.$$

(a) At the earth's surface $t = 0$, $y_2(0) \neq 0$ is given. Another boundary condition is derived from a radiation condition, requiring that only downgoing waves exist for $t \geq b$. Assuming that the properties of the medium are constant for $t \geq b$, this yields

$$\nu_\beta y_1(b) + \mu^{-1} y_2(b) = 0,$$

where

$$\nu_\beta = \sqrt{k^2 - (\omega/\beta)^2}.$$

Derive this boundary condition.
[Hint: The eigenvalues of A are $\pm\nu_\beta$.]

(b) Describe a multiple shooting method that would yield the exact solution (except for roundoff errors) for this BVP.

[Note that this problem has to be solved many times, for various values of k and ω, because the obtained solution is used for integrand evaluation for a double integral in k and ω, called the Hankel transform. It is therefore worthwhile to tailor a particularly good method for this simple BVP.]

7.6. Delay differential equations arise often in applications. There are some situations in which a conversion to an ODE system can be useful. Consider a problem with a single, constant delay $\tau > 0$,

$$\begin{aligned} \mathbf{z}'(t) &= \mathbf{f}(t, \mathbf{z}(t)) + A(t)\mathbf{z}(t-\tau), \qquad 0 < t < b, && (7.14a) \\ B_{01}\mathbf{z}(t) &= \mathbf{b}_1(t), \qquad -\tau \le t \le 0, && (7.14b) \\ B_{b2}\mathbf{z}(b) &= \mathbf{b}_2, && (7.14c) \end{aligned}$$

where in (7.14a) there are m equations, B_{01} is full rank $k \times m$, B_{b2} is $(m-k) \times m$,

$$B_0 = \begin{pmatrix} B_{01} \\ 0 \end{pmatrix}$$

is $m \times m$, and $A(t)$ can be written as $A(t) = \hat{A}(t)B_0$. We further assume that there is a unique, continuous solution and that $b = \tau J$ for some positive integer J.

(a) Show that the functions

$$\mathbf{y}_j(s) = \mathbf{z}(s + (j-1)\tau), \qquad j = 1, \dots, J,$$

satisfy the ODE system

$$\begin{aligned} \mathbf{y}_j'(s) &= \mathbf{f}(s + (j-1)\tau, \mathbf{y}_j(s)) + A(s + (j-1)\tau)\mathbf{y}_{j-1}(s), \\ &\qquad j = 2, \dots, J, \qquad 0 < s < \tau, \\ \mathbf{y}_1'(s) &= \mathbf{f}(s, \mathbf{y}_1(s)) + \hat{A}(s)\hat{\mathbf{b}}_1(s-\tau), \\ &\quad B_{01}\mathbf{y}_1(0) = \mathbf{b}_1(-\tau), \ B_{b2}\mathbf{y}_J(\tau) = \mathbf{b}_2, \\ &\quad \mathbf{y}_j(\tau) = \mathbf{y}_{j+1}(0), \qquad j = 1, \dots, J-1, \end{aligned}$$

where $\hat{\mathbf{b}}_1$ is just $\mathbf{b}_1$ extended by $m-k$ zeros. This is a BVP in standard form.

(b) Solve the following problem using your code of Exercise 7.1 or a library BVP package:

$$\begin{aligned} u''(t) &= -\frac{1}{16}\sin u(t) - (t+1)u(t-1) + t, \qquad 0 < t < 2, \\ u(t) &= t - \frac{1}{2}, \qquad -1 \le t \le 0, \\ u(2) &= -\frac{1}{2}. \end{aligned}$$

[Recall Section 6.4, in case you need the boundary conditions in separated form.]

(c) When $k = m$ and $B_0 = I$, (7.14) is an initial value delay ODE. Describe a method to convert this to a sequence of *initial value* ODEs of increasing size.

(d) Explain why both conversion tricks to standard BVP and to standard IVP forms lose their appeal when τ shrinks, i.e., $\tau \ll b$.

[This is a curious thing, because as $\tau \to 0$ the delay ODE (7.14a) becomes "closer to" an ODE system of size m. For more on this topic, see [85, 8, 50] and references therein.]

7.7. The well-known Newton–Kantorovich Theorem guarantees convergence of Newton's method starting at $\mathbf{c}^0$ for the nonlinear system of algebraic equations $\mathbf{h}(\mathbf{c}) = \mathbf{0}$. With the notation $J(\mathbf{c}) = \frac{\partial \mathbf{h}}{\partial \mathbf{c}}$, a sufficient condition for convergence is

$$\alpha\beta\gamma < \frac{1}{2},$$

where

$$\begin{aligned} \|J(\mathbf{c}^0)^{-1}\mathbf{h}(\mathbf{c}^0)\| &\leq \alpha, \\ \|J(\mathbf{c}^0)^{-1}\| &\leq \beta, \\ \|J(\mathbf{c}) - J(\mathbf{d})\| &\leq \gamma\|\mathbf{c} - \mathbf{d}\|. \end{aligned}$$

Show that for the simple shooting method for (7.1), (7.2), we can merely bound $\gamma \leq e^{Lb}$, where L is the Lipschitz constant of $\mathbf{f}$.

[This bound may be realized in practice, and indicates potential trouble in the convergence of Newton's method, unless $\mathbf{c}^0$ is very close to the exact solution $\mathbf{c}$ so that α is very small. The bound on γ is improved a lot when using multiple shooting with uniformly distributed shooting points [96].]

7.8. For the multiple shooting method we are faced with the challenge of solving the linear equations (7.9), where the matrix $\mathbf{A}$ may be large and sparse if there are many shooting points. A simple way of doing this involves viewing the equations (7.8) as a recursion. Thus, we write for (7.8a)

$$\begin{aligned} \mathbf{c}_{N-1} &= Y_{N-1}(t_{N-1})\mathbf{c}_{N-2} + \mathbf{v}_{N-1}(t_{N-1}) \\ &= Y_{N-1}(t_{N-1})[Y_{N-2}(t_{N-2})\mathbf{c}_{N-3} + \mathbf{v}_{N-2}(t_{N-2})] + \mathbf{v}_{N-1}(t_{N-1}) \\ &= \dots, \end{aligned}$$

until we can express $\mathbf{c}_{N-1}$ in terms of $\mathbf{c}_0$, and this is substituted in (7.8b). The linear system to be solved is then only $m \times m$ and can be solved by the usual means. This method is called *compactification* in [8].

(a) Carry out the method just outlined for finding $\mathbf{c}$, i.e., find the formula.

(b) Show that, unfortunately, this method can degrade the stability properties of the multiple shooting method to those of the simple shooting method; i.e., this method for solving the linear system (7.9) can be unstable.

(c) Discuss the application of this method to the problem of Exercise 7.5 [53].

[Note that, compared to simple shooting, the method just outlined does have improved convergence properties for nonlinear problems.]

7.9. This exercise is concerned with finding periodic solutions for given ODE problems. In each case you are required to plot the obtained solution in phase space *and* to find the length of the period accurate to five digits (so eye-balling or trial-and-error would not work well enough for the purpose of finding the period). You are allowed to use any initial value software and boundary value software you want (including your own program from Exercise 7.1).

(a) Find the period of the heavenly bodies example of Exercise 4.12 (Figure 4.8).

(b) Find the period of the solution of the predator-prey Example 1.2 (Figure 1.3). The initial value used in that example was $(80, 30)$.

(c) Find the attracting limit cycle and the period of the Van der Pol equation

$$u'' = (1 - u^2)u' - u. \tag{7.15}$$

Chapter 8

Finite Difference Methods for Boundary Value Problems

As in the previous chapter, we seek numerical methods for boundary value problems (BVPs) based on our knowledge of methods for initial value problems (IVPs). But unlike the previous chapter, here we will not integrate IVPs. Rather, we consider the suitability of the discretizations studied in Chapters 3, 4, and 5 for BVPs. Consider a system of ordinary differential equations (ODEs) of order m,

$$\mathbf{y}' = \mathbf{f}(t, \mathbf{y}), \qquad 0 < t < b, \tag{8.1}$$

subject to m two-point boundary conditions

$$\mathbf{g}(\mathbf{y}(0), \mathbf{y}(b)) = \mathbf{0}\,. \tag{8.2}$$

Define a mesh (or a sequence of steps; we refer to the entire mesh as π)

$$\pi \;=\; \{0 = t_0 < t_1 < \cdots < t_{N-1} < t_N = b\}$$

with $h_n = t_n - t_{n-1}$ the nth step size, and consider solving for

$$\mathbf{y}_0, \mathbf{y}_1, \ldots, \mathbf{y}_{N-1}, \mathbf{y}_N$$

with $\mathbf{y}_n$ the intended approximation of $\mathbf{y}(t_n)$. The following observations are straightforward.

- For BVPs, no particular $\mathbf{y}_n$ is entirely known before all other mesh values for $\mathbf{y}$ are known. Hence, no difference method can be regarded as explicit. So, using what we called in Chapter 4 an explicit Runge–Kutta method, for instance, offers no advantage over using what was referred to in the IVP context as implicit Runge–Kutta methods.

- It makes no sense to use multistep methods either, both because there are really no "past" known solution values, and because the sparsity structure of the linear system that results is adversely affected, compared to one-step methods.[32]
- Symmetric, implicit Runge–Kutta methods are natural, because like BVPs they are indifferent to the direction of integration; i.e., they act similarly for nondecreasing and for nonincreasing modes.

In the sequel we therefore concentrate, with the exception of Section 8.6, on symmetric, one-step methods. As in Chapter 3, we start with the midpoint and the trapezoidal methods.

8.1 Midpoint and Trapezoidal Methods

We consider below the midpoint method, and leave the parallel development for the trapezoidal method to Exercise 8.1. Recall that the midpoint method for the ODE system (8.1) reads

$$\frac{\mathbf{y}_n - \mathbf{y}_{n-1}}{h_n} = \mathbf{f}\left(t_{n-1/2}, \frac{1}{2}(\mathbf{y}_n + \mathbf{y}_{n-1})\right), \qquad n = 1, \ldots, N, \tag{8.3}$$

and require also that the boundary conditions be satisfied,

$$\mathbf{g}(\mathbf{y}_0, \mathbf{y}_N) = \mathbf{0}. \tag{8.4}$$

In (8.3)–(8.4) we have $m(N+1)$ algebraic equations for the $m(N+1)$ unknown mesh values (including the end values). These equations are nonlinear if $\mathbf{f}$ is nonlinear in $\mathbf{y}$, and there are many such equations—it is not unusual to get 500 equations for a small ODE system. Their solution is discussed below. Before this we consider an example.

Example 8.1
Consider again Example 7.2. To recall, this is a linear problem of the form (7.7) with $m = 3$,

$$A = \begin{pmatrix} 0 & 1 & 0 \\ 0 & 0 & 1 \\ -2\lambda^3 & \lambda^2 & 2\lambda \end{pmatrix},$$

and the exact solution is

$$\mathbf{y} = (u, u', u'')^T, \qquad u(t) = \frac{e^{\lambda(t-1)} + e^{2\lambda(t-1)} + e^{-\lambda t}}{2 + e^{-\lambda}} + \cos \pi t,$$

[32]Note that we are discussing first-order ODEs. For a second-order ODE a natural discretization stencil would involve two steps; see Exercises 8.9–8.12.

N	λ	Error	Rate	λ	Error	Rate	λ	Error	Rate
10	1	.60e-2		50	.57		500	.96	
20		.15e-2	2.0		.32	.84		.90	.09
40		.38e-3	2.0		.14e-1	1.9		.79	.19
80		.94e-4	2.0		.34e-1	1.9		.62	.35

Table 8.1: *Maximum errors for Example* 8.1 *using the midpoint method: Uniform meshes.*

N	λ	Error	Rate	λ	Error	Rate
10	50	.14		500	*	
20		.53e-1	1.4		.26e-1	
40		.14e-1	1.9		.60e-2	2.1
80		.32e-2	2.2		.16e-2	1.9

Table 8.2: *Maximum errors for Example* 8.1 *using the midpoint method: Nonuniform meshes.*

which determines the inhomogeneity vector $\mathbf{q}(t)$. For boundary conditions, $u(0), u(1)$, and $u'(1)$ are prescribed. In Tables 8.1 and 8.2 we record maximum errors in u at the mesh points for $\lambda = 1, 50$, and 500 using uniform meshes and specialized, nonuniform meshes. For the uniform meshes, $h = 1/N$. These results are for the midpoint method; similar results are obtained also for the trapezoidal method.

Note that for $\lambda = 1$ the second-order accuracy of the midpoint method is reflected in the computed results. Given the smoothness of the exact solution it is also clear that there is room for employing higher-order methods (see Table 8.3), especially if highly accurate trajectories are desired.

For $\lambda = 50$, and even more so for $\lambda = 500$, the method is much less accurate if we use a uniform mesh, and the convergence order is reduced. The reason has already been discussed in Chapter 3: $O(1)$ errors which are generated in the narrow layer regions near the interval ends propagate almost undamped throughout the interval (recall Figure 3.4). To retrieve the potential accuracy of the midpoint method in regions where the solution varies slowly, the mesh in layer regions must be dense. The nonuniform meshes used for Table 8.2 result from a primitive effort to handle the layer regions. They are given (for $N = 10$) by

$$0, \frac{1}{2\lambda}, \frac{3}{\lambda}, \frac{8}{\lambda}, .25, .5, .75, 1 - \frac{8}{\lambda}, 1 - \frac{3}{\lambda}, 1 - \frac{1}{2\lambda}, 1,$$

and the refinements obtained by successively subdividing each of the mesh elements into two to obtain the next mesh. For $\lambda = 500$ the errors are

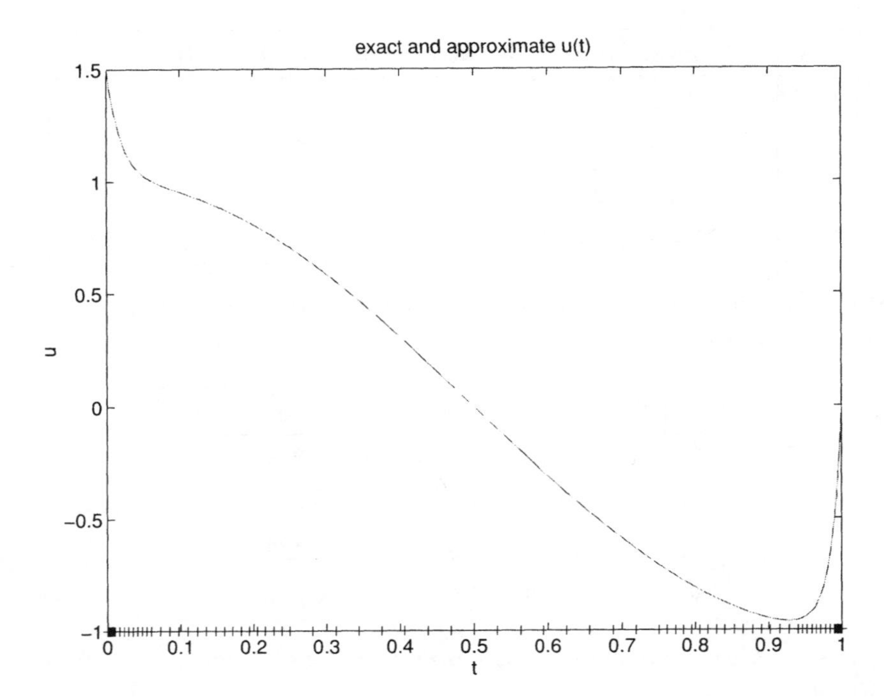

Figure 8.1: *Example* 8.1: *Exact and approximate solutions (indistinguishable) for* $\lambda = 50$, *using the indicated mesh.*

measured only at mesh points away from the layer.

Even with these simple nonuniform meshes, a significant improvement in the quality of the solution is obtained. The exact and approximate solutions for $\lambda = 50$ are plotted in Figure 8.1, together with the mesh that was used to generate these curves. This mesh corresponds to the last entry of Table 8.2 ($N = 80$). The approximate solution is in agreement with the exact one, as far as the eye can tell. It turns out that for this type of problem it is possible to construct more sophisticated meshes on which we obtain good, accurate solutions for any $\lambda \geq 1$ with N independent of λ. This is in contrast to multiple shooting techniques, where N grows linearly with λ. ♦

For solving the many nonlinear algebraic equations we again consider Newton's method, because it is basic, it is fast when it works well, and it retains the sparsity structure of the Jacobian, which is important for such a large system. As it turns out, Newton's method applied to the midpoint equations (8.3)–(8.4) is equivalent to the method of quasi-linearization coupled with the midpoint discretization for linear problems. The latter approach has the attraction of being more modular, so we describe it next.

8.1.1 Solving Nonlinear Problems: Quasi-Linearization

Newton's method for algebraic equations is obtained by expanding in Taylor series and truncating the nonlinear terms at each iteration. The quasi-linearization method does the same for the nonlinear differential system. Thus, let $\mathbf{y}^0(t)$ be an initial solution profile[33] (a guess), and write

$$\begin{aligned}
(\mathbf{y}^{\nu+1})' &= \mathbf{f}(t, \mathbf{y}^\nu) + \frac{\partial \mathbf{f}}{\partial \mathbf{y}}(t, \mathbf{y}^\nu)(\mathbf{y}^{\nu+1} - \mathbf{y}^\nu), \\
\mathbf{0} &= \mathbf{g} + \frac{\partial \mathbf{g}}{\partial \mathbf{u}}(\mathbf{y}^{\nu+1}(0) - \mathbf{y}^\nu(0)) + \frac{\partial \mathbf{g}}{\partial \mathbf{v}}(\mathbf{y}^{\nu+1}(b) - \mathbf{y}^\nu(b)),
\end{aligned}$$

where $\mathbf{y}^\nu = \mathbf{y}^\nu(t)$ is a known function at the νth iteration, and $\mathbf{g}$, $B_0 = \frac{\partial \mathbf{g}}{\partial \mathbf{u}}$, and $B_b = \frac{\partial \mathbf{g}}{\partial \mathbf{v}}$ are evaluated at the known iterate $(\mathbf{y}^\nu(0), \mathbf{y}^\nu(b))$ on the right-hand side of the last expression. Letting also

$$A(t) = \frac{\partial \mathbf{f}}{\partial \mathbf{y}}(t, \mathbf{y}^\nu(t)),$$

we obtain at the νth iteration that the next iterate $\mathbf{y}^{\nu+1} = \mathbf{y}$ satisfies the linear BVP

$$\begin{aligned}
&\mathbf{y}' = A(t)\mathbf{y} + \mathbf{q}(t), \qquad 0 < t < b, \\
&B_0\mathbf{y}(0) + B_b\mathbf{y}(b) = \mathbf{b},
\end{aligned} \tag{8.5}$$

where

$$\begin{aligned}
\mathbf{q} &= \mathbf{f}(t, \mathbf{y}^\nu(t)) - A(t)\mathbf{y}^\nu(t), \\
\mathbf{b} &= -\mathbf{g}(\mathbf{y}^\nu(0), \mathbf{y}^\nu(b)) + B_0\mathbf{y}^\nu(0) + B_b\mathbf{y}^\nu(b)\,.
\end{aligned}$$

The coefficients in the linear problem (8.5) may all depend, in general, on the current iterate $\mathbf{y}^\nu(t)$. The quasi-linearization procedure therefore defines a sequence of linear BVPs whose solutions hopefully converge to that of the given nonlinear BVP. Thus, if we know how to discretize and solve linear BVPs, then we obtain a method for nonlinear BVPs as well.

We proceed by applying the midpoint method for the linear problem. Note that the iterates $\mathbf{y}^\nu(t)$ are never really needed anywhere other than at mesh points. It is also easy to verify that the operations of linearization and discretization commute here: we obtain the same linear systems to solve as we would if we apply Newton's method directly to (8.3)–(8.4).

Example 8.2

Revisiting Example 6.2, we write the ODE in the first-order form (8.1) for $\mathbf{y}(t) = (u(t),\, u'(t))^T$,

$$\mathbf{y}' = \begin{pmatrix} y_2 \\ -e^{y_1+1} \end{pmatrix}, \qquad 0 < t < 1.$$

[33]Here and below we denote iteration number by a simple superscript, e.g., $\mathbf{y}^\nu$ for the νth iterate. This should not be confused with the notation for the νth power.

The boundary conditions are linear and homogeneous. They can be written as $B_0\mathbf{y}(0) + B_b\mathbf{y}(1) = \mathbf{0}$, with

$$B_0 = \begin{pmatrix} 1 & 0 \\ 0 & 0 \end{pmatrix}, \qquad B_b = \begin{pmatrix} 0 & 0 \\ 1 & 0 \end{pmatrix}.$$

The Jacobian matrix is apparently

$$\frac{\partial \mathbf{f}}{\partial \mathbf{y}} = \begin{pmatrix} 0 & 1 \\ -e^{y_1+1} & 0 \end{pmatrix},$$

so at the νth quasi-linearization iteration we define

$$A(t) = \begin{pmatrix} 0 & 1 \\ -e^{y_1^\nu(t)+1} & 0 \end{pmatrix}, \qquad \mathbf{q}(t) = \begin{pmatrix} y_2^\nu(t) \\ -e^{y_1^\nu(t)+1} \end{pmatrix} - A(t)\mathbf{y}^\nu(t),$$

and solve the linear system (8.5) with $\mathbf{b} = \mathbf{0}$ for $\mathbf{y} = \mathbf{y}^{\nu+1}(t)$.

Starting with the initial guess

$$u^0(t) = c_2 t(1-t), \qquad 0 \le t \le 1,$$

and employing the midpoint method with a uniform mesh of size $N = 10$, we obtain convergence after two Newton iterations to good-quality approximations of each of the two solutions depicted in Figure 6.1, upon setting $c_2 = 0.5$ and $c_2 = 10$, respectively (cf. Example 7.1). This problem is very easy to solve numerically, despite its nonunique solutions. ♦

Instead of solving in the νth quasi-linearization iteration for the next iterate $\mathbf{y}^{\nu+1}$ we can (and we prefer to) solve for the *Newton direction* at $\mathbf{y}^\nu$,

$$\boldsymbol{\eta}(t) = \mathbf{y}^{\nu+1}(t) - \mathbf{y}^\nu(t),$$

and then let

$$\mathbf{y}^{\nu+1} = \mathbf{y}^\nu + \boldsymbol{\eta}.$$

For $\boldsymbol{\eta}$ (which depends of course on ν) we have the linear problem (Exercise 8.1)

$$\begin{aligned} \boldsymbol{\eta}' &= A(t)\boldsymbol{\eta} + \mathbf{q}(t), \qquad 0 < t < b, \\ B_0\boldsymbol{\eta}(0) &+ B_b\boldsymbol{\eta}(b) = \mathbf{b}, \end{aligned} \tag{8.6}$$

where A, B_0, and B_b are as before, in (8.5), but the data simplifies to

$$\begin{aligned} \mathbf{q}(t) &= \mathbf{f}(t, \mathbf{y}^\nu) - (\mathbf{y}^\nu)', \\ \mathbf{b} &= -\mathbf{g}(\mathbf{y}^\nu(0), \mathbf{y}^\nu(b)). \end{aligned} \tag{8.7}$$

Note that in Example 8.2 we may no longer automatically set $\mathbf{b} = \mathbf{0}$ when solving for $\boldsymbol{\eta}(t)$; this depends on the initial guess $\mathbf{y}^0(t)$.

The Linear Midpoint System

The midpoint method applied to the linear problem (8.5) yields the linear equations

$$\begin{aligned}\frac{\mathbf{y}_n - \mathbf{y}_{n-1}}{h_n} &= A(t_{n-1/2})\frac{\mathbf{y}_n + \mathbf{y}_{n-1}}{2} + \mathbf{q}(t_{n-1/2}), \qquad n = 1, \ldots, N, \\ & B_0\mathbf{y}_0 + B_b\mathbf{y}_N = \mathbf{b}. \end{aligned} \tag{8.8}$$

This is a large, sparse linear system of $m(N+1)$ equations,

$$\mathbf{A}\mathbf{y}_\pi = \mathbf{r},$$

with

$$\mathbf{A} = \begin{pmatrix} S_1 & R_1 & & & \\ & S_2 & R_2 & & \\ & & \ddots & \ddots & \\ & & & S_N & R_N \\ B_0 & & & & B_b \end{pmatrix}, \tag{8.9}$$

$$\mathbf{y}_\pi = \begin{pmatrix} \mathbf{y}_0 \\ \mathbf{y}_1 \\ \vdots \\ \mathbf{y}_{N-1} \\ \mathbf{y}_N \end{pmatrix}, \qquad \mathbf{r} = \begin{pmatrix} \mathbf{q}(t_{1/2}) \\ \mathbf{q}(t_{3/2}) \\ \vdots \\ \mathbf{q}(t_{N-1/2}) \\ \mathbf{b} \end{pmatrix},$$

where

$$S_n = -\left[h_n^{-1}I + \frac{1}{2}A(t_{n-1/2})\right], \qquad R_n = \left[h_n^{-1}I - \frac{1}{2}A(t_{n-1/2})\right].$$

We see that the structure of $\mathbf{A}$ is the same as that of $\mathbf{A}$ for the multiple shooting method. In fact, to make it even more similar, we can multiply the nth block row of $\mathbf{A}$ by R_n^{-1}, obtaining block rows in the form

$$\begin{pmatrix} \cdots & R_n^{-1}S_n & I & \cdots \end{pmatrix}$$

with $R_n^{-1}S_n$ presumably approximating the fundamental solution matrix value $-Y_n(t_n)$.

To summarize, Algorithm 8.1 combines quasi-linearization with the midpoint discretization for a nonlinear BVP (8.1)–(8.2).

Algorithm 8.1. Quasi-Linearization with Midpoint

- Given
  1. $\mathbf{f}$, $\frac{\partial \mathbf{f}}{\partial \mathbf{y}}$ *for each* t *and* $\mathbf{y}$;
  2. $\mathbf{g}(\mathbf{u}, \mathbf{v})$, $\frac{\partial \mathbf{g}}{\partial \mathbf{u}}$, $\frac{\partial \mathbf{g}}{\partial \mathbf{v}}$ *for each* $\mathbf{u}$ *and* $\mathbf{v}$;
  3. *a mesh* $\pi : 0 = t_1 < \cdots < t_N = b$;
  4. *an initial guess* $\mathbf{y}^0(t)$, *or just* $\mathbf{y}_n^0 = \mathbf{y}_0(t_n)$, $n = 0, 1, \ldots, N$; *and*
  5. *a convergence tolerance* NTOL *for the nonlinear iteration.*
- *For* $\nu = 0, 1, \ldots,$ *until* $\max_{0 \le n \le N} |\mathbf{y}_n^{\nu+1} - \mathbf{y}_n^\nu| <$ NTOL*:*
  1. *For* $n = 1, \ldots, N$, form S_n, R_n, *and* $\mathbf{r}_n = \mathbf{q}(t_{n-1/2})$ *using*

$$\begin{aligned} A(t_{n-1/2}) &= \frac{\partial \mathbf{f}}{\partial \mathbf{y}} \left(t_{n-1/2}, \frac{\mathbf{y}_n^\nu + \mathbf{y}_{n-1}^\nu}{2} \right), \\ \mathbf{q}(t_{n-1/2}) &= \mathbf{f} \left(t_{n-1/2}, \frac{\mathbf{y}_n^\nu + \mathbf{y}_{n-1}^\nu}{2} \right) - \frac{\mathbf{y}_n^\nu - \mathbf{y}_{n-1}^\nu}{h_n}. \end{aligned}$$

  2. Form $\mathbf{A}$ *and* $\mathbf{r}$ *of* (8.9) *using*

$$B_0 = \frac{\partial \mathbf{g}}{\partial \mathbf{u}}(\mathbf{y}_0^\nu, \mathbf{y}_N^\nu), \quad B_b = \frac{\partial \mathbf{g}}{\partial \mathbf{v}}(\mathbf{y}_0^\nu, \mathbf{y}_N^\nu), \quad \mathbf{b} = -\mathbf{g}(\mathbf{y}_0^\nu, \mathbf{y}_N^\nu).$$

  3. Solve *the linear system of equations for* $\mathbf{y}_\pi = \boldsymbol{\eta}_\pi$.
  4. Set $\mathbf{y}_\pi^{\nu+1} = \mathbf{y}_\pi^\nu + \boldsymbol{\eta}_\pi$.

8.1.2 Consistency, 0-Stability, and Convergence

The *local truncation error*, *consistency*, and *accuracy* of a difference method are defined as in Section 3.2. There is essentially no dependence in this regard on what type of side conditions are prescribed (be they initial or boundary conditions) so long as they are approximated well. The question is still by how much the exact solution fails to satisfy the difference equations.

For the midpoint method we define

$$\mathcal{N}_\pi \mathbf{u}(t_n) \equiv \frac{\mathbf{u}(t_n) - \mathbf{u}(t_{n-1})}{h_n} - \mathbf{f} \left(t_{n-1/2}, \frac{1}{2}(\mathbf{u}(t_{n-1}) + \mathbf{u}(t_n)) \right),$$

so the numerical method is given by

$$\mathcal{N}_\pi \mathbf{y}_\pi(t_n) = \mathbf{0}$$

(with $\mathbf{g}(\mathbf{y}_0, \mathbf{y}_N) = \mathbf{0}$). By Taylor's expansion (see Exercise 3.4) we obtain that the local truncation error satisfies

$$\mathbf{d}_n = \mathcal{N}_\pi \mathbf{y}(t_n) = O(h_n^2),$$

so this is a consistent, second-order accurate method.

The definition of *convergence* is also exactly as in Section 3.2. Let

$$h = \max_{1 \le n \le N} h_n .$$

The method is *convergent of order* p if

$$\mathbf{e}_n = O(h^p)$$

for $n = 0, 1, 2, \ldots, N$, where $\mathbf{e}_n = \mathbf{y}_n - \mathbf{y}(t_n)$. We expect second-order convergence for the midpoint method.

The vehicle that carries accuracy results into convergence statements is 0-stability. For nonlinear problems we must confine ourselves to a vicinity of an exact, isolated solution (recall Chapter 6). Consider a "discrete tube" around such an exact solution $\mathbf{y}(t)$,

$$S_{\rho,\pi}(\mathbf{y}) = \{\mathbf{u}_\pi ; |\mathbf{u}_i - \mathbf{y}(t_i)| \le \rho,\ 0 \le i \le N\} \tag{8.10}$$

(the notation π is for the particular mesh considered, and $\rho > 0$ is the radius of the tube around $\mathbf{y}(t)$). The rest of the 0-stability definition is similar to the IVP case. The difference method is 0-*stable* if there are positive constants h_0, ρ, and K such that for any mesh π with $h \le h_0$ and any mesh functions $\mathbf{x}_\pi$ and $\mathbf{z}_\pi$ in $S_{\rho,\pi}(\mathbf{y})$,

$$\begin{aligned} |\mathbf{x}_n - \mathbf{z}_n| \ \le\ & K\{|\mathbf{g}(\mathbf{x}_0, \mathbf{x}_N) - \mathbf{g}(\mathbf{z}_0, \mathbf{z}_N)| \\ & + \max_{1 \le j \le N} |\mathcal{N}_\pi \mathbf{x}_\pi(t_j) - \mathcal{N}_\pi \mathbf{z}_\pi(t_j)|\}, \quad 0 \le n \le N. \end{aligned} \tag{8.11}$$

Substituting $\mathbf{x}_n \leftarrow \mathbf{y}(t_n)$ and $\mathbf{z}_n \leftarrow \mathbf{y}_n$ into (8.11) we obtain an extension of the Fundamental Theorem 3.1 to the BVP case,

$$|\mathbf{e}_n| \le K \max_j |\mathbf{d}_j| = O(h^p), \quad 0 \le n \le N . \tag{8.12}$$

In particular, the midpoint method is second-order convergent. Note also that, as in the IVP case, the bound (8.12) is useful only if K is of the order of magnitude of the stability constant of the given differential problem.

Proving 0-stability

How can we show 0-stability? Below we consider the linear case. For the nonlinear problem we consider a linearization, much in the spirit of the quasi-linearization method and the variational problem (6.5). The difference operator must satisfy certain smoothness and boundedness requirements, and then the results extend.

For the linear BVP (8.5) the midpoint method (8.8) has been cast into matrix form in (8.9). Obviously, 0-stability is obtained if there is a constant K such that for all meshes π with h small enough,

$$\|\mathbf{A}^{-1}\| \le K .$$

Indeed, then we would have for the exact solution $\mathbf{y}(t)$, written at mesh points as

$$\mathbf{y}_e = (\mathbf{y}(0), \mathbf{y}(t_1), \ldots, \mathbf{y}(t_{N-1}), \mathbf{y}(b))^T ,$$

the estimates

$$\begin{aligned} \mathbf{A}\mathbf{y}_e &= \mathbf{r} + O(h^2), \\ \mathbf{A}(\mathbf{y}_e - \mathbf{y}_\pi) &= O(h^2), \\ |\mathbf{y}_e - \mathbf{y}_\pi| &\le K\, O(h^2) . \end{aligned}$$

To show 0-stability we call upon the closeness of $\mathbf{A}$ to the multiple shooting matrix. It is not difficult to see that

$$R_n^{-1} S_n = -Y_n(t_n) + O(h_n^2) .$$

Hence, denoting the multiple shooting matrix

$$\mathbf{M} = \begin{pmatrix} -Y_1(t_1) & I & & & \\ & -Y_2(t_2) & I & & \\ & & \ddots & \ddots & \\ & & & -Y_N(t_N) & I \\ B_0 & & & & B_b \end{pmatrix}$$

(which is in a slightly, but not meaningfully, different form from $\mathbf{A}$ of (7.10)) and defining the block-diagonal scaling matrix

$$\mathbf{D} = \begin{pmatrix} R_1^{-1} & & & & \\ & R_2^{-1} & & & \\ & & \ddots & & \\ & & & R_N^{-1} & \\ & & & & I \end{pmatrix},$$

we obtain

$$\mathbf{D}\mathbf{A} = \mathbf{M} + \mathbf{E},$$

where $\mathbf{E}$ has the same zero-structure as $\mathbf{A}$ and $\|\mathbf{E}\| = O(h^2)$.

From this we have

$$\mathbf{A}^{-1} = (\mathbf{M} + \mathbf{E})^{-1}\mathbf{D} .$$

Taking norms, and capitalizing on our knowledge of the exact inverse of $\mathbf{M}$ (recall (7.11)), we readily obtain

$$\|\mathbf{A}^{-1}\| \leq \kappa + O(h) \equiv K, \tag{8.13}$$

where κ is the stability constant of the problem, defined in (6.13). For h small enough, the stability bound is therefore *quantitative*! If the BVP is stable and not stiff, and the local truncation error is small, then the global error is expected to have the order of the local truncation error times the stability constant of the given BVP.

It is important to understand that the closeness just discovered between the midpoint difference method and the multiple shooting method is mainly useful for theoretical purposes. The placement of shooting points in the latter method is done to reduce IVP instabilities, not to control truncation error (which is controlled by the initial value solver). Thus, the distance between shooting points is not necessarily small. If the number of shooting points needed is as large as what is typical for a difference method like midpoint, then the multiple shooting method becomes rather inefficient, because where one simple discretization step would do it fires up a whole IVP solver. Also, for stiff BVPs the midpoint method does not use steps so small that $R_n^{-1}S_n$ can be said to approximate $-Y_n(t_n)$ well (and wisely so). The interpretation of the above result is still valid as $h \to 0$, as the name "0-stability" indicates.

8.2 Solving the Linear Equations

Having discretized a linear BVP using, say, the midpoint method, we obtain a large, sparse linear system of algebraic equations to solve

$$\mathbf{A}\mathbf{y}_\pi = \mathbf{r}, \tag{8.14}$$

with $\mathbf{A}$ having the sparsity structure depicted in (8.9). It is important that the reader imagine the structure of this matrix for, say, $m = 3$ and $N = 100$—it is large and rather sparse (only 1818 entries out of 91,809 are possibly nonzero). In Figure 8.2 we depict this structure for more modest dimensions, where zeros are blanked.

Of particular concern is the block B_0 at the lower left corner of $\mathbf{A}$. If it was not there then we would have a *banded* system; i.e., all nonzero entries are concentrated in a narrow band around the main diagonal. Fortunately, the situation for *separated boundary conditions* is much better than for the general case, just as in Section 6.2. If

$$B_0 = \begin{pmatrix} B_{01} \\ 0 \end{pmatrix}, \quad B_b = \begin{pmatrix} 0 \\ B_{b2} \end{pmatrix},$$

where B_{01} has k rows and B_{b2} has $m - k$ rows, then we can simply permute the matrix $\mathbf{A}$, putting the rows of B_{01} at the top. The right-hand side $\mathbf{r}$ is

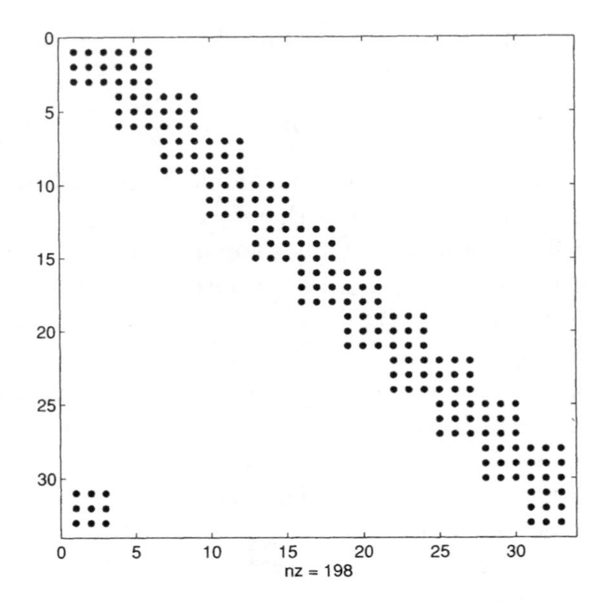

Figure 8.2: *Zero-structure of the matrix* $\mathbf{A}$, $m = 3, N = 10$. *The matrix size is* $m(N+1) = 33$.

permuted accordingly as well. This also establishes a "time" direction—the lower is the row in the permuted $\mathbf{A}$, the larger is t to which it refers. In Figure 8.3 we depict the permuted $\mathbf{A}$ corresponding to Figure 8.2, where two boundary conditions are prescribed at $t = 0$ and one at $t = b$.

A number of methods which require $O(Nm^3)$ flops (instead of the usual $O(N^3m^3)$) to solve the linear system (8.14) in the case of separated boundary conditions have been proposed in the literature. Here we describe the simplest and crudest of these and only comment on other methods.

Once permuted, the matrix $\mathbf{A}$ can be considered as banded, with $m+k-1$ diagonals below the main diagonal and $2m - k - 1$ diagonals above the main diagonal possibly having nonzero entries. Outside this total of $3m - 1$ diagonals, all entries of $\mathbf{A}$ are 0. Gaussian elimination with partial pivoting extends to the banded case in a straightforward fashion. Simply, two of the three nested loops defining the elimination process are shortened in order to not eliminate elements known to be 0 at the start. The fill-in is only within the banded structure, with the addition of a few diagonals due to the partial pivoting. It is not difficult to write a program to carry out this algorithm. Also, there exists standard software to do this, e.g., in LINPACK or LAPACK.

If you look at the band containing all nonzero elements in the matrix depicted in Figure 8.3, you will notice that there are triangles of zeros within the band for each interior mesh point. These zeros are not taken advantage of in the band method just described. Other, more sophisticated methods for solving (8.14) attempt to avoid, or at least minimize, fill-in of these triangles, thereby achieving an additional savings of up to 50% in both storage and computational efficiency.

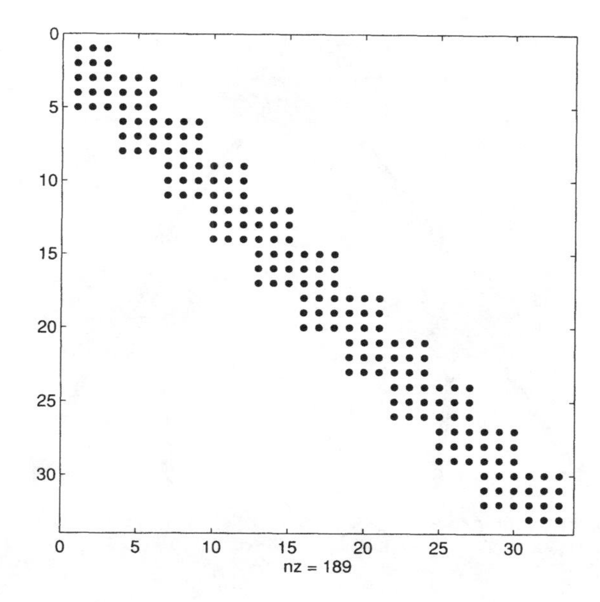

Figure 8.3: *Zero-structure of the permuted matrix* **A** *with separated boundary conditions,* $m = 3, k = 2, N = 10$.

The fact that all the nonzeros of **A** densely populate a narrow band as in Figure 8.3 is typical for boundary value ODEs, where neighboring elements (i.e., subintervals sharing unknowns) can be ordered consecutively. For boundary value partial differential equations (PDEs), on the other hand, the band is necessarily much wider and the matrix is sparse inside the band as well. Variants of Gaussian elimination become less effective then, and iterative methods like preconditioned conjugate gradients and multigrid take center stage.

8.3 Higher-Order Methods

The midpoint and trapezoidal methods may be considered as basic methods. The only problem in using them as they are for many applications is that they are only second-order accurate. There are two types of higher-order methods extending the basic methods: higher-order Runge–Kutta and acceleration techniques. The overview picture is given in Figure 8.4.

8.3.1 Collocation

One class of extensions to higher-order methods is simply higher-order implicit Runge–Kutta methods. Continuing to prefer symmetric methods, this leads to collocation methods at Gauss or Lobatto points. We have already considered the basic properties of these methods in Chapter 4, and everything else pertaining to BVPs extends in a very similar way to the treatment in the previous two sections. We summarize without repeating the proof:

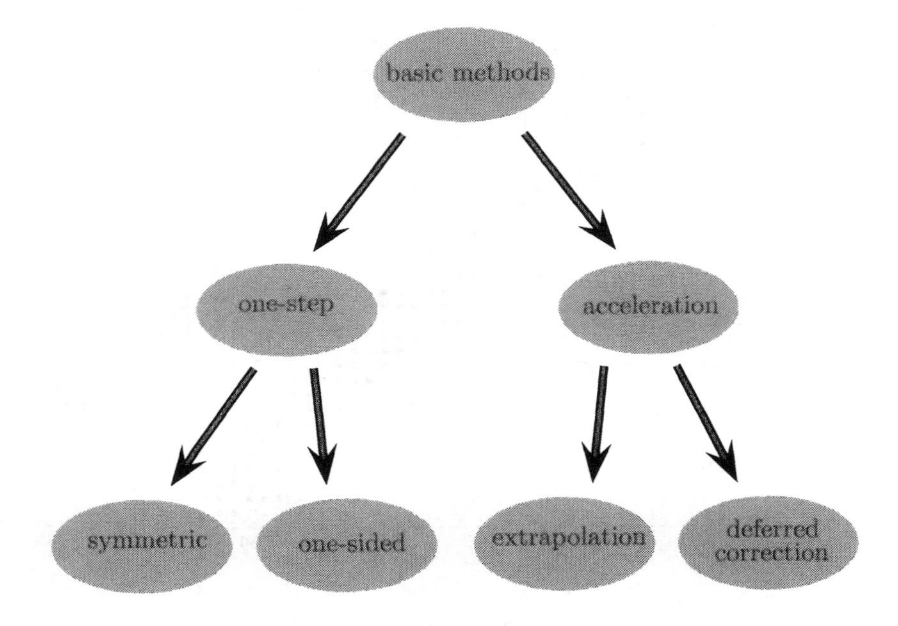

Figure 8.4: *Classes of higher-order methods.*

- Collocation at s Gaussian points is 0-stable with a numerical stability constant satisfying (8.13). It converges with the error bound

$$|\mathbf{e}_n| \leq K \max_j |\mathbf{d}_j| = O(h^{2s}), \qquad 0 \leq n \leq N. \tag{8.15}$$

Example 8.3
We repeat the computations of Example 8.1 using collocation at three Gaussian points per mesh element. The results are recorded in Tables 8.3 and 8.4.

The errors for $\lambda = 1$ in Table 8.3 reflect the fact that this method is of order 6 and has a nice error constant to boot. For $\lambda = 50$ the errors are fairly good even on the uniform mesh, although they are better on the nonuniform meshes (whose construction is discussed in Example 8.1). For $\lambda = 500$ a nonuniform mesh is certainly needed; see Table 8.4. In fact, a better layer mesh can become useful as well in order to retrieve the full convergence order (which turns out to be 4 for $h|\lambda| \gg 1$) that this method has outside the layer regions. ♦

Applying quasi-linearization to nonlinear problems and considering collocation for the resulting linear ones, we obtain a linear system of equations for the mesh unknowns as well as the internal stages (see (4.8) or (4.10)).

N	λ	Error	Rate	λ	Error	Rate	λ	Error	Rate
10	1	.60e-8		50	.54e-1		500	.71	
20		.94e-10	6.0		.66e-2	3.0		.50	.50
40		.15e-11	6.0		.32e-3	4.4		.27	.91
80		.24e-14	5.9		.73e-5	5.5		.89e-1	1.6

Table 8.3: *Maximum errors for Example* 8.1 *using collocation at three Gaussian points: Uniform meshes.*

N	λ	Error	Rate	λ	Error	Rate
10	50	.25e-2		500	.54e-3	
20		.12e-3	4.4		.14e-3	1.9
40		.27e-5	5.5		.75e-4	.90
80		.40e-7	6.1		.33e-4	1.2

Table 8.4: *Maximum errors for Example* 8.1 *using collocation at three Gaussian points: Nonuniform meshes.*

Since, unlike in the IVP case, the solution is not known anywhere in its entirety until it is known at all mesh points, one approach is to solve for all mesh values *and* internal stages simultaneously. This alters the structure of the linear system (8.14), but $\mathbf{A}$ is still in a block form and is banded independently of N.

Alternatively, we eliminate the internal stages locally, in each mesh subinterval n, in terms of the mesh values $\mathbf{y}_{n-1}$ and $\mathbf{y}_n$. This IVP-style approach is called *local elimination,* or *parameter condensation,* in the finite element literature. The remaining global system to be solved, (8.14), has the *almost block diagonal* form (8.9) independently of s, which adds an attractive modularity to the process. However, the partial decompositions used for the local elimination stage have to be stored from one nonlinear iteration to the next, so the advantage here is in elegance, not in storage or computational efficiency.

8.3.2 Acceleration Techniques

The other possibility for extending basic methods to higher order is to stay with the midpoint or the trapezoidal method as the basic discretization method and to accelerate its convergence by applying it more than once.

Extrapolation

One way of accelerating convergence is *extrapolation*, where the method is applied on more than one mesh and the results are combined to kill off the lower-order terms in the error expansion. For instance, if the global error on a given mesh π has the form

$$\mathbf{e}_n = \mathbf{y}(t_n) - \mathbf{y}_n = \mathbf{c}h_n^2 + O(h^4),$$

where $\mathbf{c}$ may vary (slowly) in t but is independent of h, then subdividing each mesh subinterval into two and applying the same method again yields for the solution $\tilde{\mathbf{y}}_{2n}$

$$\mathbf{y}(t_n) - \tilde{\mathbf{y}}_{2n} = \frac{1}{4}\mathbf{c}h_n^2 + O(h^4),$$

so $\frac{4\tilde{\mathbf{y}}_{2n} - \mathbf{y}_n}{3}$ is a fourth-order accurate approximate solution. This process can be repeated to obtain even higher-order methods (Exercise 8.4).

Deferred Correction

Another possibility for accelerating convergence is *deferred correction*, where the discretization on the same mesh is applied a few times, at each instance using the previous approximation to correct the right-hand side by better approximating the local truncation error.

Unlike extrapolation, which uses the expansion in powers of h for the global error, deferred correction uses the corresponding expansion for the local truncation error. For instance, applying the trapezoidal method to (8.1) we obtain (Exercise 8.8)

$$\mathbf{d}_n = \sum_{j=1}^{r} h_n^{2j} T_j[\mathbf{y}(t_{n-1/2})] + O(h_n^{2r+2}), \tag{8.16}$$

where

$$T_j[\mathbf{z}(t)] = \frac{-1}{2^{2j-1}(2j+1)!}\mathbf{f}^{(2j)}(t, \mathbf{z}(t)) \tag{8.17}$$

if $\mathbf{f}$ has continuous partial derivatives up to order $2r+2$ for some positive integer r. Now, let $\mathbf{y}_\pi = \{\mathbf{y}_n\}_{n=0}^N$ be the obtained solution on a given mesh using the trapezoidal method to discretize the stable BVP (8.1)–(8.2), and denote $\mathbf{f}_n = \mathbf{f}(t_n, \mathbf{y}_n)$, as in Chapter 5. Then we can use these values to approximate T_1 up to $O(h_n^2)$, e.g.,

$$\begin{aligned} T_1[\mathbf{y}(t_{n-1/2})] &\approx T_{1,n-1/2} \\ &= \frac{1}{24h_n^2}(-\mathbf{f}_{n-2} + \mathbf{f}_{n-1} + \mathbf{f}_n - \mathbf{f}_{n+1}), \quad 2 \le n \le N-1. \end{aligned} \tag{8.18}$$

This can be added to the right-hand side of the trapezoidal discretization; i.e., we solve

$$\begin{aligned} \frac{\tilde{\mathbf{y}}_n - \tilde{\mathbf{y}}_{n-1}}{h_n} &= \frac{1}{2}(\mathbf{f}(t_n, \tilde{\mathbf{y}}_n) + \mathbf{f}(t_{n-1}, \tilde{\mathbf{y}}_{n-1})) + h_n^2 T_{1,n-1/2}, \quad 1 \le n \le N, \\ & \mathbf{g}(\tilde{\mathbf{y}}_0, \tilde{\mathbf{y}}_N) = \mathbf{0}. \end{aligned}$$

The local truncation error is now $O(h_n^4)$, because the sum in the expression in (8.16) starts from $j = 2$. Hence the global error is also fourth order.

As in the case of extrapolation, the deferred correction process can be repeated to obtain higher-order approximations. Moreover, all approximations are solved for on the same mesh. For a linearized problem, one matrix $\mathbf{A}$ must be decomposed. Then, in the ensuing iterations which gradually increase the accuracy, only the right-hand side vectors are updated. The corresponding solution iterates are each computed by a pair of forward-backward substitutions. It may look at this point as if we are getting something from nothing! The catch, though, is in having to use more and more cumbersome and accurate approximations to the T_j's. The extrapolation method is more expensive but simpler.

These acceleration methods are useful and important in practice. They have useful counterparts for IVPs and differential-algebraic equations as well. Methods of both collocation and the acceleration types just described have been implemented in general-purpose codes. In a broad-brush comparison of the methods, it seems that they share many attributes. Methods of the acceleration type seem to be faster for simple problems, while methods of higher-order collocation at Gaussian points do a better job for stiff BVPs.

8.4 More on Solving Nonlinear Problems

Newton's method (or quasi-linearization) converges very rapidly *if* the first iterate is already a sufficiently good approximation to the (isolated) solution. This is the typical case for IVPs, even for stiff problems, where the known value of $\mathbf{y}_{n-1}$ is only $O(h_n)$ away from the sought value of $\mathbf{y}_n$. But for BVPs, no such high-quality initial iterate is generally available, and getting the nonlinear iteration to converge is a major practical challenge. This makes for one of the most important practical differences between general-purpose IVP and BVP solvers. Below we briefly discuss some useful approaches.[34]

8.4.1 Damped Newton

For the nonlinear system

$$\mathbf{h}(\mathbf{y}_\pi) = \mathbf{0},$$

Newton's (or the quasi-linearization) method at the νth iteration can be written as solving the linear system

$$\left(\frac{\partial \mathbf{h}}{\partial \mathbf{y}}(\mathbf{y}_\pi^\nu)\right) \boldsymbol{\eta}_\pi = -\mathbf{h}(\mathbf{y}_\pi^\nu)$$

[34]Throughout this section we consider a system of nonlinear algebraic equations and call the vector of unknowns $\mathbf{y}_\pi$. The somewhat cumbersome index π is there merely to remind us that we seek a mesh function, approximating the solution of the BVP. However, no special properties of the mesh function as such are utilized.

and forming

$$\mathbf{y}_\pi^{\nu+1} = \mathbf{y}_\pi^\nu + \boldsymbol{\eta}_\pi$$

(see (8.6) and Algorithm 8.1). This can be interpreted as taking a step of length 1 in the direction $\boldsymbol{\eta}_\pi$. If the model on which Newton's method is based (which can be viewed as assuming a local quadratic behavior of an appropriate objective function) is too optimistic, then a smaller step in this direction may be called for. In the *damped Newton* method we then let

$$\mathbf{y}_\pi^{\nu+1} = \mathbf{y}_\pi^\nu + \gamma\boldsymbol{\eta}_\pi, \tag{8.19}$$

where the parameter γ, $0 < \gamma \leq 1$, is chosen to ensure a decrease at each iteration in the objective function. For example, we can require

$$|\mathbf{h}(\mathbf{y}_\pi^{\nu+1})|_2 \leq (1-\delta)|\mathbf{h}(\mathbf{y}_\pi^\nu)|_2, \tag{8.20}$$

where δ ensures some minimum decrease, e.g., $\delta = 0.01$.

It can be shown theoretically that a sequence $\{\gamma_\nu\}$ can be found under certain conditions (which include nonsingularity of $\frac{\partial \mathbf{h}}{\partial \mathbf{y}}$ with a reasonable bound on the inverse) such that the damped Newton method converges *globally*, i.e., from any starting iterate $\mathbf{y}_\pi^0$. No such theorem holds for Newton's method without damping, which is assured to converge only locally, i.e., with $\mathbf{y}_\pi^0$ "close enough" to the sought solution $\mathbf{y}_\pi$ (recall, e.g., Exercise 7.7).

In practice this technique is useful on some occasions, but it is not sufficient for really tough BVPs. Typically, in such tough problems the Newton direction $\boldsymbol{\eta}_\pi$ is so polluted that it makes no sense to step in that direction for any step length. There seems to be no easy substitute for the remedy of finding better initial iterates.

8.4.2 Shooting for Initial Guesses

Often users feel more comfortable supplying only guesses for the initial values $\mathbf{y}^0(0)$, rather than an entire solution profile $\mathbf{y}^0(t)$. This is all that is required to fire up a simple shooting method. But if the stability of the BVP is such that a shooting method may indeed be used, then one can instead use an initial value code to solve the IVP *once* for the guessed initial values, obtaining an initial solution profile $\mathbf{y}^0(t)$. Then a quasi-linearization iteration for a finite difference method may be started.

This idea is a trick of convenience. It has obvious limitations. A more powerful (and more expensive) approach is to develop an appropriate initial solution gradually, solving a sequence of BVPs. The latter idea is discussed next.

8.4.3 Continuation

This approach is powerful and general. We embed the given problem in a family of problems

$$\boldsymbol{\phi}(\mathbf{y}_\pi, \mu) = \mathbf{0}, \qquad \mu_0 \leq \mu \leq \mu_1, \tag{8.21}$$

where the problem $\boldsymbol{\phi}(\mathbf{y}_\pi, \mu_0) = \mathbf{0}$ is easy to solve and $\boldsymbol{\phi}(\mathbf{y}_\pi, \mu_1) = \mathbf{h}(\mathbf{y}_\pi)$. Under suitable conditions this defines a *homotopy path* from an easy problem to the given problem, which we traverse numerically. Thus, we solve at each continuation step the problem

$$\boldsymbol{\phi}(\mathbf{y}_\pi, \mu + \Delta\mu) = \mathbf{0}$$

(call its solution $\mathbf{y}_\pi(t; \mu + \Delta\mu)$), given the solution $\mathbf{y}_\pi(t; \mu)$, where $\Delta\mu$ is a sufficiently small step size in μ. The simplest use of $\mathbf{y}_\pi(t; \mu)$ is as a first iterate $\mathbf{y}_\pi^0(t; \mu + \Delta\mu) = \mathbf{y}_\pi(t; \mu)$, but it is possible to get fancier.

This approach can be very successful in practice, although it can become expensive and it seems hard to automate for really difficult problems. The big question is defining the family of problems (8.21). The homotopy path has to somehow parameterize the problem well, and automatic choices such as a simple interpolation between μ_0 and μ_1 typically do not work well in this sense. Fortunately, there often exists a natural parameterization and embedding of this sort in applications.

Example 8.4
Often a nonlinear BVP results from the need to find a steady state solution of a time-dependent PDE in one space variable. Solving the PDE, starting from some initial solution profile, can then be considered as a continuation method for the steady state problem.

For instance, consider the diffusion problem of Example 1.3,

$$\frac{\partial u}{\partial t} = \frac{\partial}{\partial x}\left(p\frac{\partial u}{\partial x}\right) + g(x, u).$$

For a steady state solution, setting $\frac{\partial u}{\partial t} = 0$ yields the ODE in x:

$$0 = (pu')' + g(x, u),$$

where prime ($'$) denotes differentiation with respect to the independent variable x. This ODE is typically equipped with one boundary condition at each end of the interval in x.

Now, solving this nonlinear BVP numerically can be achieved by discretizing the space variable of the PDE while keeping $\frac{\partial u}{\partial t}$ in, and then applying the method of lines in t.

The time-embedding continuation method is very natural, but it can be very slow. One can often solve the steady state problem at a tiny fraction of the cost of solving the PDE. But this method has the attraction of generality: a straightforward numerical method is applied, regardless of how difficult the nonlinearity resolution is. Moreover, playing with different initial values may lead to different steady states (in cases where there is more than one such solution), perhaps in an intuitive way. ♦

The continuation technique opens the door to a variety of interesting topics such as *path following* and constructing *bifurcation diagrams*, but stepping through that door leads us outside the scope of this book, so we merely give a simple example here.

Example 8.5
The problem considered in Examples 8.2 and 6.2 certainly requires no fancy means to find its two solutions, once we have an idea that there are two such solutions and roughly what they look like. But to find them, and much more, automatically, we embed it in the family of problems

$$\begin{aligned} &u'' + \lambda e^u = 0, \\ &u(0) = u(1) = 0, \end{aligned} \tag{8.22}$$

and consider choosing the continuation parameter $\mu = \lambda$. As it turns out, continuation in λ starting from $\lambda = 0$ (where the solution is $u \equiv 0$) leads to the stabler of the two solutions for $\lambda = e$ (which is the value of λ considered in Example 8.2). Continuing with λ further, the problem becomes singular at some $\lambda = \lambda^* \approx 3.5$. What happens is that two solutions approach each other as λ increases and then cease to be isolated.

A more general continuation procedure uses *arclength* along the homotopy path for the continuation parameter μ. This is the preferred procedure for a general-purpose implementation, but it is again beyond the scope of our presentation. However, for this example there is also a simple trick: instead of doing the continuation with $\mu = \lambda$ of (8.22), use $\mu = \|u\|_2$. Thus, consider the embedding

$$\begin{aligned} &u'' + \lambda e^u = 0, \\ &\lambda' = 0, \\ &w' = u^2, \\ &u(0) = u(1) = 0, \quad w(0) = 0, \quad w(1) = \mu^2 . \end{aligned} \tag{8.23}$$

Carrying out the continuation process for this system (8.23) from $\mu = 0$ to $\mu = 8$ yields the bifurcation diagram depicted in Figure 8.5, where the computation for each μ was carried out using a standard BVP code. (Collocation at 3 Gaussian points was utilized and the problem was solved for 800 equidistant values of μ. This does not take long—less than a minute in total on an SGI Indigo2 R4400.) Figure 8.5 clearly suggests that for $\lambda < \lambda^*$ there are two solutions, for $\lambda = \lambda^*$, there is one, and for $\lambda > \lambda^*$, there are none. The type of singularity which occurs in this example at $\lambda = \lambda^*$ is called a *fold*. ♦

8.5 Error Estimation and Mesh Selection

Another key ingredient to the success of IVP solvers which is lost in BVP solvers is the ability to control local errors *locally*, as the solution process

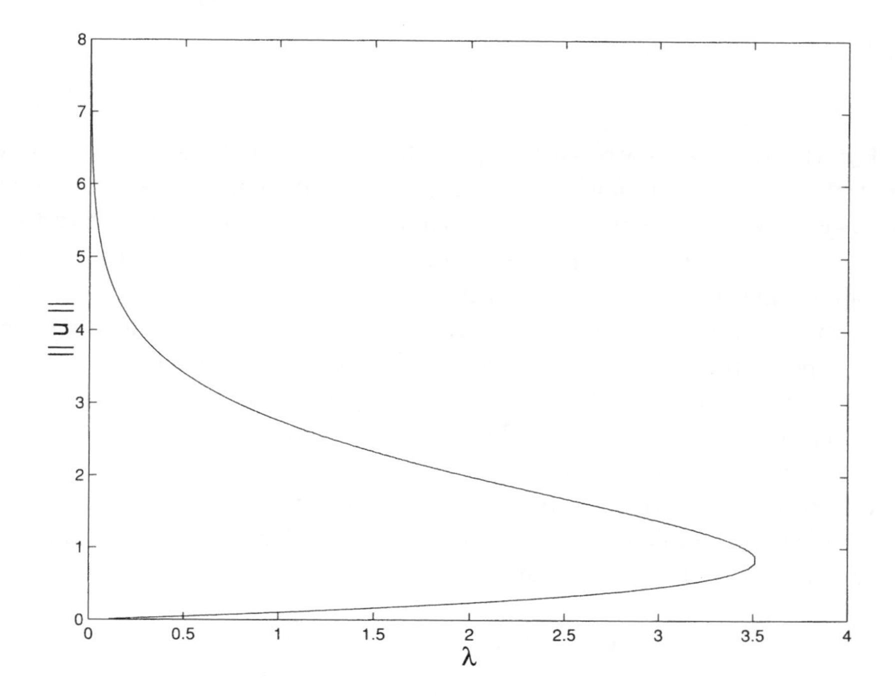

Figure 8.5: *Bifurcation diagram for Example* 8.5: $\|u\|_2$ *vs.* λ.

proceeds from $t = 0$ to $t = b$. To capitalize on this, IVP solvers often abandon controlling the global error $\mathbf{e}_n$ (although this is usually what the user may want) and control the local truncation error $\mathbf{d}_n$ instead.

In the BVP case there is no compelling reason, in general, not to estimate the global error $\mathbf{e}_n$. Such an estimate is compared against user-specified tolerances or used to select a new mesh. The process in overview for a given BVP is to discretize and solve it on a sequence of meshes, where the error in the solution on the current mesh is estimated and this information is used to decide what the next mesh should be, in case there is deemed to be a need for a next mesh. The first mesh is a guess.

The error estimation can be achieved using a process similar to the one described for extrapolation methods in Section 8.3.2. For instance, given a midpoint or a trapezoidal solution $\{\mathbf{y}_n\}$ on a mesh π and another one $\{\tilde{\mathbf{y}}_j\}$ on a mesh obtained by subdividing each element of π into two halves, we have

$$\tilde{\mathbf{y}}_{2n} - \mathbf{y}_n = \frac{3}{4}\mathbf{c}h^2 + O(h^4)\,.$$

So

$$\mathbf{e}_n = \mathbf{y}(t_n) - \mathbf{y}_n \approx \frac{4}{3}(\tilde{\mathbf{y}}_{2n} - \mathbf{y}_n)$$

and[35]

$$\tilde{\mathbf{e}}_{2n} = \mathbf{y}(t_n) - \tilde{\mathbf{y}}_{2n} \approx \frac{1}{3}(\tilde{\mathbf{y}}_{2n} - \mathbf{y}_n).$$

For the deferred correction approach the local truncation error is estimated as part of the algorithm. A global error estimate can be directly obtained as well, by solving with the truncation error as the right-hand side.

It is also possible to construct some cruder indicators for the error $\mathbf{e}_n$ on a given mesh without recomputing another solution. This can be done by taking advantage of the form of the error if it has a local leading term, or by considering arclength or other error-monitoring functions, e.g., $\hat{\mathbf{e}}_n = h_n^k |\mathbf{y}^{(k)}(t_n)|$ for some $1 \le k \le p$, with p the order of accuracy. Such monitor functions may be sufficient to select a mesh, even if they do not provide a genuine, reliable error estimate.

Given such an error estimate or indicator, the next mesh is selected based on the principle of *error equidistribution*, where one attempts to pick the mesh such that the resulting solution will satisfy

$$|\hat{\mathbf{e}}_i| \approx |\hat{\mathbf{e}}_j|, \qquad 1 \le i, j \le N.$$

This essentially minimizes $\max_n |\hat{\mathbf{e}}_n|$ for a given mesh size. The mesh size N is further selected so that

$$\max_n |\hat{\mathbf{e}}_n| \le \text{ETOL}$$

for a user-specified error tolerance.

Note: The technical level and expertise required of the reader for the next section is a touch higher than what has been required so far in this chapter, and it gets even higher in Section 8.7. But these sections are important and are worth the extra effort.

8.6 Very Stiff Problems

As in the IVP case we expect the midpoint (or trapezoidal) method to be robust when large eigenvalues with both negative and positive real parts are present, *so long as* the layers are resolved. We have already seen a demonstration in Examples 8.1 and 8.3. The properties of symmetric methods are essentially similar for the IVP and the BVP cases.

[35]Note that we do not get a good error estimate for the fourth-order extrapolated solution $\frac{1}{3}(4\tilde{\mathbf{y}}_{2n} - \mathbf{y}_n)$. The error estimates are tight only for the lower-order approximations. Similarly, using midpoint solutions on three meshes, it is possible to obtain a fourth-order approximation with an error estimate or a sixth-order approximation without a tight error estimate.

For stiff IVPs we prefer methods with stiff decay, such as BDF or collocation at Radau points. Unfortunately, it is not possible to attain this property automatically in the BVP case, because methods with stiff decay cannot be symmetric.

Upwind Discretization

For the IVP test equation

$$y' = \lambda y, \qquad \mathcal{R}\mathrm{e}(\lambda) < 0,$$

we prefer, when $h\mathcal{R}\mathrm{e}(\lambda) \ll -1$, to use a method like backward Euler,

$$y_n = y_{n-1} + h\lambda y_n .$$

Similarly, changing the direction of integration for the unstable IVP $y' = -\lambda y$ to $\tau = t_n - t$, we get

$$\frac{d\tilde{y}}{d\tau} = \lambda \tilde{y},$$

and applying backward Euler to the equation in $\tilde{y}$ readily yields the forward Euler method for the original,

$$y_n = y_{n-1} - h\lambda y_{n-1} .$$

For the system

$$\mathbf{y}' = \begin{pmatrix} \lambda & 0 \\ 0 & -\lambda \end{pmatrix} \mathbf{y},$$

it then makes sense to use *upwinding*:[36]

$$\begin{aligned} y_{1,n} &= y_{1,n-1} + h\lambda y_{1,n}, \\ y_{2,n} &= y_{2,n-1} - h\lambda y_{2,n-1}. \end{aligned}$$

For a general stiff problem, unfortunately, the increasing and decreasing modes are coupled together, and there are also slow solution modes for which a higher-order symmetric discretization method is perfectly suitable. Consider the general linearized differential system related to (8.1),

$$\mathbf{y}' = A(t)\mathbf{y} + \mathbf{q}(t),$$

where $A(t) = \frac{\partial \mathbf{f}}{\partial \mathbf{y}}$, and define the transformation for some nonsingular, sufficiently smooth matrix function $T(t)$,

$$\mathbf{w} = T^{-1}\mathbf{y}.$$

[36]The term *upwind* originates from computational fluid dynamics, where the direction of stable integration corresponds to the upwind (i.e., against the wind) direction of the flow. This type of discretization has also been called *upstream*, a name naturally arising from applications where what flows is liquid. See Exercise 8.11.

Then $\mathbf{w}(t)$ satisfies the ODE

$$\mathbf{w}' = (T^{-1}AT + T^{-1}T')\mathbf{w} + T^{-1}\mathbf{q}.$$

Now, if T is such that the transformed matrix A can be written in block form

$$T^{-1}AT + T^{-1}T' = \begin{pmatrix} B_1 & 0 & 0 \\ 0 & B_2 & 0 \\ 0 & 0 & B_3 \end{pmatrix},$$

where

- B_1 is dominated by eigenvalues with large negative real parts,
- B_2 is dominated by eigenvalues with large positive real parts,
- $\|B_3\|$ is not large,

then we can use backward Euler for $\mathbf{w}_1$, forward Euler for $\mathbf{w}_2$, and the trapezoidal method for $\mathbf{w}_3$, where $\mathbf{w}_i$ corresponds to the equations involving B_i and $\mathbf{w}^T = (\mathbf{w}_1^T, \mathbf{w}_2^T, \mathbf{w}_3^T)$. The equations resulting from such a discretization need not be solved for $\mathbf{w}$. Rather, the back transformation from $\mathbf{w}$ to $\mathbf{y}$ is used to transform them into difference equations for $\mathbf{y}$.

Example 8.6
The stable BVP

$$\mathbf{y}' = \begin{pmatrix} \cos t & \sin t \\ -\sin t & \cos t \end{pmatrix} \begin{pmatrix} \lambda & 1 \\ -1 & -\lambda \end{pmatrix} \begin{pmatrix} \cos t & -\sin t \\ \sin t & \cos t \end{pmatrix} \mathbf{y},$$
$$y_1(0) = 1, \quad y_1(1) = 2$$

is stiff when $\mathcal{R}\mathrm{e}(\lambda) \ll -1$. Applying forward Euler or backward Euler to this ODE with a step size h yields disastrous results when $h\mathcal{R}\mathrm{e}(\lambda) < -2$. But for $\mathbf{w} = T^{-1}\mathbf{y}$, where

$$T(t) = \begin{pmatrix} \cos t & \sin t \\ -\sin t & \cos t \end{pmatrix},$$

we obtain the decoupled system

$$\mathbf{w}' = \begin{pmatrix} \lambda & 0 \\ 0 & -\lambda \end{pmatrix} \mathbf{w}$$

and the upwind method described above can be applied, yielding a very stable discretization method. We write it in matrix form as

$$\begin{pmatrix} 1-h\lambda & 0 \\ 0 & 1 \end{pmatrix} \mathbf{w}_n = \begin{pmatrix} 1 & 0 \\ 0 & 1-h\lambda \end{pmatrix} \mathbf{w}_{n-1}\,.$$

Defining for each n

$$\mathbf{y}_n = T_n \mathbf{w}_n = \begin{pmatrix} \cos t_n & \sin t_n \\ -\sin t_n & \cos t_n \end{pmatrix} \mathbf{w}_n,$$

the obtained upwind method for $\mathbf{y}_\pi$ is

$$\begin{pmatrix} 1-h\lambda & 0 \\ 0 & 1 \end{pmatrix} \begin{pmatrix} \cos t_n & -\sin t_n \\ \sin t_n & \cos t_n \end{pmatrix} \mathbf{y}_n$$

$$= \begin{pmatrix} 1 & 0 \\ 0 & 1-h\lambda \end{pmatrix} \begin{pmatrix} \cos t_{n-1} & -\sin t_{n-1} \\ \sin t_{n-1} & \cos t_{n-1} \end{pmatrix} \mathbf{y}_{n-1}, \quad 1 \le n \le N,$$

with

$$y_{1,0} = 1, \quad y_{1,N} = 2.$$

In Figure 8.6 we display the approximate solution using the upwind method just described and a uniform mesh with $\lambda = -1000$, $h = 0.1$. We also display the "exact" solution, obtained using the code COLNEW.[37] Note that despite the fact that the boundary layers are totally missed (i.e., skipped over), the solution values at mesh points are approximated well by the upwind method, in analogy to the IVP case depicted in Figure 3.6. ♦

The upwind discretization method outlined above, and other similar methods of higher order, work very well for special classes of problems (e.g., gas dynamics in PDEs). Layer details can be skipped, as with the backward Euler method in Chapter 3. But finding the transformation T in general and applying it in practice are major obstacles. For linear problems there are recipes for this, but the general case involves such an additional amount of work that straightforward collocation at Gaussian points is often more efficient. Things are worse for nonlinear problems, where the entire process is difficult because the linearization is based on unknowns and a mistake in the sign of a fast mode is a disaster akin to simple shooting. (Alternatively, upon using quasi-linearization, stable but entirely wrong linear problems are solved, so the nonlinear iteration may not converge.)

[37]The COLNEW computations employed collocation at 3 Gaussian points per subinterval on a nonuniform mesh with 52 subintervals. The code selected the mesh automatically, satisfying a global error tolerance of 1.e − 5.

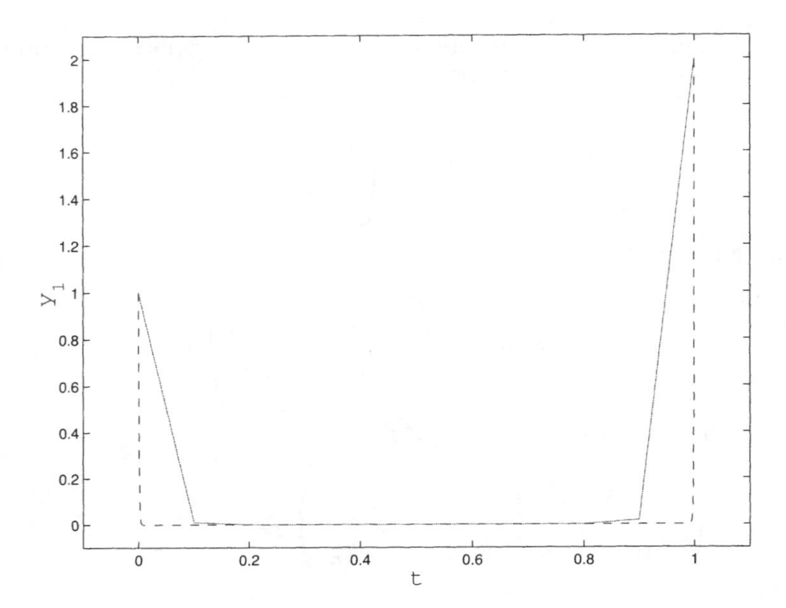

Figure 8.6: *Solution for Example* 8.6 *with* $\lambda = -1000$ *using an upwind discretization with a uniform step size* $h = 0.1$ *(solid line). The "exact" solution is also displayed (dashed line).*

Symmetric Discretization

For symmetric difference methods no explicit decoupling transformation is usually needed. But, as mentioned earlier, the lack of stiff decay can be evident in computations. For the test equation

$$y' = \lambda y$$

the midpoint method yields, to recall,

$$y_n = \frac{2 + h\lambda}{2 - h\lambda} y_{n-1}.$$

So, although $|y_n| \leq |y_{n-1}|$ precisely whenever the exact solution satisfies $|y(t_n)| \leq |y(t_{n-1})|$, as $h|\lambda| \to \infty$ we get

$$y_n \approx -y_{n-1} \, .$$

Thus, if $y(0) = 1$ and $h\mathcal{R}\mathrm{e}(\lambda) \ll -1$, then the exact solution satisfies $y(h) \approx 0$, yet for the numerical solution,

$$y_n \approx (-1)^n, \qquad n = 1, \ldots, N \, .$$

This not only necessitates covering layer regions (where the solution varies rapidly) with dense meshes, as we have already seen in Examples 3.2 and 8.1, it also means that in the very stiff case local errors propagate through smooth solution regions (where h is relatively large) almost undamped, yielding nonlocal error effects.

There are some exotic examples where a symmetric method may even blow up when approximating a stable BVP or even a stable IVP (see Exercise 8.14). These seem to be rare in practice, though. Moreover, and perhaps more importantly, since there is almost no quadrature effect when $0 < |\lambda^{-1}| \ll 1$ in a stable problem of the form

$$y' = \lambda(y - q(t))$$

(note that the solution is $y(t) = q(t) + O(|\lambda^{-1}|)$ and the quadrature precision affects only the $O(|\lambda^{-1}|)$ term), methods which are based on high-precision quadrature may experience *order reduction* away from collocation points; see Section 4.7.3 and Exercise 8.13. For collocation at s Gaussian points we can get, assuming no layer errors and $0 < |\mathcal{R}\mathrm{e}(\lambda)|^{-1} \ll h \ll 1$, the error bound

$$\mathbf{y}(t_n) - \mathbf{y}_n = O(h^s)\,.$$

This error estimate improves to

$$\mathbf{y}(t_n) - \mathbf{y}_n = O(h^{s+1})$$

if s is odd and some mesh restrictions apply. But this still falls short of the usual nonstiff order $O(h^{2s})$ when $s > 1$. For Example 8.3, the effective order is $s + 1 = 4$ (instead of $2s = 6$) when $h\lambda$ is very large.

Despite their limitations, as candidates for constructing a general-purpose solver, symmetric discretizations seem to win. It appears that upwinding techniques should be reserved for special problems where the explicit decoupling of modes can proceed with relative ease; see Exercises 8.11 and 8.12.

8.7 Decoupling

The concept of decoupling of modes of distinctly different types is fundamental to the understanding of numerical discretizations for differential equations. But it seems to be particularly important in the context of boundary value ODEs, so we briefly discuss it here. This section can be viewed as a road map for numerical methods for BVPs.

As we have seen in Section 6.2 a stable linear BVP must have a dichotomy (i.e., a certain number of its fundamental modes are nonincreasing and the rest are nondecreasing) throughout the interval $[0, b]$ on which the differential problem is defined. Slow modes can be grouped either with the rapidly increasing or with the rapidly decreasing modes, but this is possible only locally; i.e., a mode can change from fast to slow and vice versa in different subintervals of $[0, b]$.

What must be avoided then is a numerical integration of fast modes in the direction of their increase. When the modes are decoupled, e.g., in the system

$$\mathbf{y}' = \begin{pmatrix} \lambda & \mu \\ 0 & -\lambda \end{pmatrix} \mathbf{y}, \tag{8.24}$$

where $\mathcal{R}e(\lambda) > 0$ and $|\mu|$ is not much larger than $|\lambda|$, then the modes can be integrated in suitable directions—for (8.24) the second ODE is integrated from 0 to b and then the first is integrated from b to 0. In the general case where the modes are not decoupled, some decoupling must be applied. The simple shooting method does not apply any decoupling, hence it is unstable in the presence of fast (not even very fast) increasing and decreasing modes.

For the multiple shooting method discussed in Section 7.2 and the finite difference methods discussed in the early sections of this chapter, stability is proved provided that sufficiently many shooting points or mesh points are used. This is the limit where these methods are all similar to one another and bounds like (8.13) apply, indicating that the discretization follows the continuous system closely. The decoupling of modes is then achieved implicitly, *through the solution of the linear system of algebraic equations.* Recall from Section 8.2 that this system has an almost block diagonal form (Figure 8.3) corresponding to the sequential ordering of the mesh points. So, when performing *LU* decomposition and forward and backward substitutions in order to solve a system like (8.14), we are in effect sweeping forward and then backward along the interval of integration. The *LU* decomposition itself can be seen to correspond to a decoupling transformation along the lines given by the dichotomy bound (6.16).

The great robustness of symmetric difference methods arises from the possibility of achieving the decoupling effect implicitly, i.e., without an explicit transformation, even for stiff BVPs. But for very stiff problems a method like midpoint also tends to transform fast increasing and decreasing modes into slower ones, when the step size is not very small.

Some unfortunate effects may also result when fast and slow modes are not suitably decoupled by the numerical method. This may occur already for stable IVPs (Exercise 8.14 is a case in point), but such problems are rarer in practice and, moreover, the decoupling must be done locally, hence explicitly, as described in Section 8.6, which is not very practical for many applications. Trouble can happen also for DAEs when different solution components are not properly decoupled, as we will see in the next two chapters.

8.8 Software, Notes, and References

8.8.1 Notes

Much of the early theoretical development of the theory of numerical methods for BVPs was done by H. Keller and appears in [59] as well as in the more modern reference book [8].

A lot of work was done on the numerical solution of second-order two-point BVPs. Often a single ODE is considered in this context. Many papers on numerical methods for stiff BVPs of this sort have appeared, probably both because of the relevance to advection-diffusion PDEs where advection dominates and because of the relative tractability of these problems compared to the general stiff system case. We have devoted a series of exercises to this (Exercises 8.9–8.12), and refer for more to [8] and the references therein.

But for our main exposition we consider the general ODE system case, which naturally extends our discussion in the previous IVP and shooting chapters. All of the material covered in this chapter, including proofs and references which we have omitted here (plus, be warned, much more!) can be found in [8].

The linear system solvers used in all the leading software are more sophisticated than the band solver that we have described. See Chapter 7 of [8] and references therein. For the parallel solution of such systems, see Wright [97].

A thorough treatment of discretization methods and their asymptotic expansions can be found in [89]. See also the early book [41]. V. Pereyra made fundamental contributions to the theory of deferred corrections. An important work in the early development of collocation at Gaussian points is de Boor and Swartz [36], although the later treatment in [8] is cleaner.

The earliest uses of the principle of error equidistribution seem to have been made in de Boor [35]; see also [70]. (The apparent addition of the word "equidistribution" to the English language is due to M. Lentini and V. Pereyra.)

Major contributions on decoupling in BVPs were made in the 1980s by R. Mattheij and appear in [8].

In Chapter 11 of [8] there is a brief description plus relevant references of a number of topics which we have omitted here, except in the occasional exercise. These include eigenvalue problems, singular BVPs, BVPs on infinite intervals, singular points, bifurcation, arclength continuation, and highly oscillatory BVPs.

Finally, while we have treated finite difference methods exclusively in this book, there has been much theoretical development on finite element methods as well (see, e.g., [90, 20]). The power of the latter methods, however, appears to be more pronounced in the PDE context.

8.8.2 Software

Most general-purpose codes for BVPs which are publicly available use the methods described in this chapter.

- The code COLSYS by Ascher, Christiansen, and Russell [5] and its newer version COLNEW by Bader and Ascher [13] use collocation at

Gaussian points. This code is available from NETLIB.

- Also available from NETLIB is the code TWPBVP by Cash and Wright [30], which uses deferred correction in combination with certain non-collocation Runge–Kutta methods called mono-implicit, which we have not covered in this book.

- The NAG library contains the code PASVAR by Lentini and Pereyra [63], which also uses deferred correction. This code has been influential for many years.

- The code AUTO, by Doedel and Kernevez [38], which does bifurcation analysis and finds periodic solutions, is based on Gauss collocation.

8.9 Exercises

8.1. Show that the formulation of the quasi-linearization method using $\boldsymbol{\eta}$ as defined in (8.6)–(8.7) is equivalent to the formulation using (8.5).

8.2. Carry out the development of theory and practice as in Section 8.1 for the trapezoidal method (3.32) instead of the midpoint method.

8.3. (a) Write down the quasi-linearization (or the linearization) problem (8.6) for the BVP (8.23).

(b) Show that this linearized problem is singular (i.e., it does not have a unique solution) when it is carried out about $u \equiv 0$. Conclude that starting the quasi-linearization iteration with the initial guess $u^0 \equiv 0$ is unwise in this example.

8.4. It can be shown that the error when applying the trapezoidal method to a sufficiently smooth BVP (8.1)–(8.2) has the expansion

$$\mathbf{e}_n = \mathbf{y}(t_n) - \mathbf{y}_n = \sum_{j=1}^{l} \mathbf{c}_j h_n^{2j} + O(h^{2l+1}), \tag{8.25}$$

where $h = \max_n h_n$ on a mesh π which satisfies

$$h / \min_n h_n \leq constant.$$

The functions $\mathbf{c}_j$ are independent of the mesh π. Just how large l is depends on the smoothness of the problem, and we assume $l \geq 3$.

(a) Construct a method of order 6 using extrapolation, based on the trapezoidal method.

(b) Apply this extrapolation method to the problem of Examples 8.1 and 8.3, using the same parameter values and meshes. Compare with collocation at three Gaussian points (Example 8.3). What are your conclusions?

8.5. Use your code from the previous exercise, or any available software based on the methods discussed in this chapter, to solve the following problems to about five-digit accuracy.

(a) Find a nontrivial solution for the problem (7.13) of Exercise 7.4.

(b) Find the attracting limit cycle and the period of the Van der Pol equation (7.15).

(c) Solve (8.23) for $\mu = 1$. What is the corresponding value of λ?

8.6. The injected fluid flow through a long, vertical channel gives rise to the BVP

$$\begin{aligned} u'''' &= R\,(u'u'' - uu'''), \\ &\quad u(0) = u'(0) = 0, \\ &\quad u(1) = 1, \quad u'(1) = 0, \end{aligned}$$

where u is a potential function and R is a given (constant) Reynolds number.

Use your code from the previous exercise, or any available software (we suggest that it be based on the methods discussed in this chapter), to solve this problem for four values of R: $R = 10,\ 100,\ 1000,$ and 10,000. Observe the increased difficulty, due to a boundary layer near the left boundary, as R increases.

8.7. Consider the following particle diffusion and reaction system:

$$\begin{aligned} T'' + \frac{2}{t}T' &= -\phi^2 \beta C e^{\gamma(1-T^{-1})}, \\ C'' + \frac{2}{t}C' &= \phi^2 C e^{\gamma(1-T^{-1})}, \end{aligned}$$

where $C(t)$ is the concentration and $T(t)$ is the temperature. Representative values for the constants are $\gamma = 20,\ \beta = 0.02,\ \phi = 14.44$. The boundary conditions at $t = 0$ are

$$T'(0) = C'(0) = 0.$$

Use any available software (we suggest that it be based on the methods discussed in this chapter) to solve this problem for the following sets of additional boundary conditions:

(a) $T'(1) = C'(1) = 1$.

(b) $-T'(1) = \alpha(T(1) - 1),\ -C'(1) = \hat{\alpha}(C(1) - 1)$, with $\alpha = 5,\ \hat{\alpha} = 250$.

[This case may cause you more grief. Note that there is a thin boundary layer near $t = 1$.]

8.8. (a) Show that the error expansion (8.16)–(8.17) holds for the trapezoidal method.

(b) The centered approximation (8.18) is not good near the boundary, e.g., for $n = 1$. Construct one-sided, second-order accurate difference methods for near-boundary points.

(c) To what order should T_1 and T_2 be approximated in order to achieve a sixth-order deferred correction method? Explain why you need the fourth-order $\tilde{\mathbf{y}}_n$, not just the second-order $\mathbf{y}_n$, to construct such higher-order approximations for the truncation error terms.

8.9. Consider the scalar ODE of order 2:

$$
\begin{aligned}
&-(a(t)u')' + b(t)u = q(t), \\
&u(0) = b_1, \quad u(1) = b_2,
\end{aligned} \tag{8.26}
$$

where $a > 0, b \geq 0 \, \forall t$. We convert this ODE into a first-order system without differentiating a by

$$
\begin{aligned}
u' &= a^{-1}v, \\
v' &= bu - q.
\end{aligned}
$$

(a) Show that if we discretize the first-order system by the midpoint method we obtain a five-diagonal matrix $\mathbf{A}$.

(b) Consider instead a *staggered midpoint method*: on a uniform mesh, the equation for u is centered at $t_{n-1/2}$ and the equation for v is centered at t_n, with u_π defined at mesh points and v_π at midpoints:

$$
\begin{aligned}
(u_n - u_{n-1})/h &= a^{-1}(t_{n-1/2})v_{n-1/2}, \\
(v_{n+1/2} - v_{n-1/2})/h &= b(t_n)u_n - q(t_n).
\end{aligned}
$$

Show that by eliminating the v-values we obtain for the mesh values in u a *tridiagonal* matrix $\mathbf{A}$. Under what condition are we assured that it is diagonally dominant?

(c) The usual three-point formula for discretizing (8.26) becomes first order if the mesh is no longer uniform. Generalize the staggered midpoint method developed above to obtain a second-order accurate, three-point method for u on an arbitrary mesh.

[Hint: You can use quadratic interpolation of three adjacent mesh values of u without changing the sparsity structure of $\mathbf{A}$.]

(d) Try your method on the problem given by $a = 1+t^2$, $b = 1$, $u(t) = \sin(t)$ (calculate the appropriate $q(t)$ and $\mathbf{b}$ required for this exact solution). Compute maximum errors on three meshes:

- a uniform mesh with $h = .01$,

- a uniform mesh with $h = .02$,
- a nonuniform mesh with 100 subintervals. The step sizes are to be chosen by a random number generator, scaled, and translated to lie between .01 and .02.

What are your observations?

8.10. For the second-order ODE system

$$\mathbf{y}'' = \mathbf{f}(t, \mathbf{y}),$$

we can consider the linear three-point methods

$$\alpha_0 \mathbf{y}_{n+1} + \alpha_1 \mathbf{y}_n + \alpha_2 \mathbf{y}_{n-1} = h^2(\beta_0 \mathbf{f}_{n+1} + \beta_1 \mathbf{f}_n + \beta_2 \mathbf{f}_{n-1}), \quad (8.27)$$

where we use the notational convention of Chapter 5 and set $\alpha_0 = 1$. (Note that these methods are *compact*: the order of the difference equation is 2, just like the order of the ODE, so there are no parasitic roots for the stability polynomial here.)

(a) Derive order conditions (as in Section 5.2) for (8.27).

(b) Show that to obtain a consistent method (with a constant h) we must set $\alpha_1 = -2, \alpha_2 = 1$, as in the usual discretization for $\mathbf{y}''$, and

$$\beta_0 + \beta_1 + \beta_2 = 1\,.$$

(c) Show that to obtain a second-order method we must set in addition

$$\beta_0 = \beta_2\,.$$

In particular, the usual formula with $\beta_0 = \beta_2 = 0$ and $\beta_1 = 1$ is second-order accurate.

(d) Show that Cowell's method

$$\mathbf{y}_{n+1} - 2\mathbf{y}_n + \mathbf{y}_{n-1} = \frac{h^2}{12}(\mathbf{f}_{n+1} + 10\mathbf{f}_n + \mathbf{f}_{n-1}) \quad (8.28)$$

is fourth-order accurate.

(e) Describe in detail an implementation of the method (8.28) for the Dirichlet BVP, where $\mathbf{y}(0)$ and $\mathbf{y}(b)$ are given.

8.11. Consider the scalar Dirichlet problem

$$-\varepsilon u'' + a u' = q(t),$$
$$u(0) = b_1, \quad u(1) = b_2,$$

where $a \neq 0$ is a real constant and $0 < \varepsilon \ll 1$. Assume for simplicity that there are no boundary layers (i.e., the values b_1 and b_2 agree with the reduced solution satisfying $u' = q/a$), and consider three-point discretizations on a uniform mesh with step size $h = \frac{1}{N+1}$,

$$\alpha_n u_n = \beta_n u_{n-1} + \gamma_n u_{n+1} + q_n, \quad 1 \le n \le N,$$
$$u_0 = b_1, \quad u_{N+1} = b_2 .$$

The solution for $\mathbf{u}_\pi = (u_1, \ldots, u_N)^T$ requires solving a linear tridiagonal system with the matrix

$$\mathbf{A} = \begin{pmatrix} \alpha_1 & -\gamma_1 & & & \\ -\beta_2 & \alpha_2 & -\gamma_2 & & \\ & \ddots & \ddots & \ddots & \\ & & -\beta_{N-1} & \alpha_{N-1} & -\gamma_{N-1} \\ & & & -\beta_N & \alpha_N \end{pmatrix}.$$

It is desirable that $\mathbf{A}$ be diagonally dominant. A related, important requirement is that the method be *positive*:

$$\alpha_n > 0, \quad \beta_n \ge 0, \quad \gamma_n \ge 0 \ \forall n.$$

(This implies a discrete maximum principle which yields stability.)

(a) A *symmetric*, or *centered* second-order discretization is given by

$$\frac{\varepsilon}{h^2}(-u_{n-1} + 2u_n - u_{n+1}) + \frac{a}{2h}(u_{n+1} - u_{n-1}) = q(t_n) .$$

Show that $\mathbf{A}$ is diagonally dominant and the method is positive if and only if

$$R = \frac{|a|h}{\varepsilon} \le 2.$$

(R is called the *mesh Reynolds number*.)

(b) An *upwind* method is obtained by replacing the discretization of u' with forward or backward Euler, depending on sign(a):

$$\frac{\varepsilon}{h^2}(-u_{n-1} + 2u_n - u_{n+1}) + \frac{a}{h}\phi_n = q(t_n),$$

$$\phi_n = \begin{cases} u_{n+1} - u_n, & a < 0, \\ u_n - u_{n-1}, & a \ge 0. \end{cases}$$

Show that this method is positive and $\mathbf{A}$ is diagonally dominant for all $R \ge 0$. It is also only first-order accurate.

(c) Show that

$$\frac{a}{h}\phi_n = a\frac{u_{n+1} - u_{n-1}}{2h} + \psi_n,$$

where ψ_n is the three-point discretization of $-h|a|u''$. The upwind method can therefore be viewed as adding an $O(h)$ *artificial diffusion* term to the centered discretization.

8.12. This exercise continues the previous one.

(a) Extend the definitions of the centered and upwind three-point discretizations to the ODE

$$-\varepsilon u'' + a(t)u' + b(t)u = q(t), \qquad 0 < t < 1,$$

where $a(t)$ varies smoothly and can even change sign on $(0, 1)$, and $b(t)$ is a smooth, bounded function. What happens when $a(t) = t - \frac{1}{2}$ and $0 < \varepsilon \ll h \ll 1$? What happens when there are boundary or turning-point layers?

(b) Extend the definitions of centered and upwind three-point discretizations to the nonlinear problem

$$-\varepsilon u'' + uu' + b(t)u = q(t), \qquad 0 < t < 1,$$
$$u(0) = -1, \quad u(1) = 1.$$

(c) When $R < 2$ the centered method is preferred because of its accuracy. When $R > 2$ the upwind method has superior stability properties. Design a method which mixes the two and gradually switches between them, adding at each mesh point just enough artificial diffusion to achieve positivity, for any values of ε, h, and a or u.

8.13. (a) Write down the midpoint method, on an arbitrary mesh π, for the scalar ODE

$$y' = \lambda(y - q(t)), \quad 0 < t < b,$$

where $q(t)$ is a smooth, bounded function and $\lambda \ll -1$. Consider the IVP case with $\lambda h_n \ll -1$, where h is the maximum step size ($h = \max_{1 \le n \le N} h_n$). Assume no initial layer, so $|y''|$ and higher solution derivatives are bounded independently of λ.

(b) Show that the local truncation error satisfies

$$d_n = -\frac{z_n h_n}{8}(y''(t_n) + O(h_n)) + O(h_n^2),$$

where $z_n = \lambda h_n$, and that the global error $e_n = y(t_n) - y_n$ satisfies

$$e_n = (1 - z_n/2)^{-1}(1 + z_n/2)e_{n-1} + (1 - z_n/2)^{-1}h_n d_n, \quad 1 \le n \le N.$$

(c) Letting $z_n \to -\infty$, $1 \le n \le N$, show that the global error satisfies

$$e_n = \sum_{j=0}^{n}(-1)^j \frac{h_j^2}{4}(y''(t_j) + O(h_j)).$$

This is the leading error term for the case $z_n \ll -1$, $1 \le n \le N$.

(d) Conclude that (for $b = O(1)$) the error for the midpoint method in the very stiff limit reduces to $O(h)$. However, if the mesh is *locally almost uniform*, i.e., the steps can be paired such that for each odd j

$$h_{j+1} = h_j(1 + O(h_j)),$$

then the convergence order is restored to $O(h^2)$.

[This mesh restriction is mild: take any mesh, and double it as for extrapolation by replacing each element by its two halves. The resulting mesh is locally almost uniform. Note, on the other hand, that even when second-order accuracy is thus restored, there is no error expansion of the type utilized in Section 8.3.2.]

(e) Can you guess why we have included this exercise here, rather than in Chapters 3 or 4?

8.14. Consider the IVP

$$\begin{pmatrix} 1 & -t \\ 0 & \varepsilon \end{pmatrix} \mathbf{y}' = \begin{pmatrix} -1 & 1+t \\ \beta & -(1+\beta t) \end{pmatrix} \mathbf{y} + \begin{pmatrix} 0 \\ \sin t \end{pmatrix}, \tag{8.29}$$

$$\mathbf{y}(0) = (1, \beta)^T,$$

where the two parameters β and ε are real and $0 < \varepsilon \ll 1$.

(a) Apply the transformation

$$\begin{pmatrix} 1 & -t \\ 0 & 1 \end{pmatrix} \mathbf{y} = \mathbf{w}$$

to show that this problem is stable and to find the exact solution.

(b) Let $\varepsilon = 10^{-10}$. There is no initial layer, so consider applying the midpoint method with a uniform step size

$$h = .1/\max(|\beta|, 1)$$

to (8.29). Calculate maximum errors in y_1 for $\beta = 1,\ 100,\ -100$. What are your observations?

(c) Attempt to explain the observed results. [This may not be easy.]

Part IV: Differential-Algebraic Equations

Chapter 9

More on Differential-Algebraic Equations

In this chapter and the next we study differential-algebraic equations (DAEs), already introduced in Section 1.3. Here we consider the mathematical structure of such systems and some essential analytical transformations. Numerical approaches and discretizations are discussed in the next chapter. But here, too, our motivation remains finding practical computer solutions. Compared to Chapters 2 and 6, this chapter is unusually long. One reason is that DAE theory is much more recent than ordinary differential equation (ODE) theory. As a result, DAE theory is more in a state of flux, and good expositions are scarce. More importantly, understanding the principles highlighted here is essential for, and will get you a long way towards, constructing good numerical algorithms.

To get a taste of the similarity and the difference between DAEs and ODEs, consider two functions $y(t)$ and $z(t)$ which are related on some interval $[0, b]$ by

$$y'(t) = z(t), \qquad 0 \leq t \leq b, \tag{9.1}$$

and the task of recovering one of these functions from the other. To recover z from y one needs to differentiate $y(t)$—an automatic process familiar to us from a first calculus course. To recover y from z one needs to integrate $z(t)$—a less automatic process necessitating an additional side condition (such as the value of $y(0)$).

This would suggest that differentiation is a simpler, more straightforward process than integration. On the other hand, though, note that $y(t)$ is generally a smoother function than $z(t)$. For instance, if $z(t)$ is bounded but has jump discontinuities, then $y(t)$ is once differentiable; see Figure 9.1.

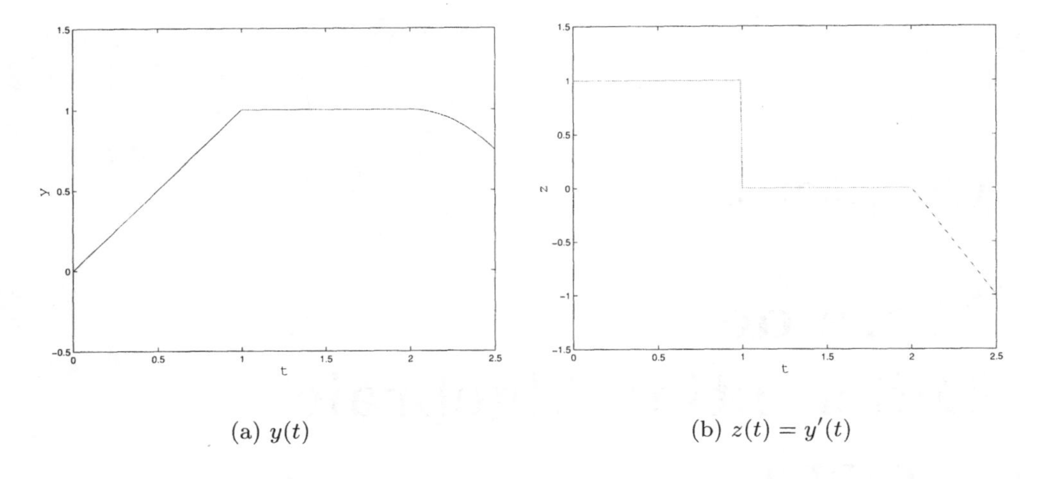

(a) $y(t)$ (b) $z(t) = y'(t)$

Figure 9.1: *A function and its less smooth derivative.*

Thus, integration is a *smoothing* process, while differentiation is an *antismoothing* process. The differentiation process is in a sense unstable,[38] although it is often very simple to carry out analytically.

A differential equation involves integration, hence smoothing: the solution $\mathbf{y}(t)$ of the linear system $\mathbf{y}' = A\mathbf{y} + \mathbf{q}(t)$ is one derivative smoother than $\mathbf{q}(t)$. A DAE, on the other hand, involves both differentiations and integrations. The class of DAEs contains all ODEs, as well as the problems in Example 9.1 below. But it also contains problems where both differentiations and integrations are intertwined in a complex manner, and that's when the fun really starts: simple differentiations may no longer be possible, but their effect complicates the numerical integration process, potentially well beyond what we have seen so far in this book.

9.1 Index and Mathematical Structure

Since a DAE involves a mixture of differentiations and integrations, one may hope that applying analytical differentiations to a given system and eliminating as needed, repeatedly if necessary, will yield an explicit ODE system for all the unknowns. This turns out to be true unless the problem is singular. The number of differentiations needed for this transformation is called the *index* of the DAE. Thus, ODEs have index 0. We will refine this definition later, but first let us consider some simple examples.

Example 9.1
Let $q(t)$ be a given, smooth function, and consider the following problems for $\mathbf{y}(t)$.

[38] If we add to $y(t)$ a small perturbation $\epsilon \cos \omega t$, where $|\epsilon| \ll 1$ and $\omega > |\epsilon^{-1}|$, then $z(t)$ is perturbed by a large amount $|\omega\epsilon|$.

- The scalar equation

$$y = q(t) \tag{9.2}$$

is a (trivial) index-1 DAE, because it takes one differentiation to obtain an ODE for y.

- For the system

$$\begin{aligned} y_1 &= q(t), \\ y_2 &= y_1', \end{aligned} \tag{9.3}$$

we differentiate the first equation to get

$$y_2 = y_1' = q'(t)$$

and then

$$y_2' = y_1'' = q''(t)\,.$$

The index is 2 because two differentiations of $q(t)$ were needed.

- A similar treatment for the system

$$\begin{aligned} u &= q(t), \\ y_3 &= u'', \end{aligned} \tag{9.4}$$

necessitates three differentiations to obtain an ODE for y_3, hence the index is 3. ♦

Note that whereas m initial or boundary conditions must be given to specify the solution of an ODE of size m, for the simple DAEs of Example 9.1 the solution is completely determined by the right-hand side. More complicated DAE systems will usually also include some ODE subsystems. Thus, the DAE system will in general have l *degrees of freedom*, where l is anywhere between 0 and m.

In general it may be difficult, or at least not immediately obvious, to determine which l pieces of information are needed to determine the DAE solution. Often the entire initial solution vector is known. Initial or boundary conditions which are specified for the DAE must be *consistent*. In other words, they must satisfy the constraints of the system. For example, an initial condition on the index-1 system (9.2) (which is needed if we write it as an ODE) must satisfy $y_1(0) = q(0)$. For the index-2 system (9.3), the situation is somewhat more complicated. Not only must any solution satisfy the obvious constraint $y_1 = q(t)$, there is also a *hidden constraint* $y_2 = q'(t)$, which the solution must satisfy at any point t, so the only consistent initial conditions are $y_1(0) = q(0)$, $y_2(0) = q'(0)$. This is an important difference between index-1 and *higher-index* (index greater than 1) DAEs: *higher-index DAEs include some hidden constraints.* These hidden constraints are the derivatives of the explicitly stated constraints in the system. Index-2 systems include hidden constraints which are the first derivative of explicitly stated constraints. Higher-index systems include hidden constraints which correspond to higher-order derivatives; for example, solutions to the index-3 system (9.4) must satisfy the hidden constraints $u' = q'(t)$ and $y_3 = q''(t)$.

The most general form of a DAE is given by

$$\mathbf{F}(t, \mathbf{y}, \mathbf{y}') = \mathbf{0}, \tag{9.5}$$

where $\partial\mathbf{F}/\partial\mathbf{y}'$ may be singular. The rank and structure of this Jacobian matrix may depend, in general, on the solution $\mathbf{y}(t)$, and for simplicity we will always assume that it is independent of t. Recall also from Section 1.3 the important special case of a *semi-explicit* DAE, or an *ODE with constraints*,

$$\begin{aligned} \mathbf{x}' &= \mathbf{f}(t, \mathbf{x}, \mathbf{z}), && (9.6a) \\ \mathbf{0} &= \mathbf{g}(t, \mathbf{x}, \mathbf{z}). && (9.6b) \end{aligned}$$

This is a special case of (9.5). The index is 1 if $\partial\mathbf{g}/\partial\mathbf{z}$ is nonsingular, because then one differentiation of (9.6b) yields $\mathbf{z}'$ in principle.[39] For the semi-explicit index-1 DAE we can distinguish between *differential variables* $\mathbf{x}(t)$ and *algebraic variables* $\mathbf{z}(t)$. The algebraic variables may be less smooth than the differential variables by one derivative (e.g., the algebraic variables may be nondifferentiable).

In the general case, each component of $\mathbf{y}$ may contain a mix of differential and algebraic components, which makes the numerical solution of such high-index problems much harder and riskier. The semi-explicit form is decoupled in this sense. On the other hand, any DAE (9.5) can be written in the semi-explicit form (9.6) but with the index increased by 1, upon defining $\mathbf{y}' = \mathbf{z}$, which gives

$$\begin{aligned} \mathbf{y}' &= \mathbf{z}, && (9.7a) \\ \mathbf{0} &= \mathbf{F}(t, \mathbf{y}, \mathbf{z}). && (9.7b) \end{aligned}$$

Needless to say, this rewriting alone does not make the problem easier to solve. The converse transformation is also possible: given a semi-explicit index-2 DAE system (9.6), let $\mathbf{w}' = \mathbf{z}$. It is easily shown that the system

$$\begin{aligned} \mathbf{x}' &= \mathbf{f}(t, \mathbf{x}, \mathbf{w}'), && (9.8a) \\ \mathbf{0} &= \mathbf{g}(t, \mathbf{x}, \mathbf{w}') && (9.8b) \end{aligned}$$

is an index-1 DAE and yields exactly the same solution for $\mathbf{x}$ as (9.6). The classes of fully implicit index-1 DAEs of the form (9.5) and semi-explicit index-2 DAEs of the form (9.6) are therefore equivalent.

[39] Note that a differentiation of a vector function counts as *one* differentiation.

It is important to note, as the following example illustrates, that in general the index depends on the solution and not only on the form of the DAE. This is because the local linearization, hence the partial derivative matrices, depend on the solution.

Example 9.2
Consider the DAE system for $\mathbf{y} = (y_1, y_2, y_3)^T$,

$$\begin{aligned} y_1' &= y_3, \\ 0 &= y_2(1 - y_2), \\ 0 &= y_1 y_2 + y_3(1 - y_2) - t. \end{aligned}$$

The second equation has two solutions $y_2 = 0$ and $y_2 = 1$, and it is given that $y_2(t)$ does not switch arbitrarily between these two values (e.g., another equation involving y_2' and y_4' is prescribed with $y_4(0)$ given, implying continuity of $y_2(t)$).

1. Setting $y_2 = 0$, we get from the third equation $y_3 = t$. Then from the first equation, $y_1 = y_1(0) + t^2/2$. The system has index 1 and the solution is
$$\mathbf{y}(t) = (y_1(0) + t^2/2,\, 0,\, t)^T .$$
Note that this is an index-1 system in semi-explicit form.

2. Setting $y_2 = 1$, the third equation reads $y_1 = t$. Then, upon differentiating the first equation, $y_3 = 1$. The system has index 2 and the solution is
$$\mathbf{y}(t) = (t,\, 1,\, 1)^T .$$
Note that, unlike in the index-1 case, no initial value is required.

If we replace the algebraic equation involving y_2 by its derivative and simplify, we obtain the DAE

$$\begin{aligned} y_1' &= y_3, \\ y_2' &= 0, \\ 0 &= y_1 y_2 + y_3(1 - y_2) - t. \end{aligned} \tag{9.9}$$

Now the index depends on the initial conditions. If $y_2(0) = 0$ the index is 1, and if $y_2(0) = 1$ the index equals 2. ♦

We are ready to define the index of a DAE.

For general DAE systems (9.5), the *index* along a solution $\mathbf{y}(t)$ is the minimum number of differentiations of the system which would be required to solve for $\mathbf{y}'$ uniquely in terms of $\mathbf{y}$ and t (i.e., to define an ODE for $\mathbf{y}$). Thus, the index is defined in terms of the overdetermined system

$$\begin{aligned} \mathbf{F}(t,\mathbf{y},\mathbf{y}') &= \mathbf{0}, \\ \frac{d\mathbf{F}}{dt}(t,\mathbf{y},\mathbf{y}',\mathbf{y}'') &= \mathbf{0}, \\ &\vdots \\ \frac{d^p\mathbf{F}}{dt^p}(t,\mathbf{y},\mathbf{y}',\dots,\mathbf{y}^{(p+1)}) &= \mathbf{0} \end{aligned} \tag{9.10}$$

to be the smallest integer p so that $\mathbf{y}'$ in (9.10) can be solved for in terms of $\mathbf{y}$ and t.

We note that in practice, differentiation of the system as in (9.10) is rarely done in a computation. However, such a definition is very useful in understanding the underlying mathematical structure of the DAE system, and hence in selecting an appropriate numerical method.

Example 9.3

The computer-aided design of electrical networks involves simulations of the behavior of such networks in time. Electric circuits are assembled from basic elements such as resistors, diodes, inductors, capacitors, and sources. Large circuits can lead to large DAE systems.

A circuit is characterized by the type of elements it has and by its network's topology. For each element there is a relationship between the voltage drop between the nodes of the element and the current. For instance, a linear resistor satisfies, by Ohm's law,

$$U = RI,$$

where U is the potential drop, $I = Q'$ is the current (Q is the charge), and R is the resistance; for a linear inductor,

$$U = LI',$$

where L is the inductance; and for a linear capacitor,

$$I = CU',$$

where C is the capacitance. There are nonlinear versions of these too, e.g., $L = L(I)$ for a current-controlled inductor or $C = C(U)$ for a voltage-controlled capacitor.

The network consists of nodes and branches (it is a directed graph) and its topology can be encoded in an *incidence matrix* A. The (i, j)th entry of A is 1 if current flows from node i into branch j, -1 if current flows in branch j towards node i, and 0 if node i and branch j are not adjacent. Thus, A is typically large and very sparse. Let $\mathbf{u}_N$ be the vector function of all node potentials, $\mathbf{u}_B$ the branch potentials, and $\mathbf{i}_B$ the (branch) currents. *Kirchoff's current law* states that

$$A\mathbf{i}_B = 0, \tag{9.11a}$$

and *Kirchoff's voltage law* states that

$$\mathbf{u}_B = A^T\mathbf{u}_N. \tag{9.11b}$$

Adding to this the characteristic element equations as described earlier,

$$\phi(\mathbf{i}_B, \mathbf{u}_B, \mathbf{i}'_B, \mathbf{u}'_B) = 0, \tag{9.11c}$$

we obtain a typically very large, sparse DAE.

The sparse tableau approach leading to the DAE (9.11) is general, and software can be written to generate the equations from a given functional description of a circuit, but it is not favored in practice because it leads to too much redundancy in the unknowns. Instead, the *modified nodal analysis* eliminates $\mathbf{u}_B$ (via (9.11b)) and the currents $\mathbf{i}_B$, except for those currents through voltage-controlled elements (inductors and voltage sources). This leads to a large, sparse, but smaller DAE of the form

$$M(\mathbf{y})\mathbf{y}' + \mathbf{f}(\mathbf{y}) = \mathbf{q}(t), \tag{9.12}$$

where the possibly singular and still quite sparse M describes the dynamic elements, $\mathbf{f}$ corresponds to the other elements, and $\mathbf{q}$ are the independent sources.

The index of (9.12) depends on the type of circuit considered. In practical applications it often equals 0 or 1, but it may be higher. This index is often lower than that of (9.11), because some constraints are eliminated. Standard software exists which generates (9.12) from a functional description. However, a further reduction to an explicit ODE in the case when M is singular is not a practical option for most large circuits, because the sparsity of M is destroyed by the necessary matrix decomposition (such as (9.27) below).

A specific instance of a circuit is given in Example 10.3. ♦

For initial value ODEs, Theorem 1.1 guarantees solution existence, uniqueness, and continuous dependence on initial data for a large class of problems. No corresponding theorem holds in such generality for boundary value ODEs (see Chapter 1). No corresponding theorem holds for general DAEs either, although there are some weaker results of this type. Boundary value DAEs are of course no less complex than boundary value ODEs and will not be considered further in this chapter.

9.1.1 Special DAE Forms

The general DAE system (9.5) can include problems which are not well defined in a mathematical sense, as well as problems which will result in failure for any direct discretization method (i.e., a method based on discretization of $\mathbf{y}$ and $\mathbf{y}'$ without first reformulating the equations). Fortunately, most of the higher-index problems encountered in practice can be expressed as a combination of more restrictive structures of ODEs coupled with constraints. In such systems the algebraic and differential variables are explicitly identified for higher-index DAEs as well, and the algebraic variables may all be eliminated (in principle) using the same number of differentiations. These are called *Hessenberg forms* of the DAE and are given below.

Hessenberg Index-1

$$\mathbf{x}' = \mathbf{f}(t, \mathbf{x}, \mathbf{z}), \tag{9.13a}$$

$$\mathbf{0} = \mathbf{g}(t, \mathbf{x}, \mathbf{z}). \tag{9.13b}$$

Here the Jacobian matrix function $\mathbf{g_z}$ is assumed to be nonsingular for all t. This is also often referred to as a *semi-explicit index*-1 system. Semi-explicit index-1 DAEs are very closely related to implicit ODEs. Using the implicit function theorem, we can in principle solve for $\mathbf{z}$ in (9.13b). Substituting $\mathbf{z}$ into (9.13a) yields an ODE in $\mathbf{x}$ (although no uniqueness is guaranteed; see Exercise 9.5). For various reasons, this procedure is not always recommended for numerical solution.

Hessenberg Index-2

$$\mathbf{x}' = \mathbf{f}(t, \mathbf{x}, \mathbf{z}), \tag{9.14a}$$

$$\mathbf{0} = \mathbf{g}(t, \mathbf{x}). \tag{9.14b}$$

Here the product of Jacobians $\mathbf{g_x f_z}$ is nonsingular for all t. Note the absence of the algebraic variables $\mathbf{z}$ from the constraints (9.14b). This is a *pure* index-2 DAE, and all algebraic variables play the role of index-2 variables.[40]

Example 9.4
A practical example of a pure index-2 system arises from modeling the flow of an incompressible fluid by the Navier–Stokes equations

$$u_t + uu_x + vu_y + p_x - \nu(u_{xx} + u_{yy}) = 0, \tag{9.15a}$$

$$v_t + uv_x + vv_y + p_y - \nu(v_{xx} + v_{yy}) = 0, \tag{9.15b}$$

$$u_x + v_y = 0, \tag{9.15c}$$

[40]Whether a DAE is Hessenberg index-1 or index-2 may depend on the solution (Example 9.2) but usually doesn't in practice.

where subscripts denote partial derivatives, x, y are spatial variables and t is time, u, v are the velocities in the x- and y-directions, respectively, p is the scalar pressure, and ν is the (known) kinematic viscosity. Equations (9.15a)–(9.15b) are the momentum equations, and (9.15c) is the incompressibility condition. The extension to three spatial variables is straightforward. After a careful spatial discretization of (9.15) with a finite difference, finite volume, or finite element method, the vectors $\mathbf{u}(t)$ and $\mathbf{p}(t)$ approximating $(u(t,x,y), v(t,x,y))$ and $p(t,x,y)$ in the domain of interest satisfy

$$M\mathbf{u}' + (K + N(\mathbf{u}))\mathbf{u} + C\mathbf{p} = \mathbf{f}, \tag{9.16a}$$

$$C^T\mathbf{u} = \mathbf{0}. \tag{9.16b}$$

In this DAE the mass matrix M is symmetric positive definite. Skipping some nontrivial details of the spatial discretization, we assume not only that the same matrix C appears in (9.16a) and (9.16b), but also that $C^T M^{-1} C$ is a nonsingular matrix with a bounded inverse. This yields an index-2 DAE in Hessenberg form. The DAE could be made semi-explicit upon multiplying by M^{-1}, but the sparsity of the coefficient matrices of the DAE would be lost, unless M is block diagonal. The forcing function $\mathbf{f}$ comes from the (spatial) boundary conditions.

It is well known that obtaining an accurate solution for the pressure in (9.15) can be problematic. Often this variable is treated in a different way by discretization methods. For instance, a staggered grid may be used in space, where the pressure values are considered at midcells and the velocity values "live" on cell edges. Part of the reason for this is that the pressure in (9.15) is an index-2 variable. It has the same order of (temporal) smoothness as the derivative of the velocity. The pressure in (9.16) is playing the role of the index-2 variable $\mathbf{z}$ in (9.14).

One can consider differentiating (9.15c) with respect to time and substituting into (9.15a), (9.15b) to obtain a Poisson equation for p with the right-hand side being a function of u and v. This is called the pressure-Poisson equation—the matrix $C^T M^{-1} C$ above can in fact be viewed as a discretization of the Laplace operator plus suitable boundary conditions—and the obtained system has index 1. For the index-1 system the discretization in space need no longer be staggered, but some difficulties with boundary conditions may arise. ♦

Another way to look at index-2 variables like the pressure in (9.16) derives from the observation that these DAEs are closely related to constrained optimization problems. From this point of view, $\mathbf{p}$ in (9.16) plays the role of a *Lagrange multiplier*: it forces the velocity $\mathbf{u}$ to lie in the *constraint manifold* defined by (9.16b). The relationship between higher-index DAEs and constrained optimization problems is no accident; many of these DAEs, including the incompressible Navier–Stokes equations, arise from constrained variational problems.

Example 9.5
Consider the DAE

$$\begin{aligned} y_1' &= \lambda y_1 - y_4, \\ y_2' + y_3' &= (2\lambda - \sin^2 t)(y_2 + y_3) + 1/2(y_2 - y_3)^2, \\ 0 &= y_2 - y_3 - 2(\sin t)(y_1 - 1), \\ 0 &= y_2 + y_3 - 2(y_1 - 1)^2, \end{aligned} \tag{9.17}$$

where λ is a parameter and $y_1(0) = 2, y_2(0) = 1$ are prescribed.

This DAE is not in semi-explicit form. We can, however, easily convert it to that form by the *constant, nonsingular* transformation

$$x_1 = y_1, \; x_2 = \frac{1}{2}(y_2 + y_3), \; z_1 = \frac{1}{2}(y_2 - y_3), \; z_2 = y_4,$$

yielding

$$x_1' = \lambda x_1 - z_2, \tag{9.18a}$$
$$x_2' = (2\lambda - \sin^2 t)x_2 + z_1^2, \tag{9.18b}$$
$$0 = z_1 - (\sin t)(x_1 - 1), \tag{9.18c}$$
$$0 = x_2 - (x_1 - 1)^2. \tag{9.18d}$$

The DAE is now in the semi-explicit form (9.6), but it is not in Hessenberg form. In particular, (9.18c) yields $z_1 = z_1(\mathbf{x})$, so z_1 is an index-1 algebraic variable, whereas z_2 cannot be eliminated without differentiation. A differentiation of (9.18d) and a substitution into (9.18a) confirm that, for the given initial conditions, z_2 can be subsequently eliminated. Hence the DAE is index-2 and z_2 is an index-2 algebraic variable.

Note that if we further carry out the substitution for z_1, then the resulting DAE

$$\begin{aligned} x_1' &= \lambda x_1 - z_2, \\ x_2' &= (2\lambda - \sin^2 t)x_2 + (\sin^2 t)(x_1 - 1)^2, \\ 0 &= x_2 - (x_1 - 1)^2, \end{aligned} \tag{9.19}$$

is Hessenberg index-2. ♦

Hessenberg Index-3

$$\mathbf{x}' = \mathbf{f}(t, \mathbf{x}, \mathbf{y}, \mathbf{z}), \tag{9.20a}$$
$$\mathbf{y}' = \mathbf{g}(t, \mathbf{x}, \mathbf{y}), \tag{9.20b}$$
$$\mathbf{0} = \mathbf{h}(t, \mathbf{y}). \tag{9.20c}$$

Here the product of three matrix functions $\mathbf{h_y g_x f_z}$ is nonsingular.

Example 9.6
The mechanical systems with holonomic constraints described in Example 1.6 are Hessenberg index-3. This type of DAE often arises from second-order ODEs subject to constraints.

Indeed, the ODEs describe Newton's second law of motion relating body accelerations to forces. Since accelerations are second derivatives of positions, constraints imposed on the positions imply that two differentiations must be buried in the system of ODEs with constraints. ♦

The index of a Hessenberg DAE is found, as in the general case, by differentiation. However, here only algebraic constraints must be differentiated.

Example 9.7
To illustrate, we find the index of a simple mechanical system, the pendulum in Cartesian coordinates from Example 1.5. We use the notation $\mathbf{q}$ for the position coordinates and $\mathbf{v} = \mathbf{q}'$ for the velocities. First, the DAE is written as a first-order system

$$\begin{aligned} q_1' &= v_1, & (9.21a)\\ q_2' &= v_2, & (9.21b)\\ v_1' &= -\lambda q_1, & (9.21c)\\ v_2' &= -\lambda q_2 - g, & (9.21d)\\ 0 &= q_1^2 + q_2^2 - 1\,. & (9.21e) \end{aligned}$$

(Note that $\lambda = \lambda(t)$ is an unknown function and g is the known, scaled constant of gravity.) Then the *position constraint* (9.21e) is differentiated once, to obtain

$$q_1 q_1' + q_2 q_2' = 0\,.$$

Substituting for $\mathbf{q}'$ from (9.21a) and (9.21b) yields the *velocity constraint*

$$\mathbf{q}^T\mathbf{v} = q_1 v_1 + q_2 v_2 = 0\,. \qquad (9.22)$$

Differentiating the velocity constraint (9.22) and substituting for $\mathbf{q}'$ yields

$$q_1 v_1' + q_2 v_2' + v_1^2 + v_2^2 = 0.$$

Substituting for $\mathbf{v}'$ from (9.21c) and (9.21d), and simplifying using the position constraint, yields the *acceleration constraint*

$$-\lambda - q_2 g + v_1^2 + v_2^2 = 0. \qquad (9.23)$$

This yields λ, which can be substituted into (9.21c) and (9.21d) to obtain an ODE for $\mathbf{q}$ and $\mathbf{v}$. To obtain a differential equation for *all* the unknowns, however, we need to differentiate (9.23) one more time, obtaining an ODE for λ as well. In the process of getting to the explicit ODE system, the position constraints were differentiated three times. Hence, the index of this system is 3. ♦

The index has proven to be a useful concept for classifying DAEs, in order to construct and identify appropriate numerical methods. It is often not necessary to perform the differentiations in order to find the index, because most physical systems can be readily seen to result in systems of Hessenberg structure or in simple combinations of Hessenberg structures.

Example 9.8
Consider a tiny ball of mass 1 attached to the end of a spring of length 1 at rest with a spring constant ε^{-1}, $\varepsilon > 0$. At its other end the spring's position is fixed at the origin of a planar coordinate system (see Figure 1.2 and imagine the rod in the simple pendulum being replaced by a spring).

The sum of kinetic and potential energies in this system is

$$e(\mathbf{q}, \mathbf{v}) = \frac{1}{2}[\mathbf{v}^T\mathbf{v} + \varepsilon^{-1}(r-1)^2] + gq_2,$$

where $\mathbf{q} = (q_1, q_2)^T$ are the Cartesian coordinates, $\mathbf{v} = (v_1, v_2)^T$ are the velocities (which equal the momenta $\mathbf{p}$ in our scaled, dimensionless notation), $r = \sqrt{q_1^2 + q_2^2} = |\mathbf{q}|_2$ is the length of the spring at any given time, and g is the scaled constant of gravity. The equations of motion are (recall Section 2.5)

$$\begin{aligned} \mathbf{q}' &= e_{\mathbf{v}} = \mathbf{v}, \\ \mathbf{v}' &= -e_{\mathbf{q}} = -\varepsilon^{-1}\frac{r-1}{r}\mathbf{q} - \begin{pmatrix} 0 \\ g \end{pmatrix}. \end{aligned}$$

This is an ODE. Let us next write the same system as a DAE. Defining $\lambda = \varepsilon^{-1}(r-1)$ we get

$$\begin{aligned} \mathbf{q}'' &= -\frac{1}{r}\mathbf{q}\lambda - \begin{pmatrix} 0 \\ g \end{pmatrix}, \\ \varepsilon\lambda &= r - 1. \end{aligned}$$

This DAE is semi-explicit index-1. It is not really different from the ODE in a meaningful way (although it may suggest controlling the error also in λ in a numerical approximation).

Next, consider what happens when the spring is very stiff, almost rigid, i.e., $\varepsilon \ll 1$. We then expect the radius r to oscillate rapidly about its rest value, while the angle θ varies slowly. This is depicted in Figure 9.2.

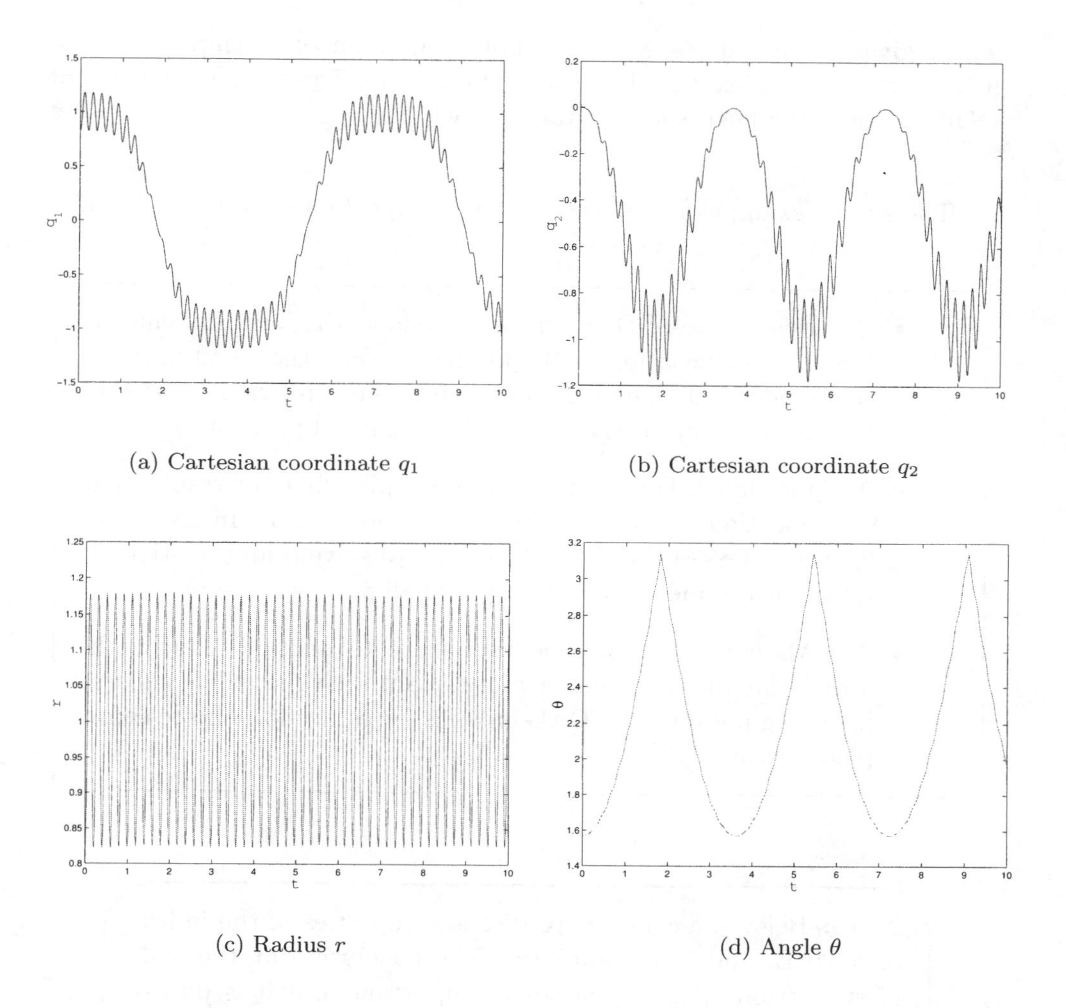

(a) Cartesian coordinate q_1

(b) Cartesian coordinate q_2

(c) Radius r

(d) Angle θ

Figure 9.2: *Stiff spring pendulum*, $\varepsilon = 10^{-3}$, *initial conditions* $\mathbf{q}(0) = (1 - \varepsilon^{1/4}, 0)^T, \mathbf{v}(0) = \mathbf{0}$.

Provided that the initial conditions yield

$$r(t) = 1 + O(\varepsilon),$$

we have $\lambda(t) = O(1)$ to balance the constraint equation in the index-1 formulation. The passage to the limit $\varepsilon \to 0$ is simple then, and we obtain the DAE

$$\begin{aligned} \mathbf{q}'' &= -\mathbf{q}\lambda - \begin{pmatrix} 0 \\ g \end{pmatrix}, \\ 0 &= r - 1, \end{aligned}$$

which gives the equations for the simple pendulum of Examples 1.5 and 9.7.[41] This is an index-3 DAE in Hessenberg form. Unlike the ε-dependent ODE solution, the DAE solution varies slowly! ♦

The simple example above leads to some important observations.

- One rich source of DAEs in practice is as the limit systems of singular perturbation ODE problems, when the small parameter tends to 0. The solution then is often referred to as the *reduced solution* for the singularly perturbed problem.
- A higher-index DAE can often be simpler than, or result as a simplification of, an ODE or a lower-index DAE. In Example 9.8 the index-3 DAE is much simpler to solve than the original ODE (or the index-1 DAE) for a small ε.
- A DAE can in a sense be very close to another DAE with a different index. Thus, a more quantitative stability theory involving not only the DAE index is necessary for a more complete picture.

Note: Below we continue to discuss properties of the index of a DAE, and DAE stability. The conclusion at the end of subsection 9.1.2 is practically important, but it is possible to skip the discussion, at least on first reading, and still understand the material in Section 9.2.

9.1.2 DAE Stability

Example 9.2 suggests that the index is a local quantity, to be measured about an isolated exact solution. Thus, we next consider perturbations in linear DAEs and their relationship to the index and to stability constants. For a nonlinear system we form the variational problem about an exact solution and its perturbations, and define the index locally based on the index of this linear problem. As in Section 6.2 we note that for a linear problem the objective is to bound the solution in terms of the data (the inhomogeneities). The same bound then holds for a perturbation to the

[41] In Figure 9.2 the initial conditions are such that $|\lambda| \gg 1$, so that the oscillations in $\mathbf{q}$ can be seen by the naked eye, but the limit DAE turns out to be the same.

solution when the inhomogeneities are replaced by their perturbations. (See the exposition following (6.14).)

For the linear ODE system

$$\mathbf{y}' = A(t)\mathbf{y} + \mathbf{q}(t), \qquad 0 < t < b,$$

subject to homogeneous initial or boundary conditions, we can transform the independent variable by $\tau = t/b$ for any large b. Let us assume that this has been done and take $b = 1$. We have seen in Sections 2.3 and 6.2 that the following stability bound holds:

$$\|\mathbf{y}\| = \max_{0 \le t \le 1} |\mathbf{y}(t)| \le \kappa \int_0^1 |\mathbf{q}(s)| ds = \kappa \|\mathbf{q}\|_1.^{42} \tag{9.24}$$

For the trivial index-1 DAE

$$\mathbf{y} = \mathbf{q}(t),$$

we have a slightly weaker bound than (9.24), namely,

$$\|\mathbf{y}\| \le \|\mathbf{q}\|$$

(weaker because the maximum norm, rather than the L_1-norm, must be used for $\mathbf{q}$). For the semi-explicit index-1 DAE

$$\mathbf{x}' = A\mathbf{x} + B\mathbf{z} + \mathbf{q}_1(t), \tag{9.25a}$$

$$\mathbf{0} = C\mathbf{x} + D\mathbf{z} + \mathbf{q}_2(t), \tag{9.25b}$$

where A, B, C, D are bounded functions of t and with D boundedly invertible, we get a similar result,

$$\|\mathbf{y}\| \le \kappa \|\mathbf{q}\|,$$

where $\mathbf{y}^T = (\mathbf{x}^T, \mathbf{z}^T)$, $\mathbf{q}^T = (\mathbf{q}_1^T, \mathbf{q}_2^T)$. The generic stability constant κ involves bounds on D^{-1}, as well as the stability constant of the *underlying ODE* for $\mathbf{x}$, once $\mathbf{z}$ given by (9.25b) has been substituted for in (9.25a). This bound can actually be refined to

$$\|\mathbf{z}\| \le \kappa \|\mathbf{q}\|, \quad \|\mathbf{x}\| \le \kappa \|\mathbf{q}\|_1 .$$

For the general index-1 linear DAE[43]

$$E(t)\mathbf{y}' = A(t)\mathbf{y} + \mathbf{q}(t) \tag{9.26}$$

still with homogeneous initial or boundary conditions, we can decompose $E(t)$ into

$$E(t) = S(t) \begin{pmatrix} I & 0 \\ 0 & 0 \end{pmatrix} T^{-1}(t), \tag{9.27}$$

[42] In (9.24) we have defined the L_1 norm. Note that $\|\mathbf{q}\|_1 = \int_0^1 |\mathbf{q}(s)| ds \le \max_{0 \le t \le 1} |\mathbf{q}(t)| = \|\mathbf{q}\|$.

[43] We assume here that the system is strictly index-1, i.e., not tending arbitrarily closely to a higher-index or singular system.

where T and S are nonsingular matrix functions with uniformly bounded condition numbers. Then a change of variables

$$\begin{pmatrix} \mathbf{x} \\ \mathbf{z} \end{pmatrix} = T^{-1}\mathbf{y},$$

where $\mathbf{x}$ has the dimension of the identity block in (9.27), yields a semi-explicit system (9.25). Hence we obtain again (assuming of course that the underlying ODE problem is stable) an estimate

$$\|\mathbf{y}\| \leq \kappa\|\mathbf{q}\|,$$

where now the condition numbers of the transformations are also lumped into the stability constant κ.

In short, for a linear index-1 problem, if

- it can be transformed (without differentiations) into a semi-explicit system, and from there to an ODE by eliminating the algebraic variables,
- the transformations are all suitably well conditioned,
- the obtained ODE problem is stable,

then the index-1 DAE problem is also stable in the usual sense. Exercise 9.7 makes this statement precise.

For higher-index problems we must differentiate at least some of the equations. For an index-p DAE we need $p-1$ differentiations to obtain an index-1 DAE, hence all we can hope for is a "stability" bound of the form

$$\|\mathbf{y}\| \leq \kappa \sum_{j=1}^{p} \|\mathbf{q}^{(j-1)}\|. \tag{9.28}$$

Fortunately, for a DAE in Hessenberg form this can be somewhat improved upon. In particular, for an index-2 Hessenberg DAE of the form (9.25) with $D \equiv 0$ and CB nonsingular, we have

$$\begin{aligned} \|\mathbf{x}\| &\leq \kappa\|\mathbf{q}\|, \\ \|\mathbf{z}\| &\leq \kappa\|\mathbf{q}'\|. \end{aligned} \tag{9.29}$$

All this suggests that a direct numerical discretization of nontrivial higher-index DAEs other than Hessenberg index-2 may encounter serious difficulties. We will see in the next chapter that this is indeed true.

9.2 Index Reduction and Stabilization: ODE with Invariant

Often, the best way to solve a high-index DAE problem is to first convert it to a lower-index system by carrying out differentiations analytically. In this section we describe some of the techniques which are available for reformulation of a higher-index, semi-explicit DAE (9.6), where differentiations are applied to the constraint equations (9.6b). The essential concept here is that the DAE is equivalent to an *ODE with an invariant.* For an index-$(p+1)$ DAE in Hessenberg form with m ODEs and l constraints, recall that we need p differentiations in order to eliminate the algebraic variables and obtain an ODE system of size m in closed form. The equations (9.6b), together with their first $p-1$ derivatives (with $\mathbf{z}(t)$ eliminated), form an invariant set defined by pl algebraic constraints. One can consider using these algebraic constraints at each t in order to define a smaller set of $m-pl$ unknowns. The differential equations for the smaller set of unknowns then describe the dynamics while enforcing the constraints. This yields an *ODE on a manifold* and is further discussed in Section 9.2.3. Since the dimension of the constraint manifold is pl, the true dimension (i.e., the number of degrees of freedom) of the entire system is $m-pl$, as discussed in the previous section.

In the presentation that follows we use constrained mechanical systems as a case study for higher-index DAEs in Hessenberg form. Problems from this important class are often solved in practice using the techniques of this section. The general principles of reformulation of DAE systems are also useful in a wide variety of other applications.

9.2.1 Reformulation of Higher-Index DAEs

Recall the mechanical systems from Example 1.6,

$$\mathbf{q}' = \mathbf{v}, \tag{9.30a}$$

$$M(\mathbf{q})\mathbf{v}' = \mathbf{f}(\mathbf{q},\mathbf{v}) - G^T(\mathbf{q})\boldsymbol{\lambda}, \tag{9.30b}$$

$$\mathbf{0} = \mathbf{g}(\mathbf{q}), \tag{9.30c}$$

where $\mathbf{q}$ are generalized body positions, $\mathbf{v}$ are generalized velocities, $\boldsymbol{\lambda} \in \Re^l$ are Lagrange multiplier functions, $\mathbf{g}(\mathbf{q}) \in \Re^l$ defines the holonomic constraints, $G = \mathbf{g_q}$ is assumed to have full row rank at each t, M is a positive definite generalized mass matrix, and $\mathbf{f}$ are the applied forces. Any explicit dependence on t is omitted for notational simplicity, but of course all the quantities above are functions of t. We also denote

$$\mathbf{x} = \begin{pmatrix} \mathbf{q} \\ \mathbf{v} \end{pmatrix} \in \Re^m,$$

corresponding to the notation in (9.6).

We now apply two differentiations to the position constraints (9.30c). The first yields the constraints on the velocity level

$$\mathbf{0} = G\mathbf{v} \qquad (= \mathbf{g}'), \tag{9.31}$$

and the second differentiation yields the constraints on the acceleration level

$$\mathbf{0} = G\mathbf{v}' + \frac{\partial(G\mathbf{v})}{\partial \mathbf{q}}\mathbf{v} \qquad (= \mathbf{g}'') . \tag{9.32}$$

Next, multiply (9.30b) by GM^{-1} and substitute from (9.32) to eliminate $\boldsymbol{\lambda}$:

$$\boldsymbol{\lambda}(\mathbf{q}, \mathbf{v}) = (GM^{-1}G^T)^{-1}\left(GM^{-1}\mathbf{f} + \frac{\partial(G\mathbf{v})}{\partial \mathbf{q}}\mathbf{v}\right). \tag{9.33}$$

Finally, $\boldsymbol{\lambda}$ from (9.33) can be substituted into (9.30b) to yield an ODE for $\mathbf{x}$,

$$\mathbf{q}' = \mathbf{v}, \tag{9.34a}$$

$$M\mathbf{v}' = \mathbf{f} - G^T(GM^{-1}G^T)^{-1}\left(GM^{-1}\mathbf{f} + \frac{\partial(G\mathbf{v})}{\partial \mathbf{q}}\mathbf{v}\right). \tag{9.34b}$$

In practice we may want to keep (9.34b) in the equivalent form (9.30b), (9.32) as long as possible and never evaluate the matrix function $\frac{\partial(G\mathbf{v})}{\partial \mathbf{q}}$ (i.e., evaluate only its product with $\mathbf{v}$).

The ODE system (9.34) has dimension m and is the result of an *unstabilized index reduction.* The constraints on the position and the velocity levels, which are now additional to this ODE, define an invariant set of dimension $2l$,

$$\mathbf{h}(\mathbf{x}) \equiv \begin{pmatrix} \mathbf{g}(\mathbf{q}) \\ G(\mathbf{q})\mathbf{v} \end{pmatrix} = \mathbf{0}. \tag{9.35}$$

Thus, any solution of the larger ODE system (9.34) with consistent initial values, i.e., with initial values satisfying $\mathbf{h}(\mathbf{x}(0)) = \mathbf{0}$, satisfies $\mathbf{h}(\mathbf{x}(t)) = \mathbf{0}$ at all later times. We denote the constraint Jacobian matrix

$$H = \mathbf{h}_{\mathbf{x}} \tag{9.36}$$

and note that for the mechanical system (9.30),

$$H = \begin{pmatrix} G & 0 \\ \frac{\partial(G\mathbf{v})}{\partial \mathbf{q}} & G \end{pmatrix} \tag{9.37}$$

has full row rank $2l$. Restricted to the constraint manifold, the ODE has dimension $m - 2l$, which is the correct dimension of the DAE (9.30).

Example 9.9
For the DAE (9.21) of Example 9.7 we substitute

$$-\lambda = q_2 g - v_1^2 - v_2^2$$

to obtain the ODE corresponding to (9.34),

$$\begin{aligned} q_1' &= v_1, \\ q_2' &= v_2, \\ v_1' &= -(v_1^2 + v_2^2 - q_2 g)q_1, \\ v_2' &= -(v_1^2 + v_2^2 - q_2 g)q_2 - g, \end{aligned}$$

and the invariant equations corresponding to (9.35),

$$\begin{aligned} 0 &= q_1^2 + q_2^2 - 1, \\ 0 &= q_1 v_1 + q_2 v_2. \end{aligned}$$

♦

9.2.2 ODEs with Invariants

Differential systems with invariants arise frequently in various applications, not only as a result of index reduction in DAEs. The invariant might represent conservation of energy, momentum, or mass in a physical system. The ODE system in Example 9.8 has the invariant that the energy is constant in t, as is typical for Hamiltonian systems. Recall also Exercises 4.16–4.17.

The relationship between DAEs and ODEs with invariants goes both ways. Not only does index reduction of a DAE lead to an ODE with an invariant, but also an ODE with an invariant

$$\mathbf{x}' = \hat{\mathbf{f}}(\mathbf{x}), \tag{9.38a}$$
$$\mathbf{h}(\mathbf{x}) = \mathbf{0} \tag{9.38b}$$

is equivalent to the Hessenberg index-2 DAE

$$\mathbf{x}' = \hat{\mathbf{f}}(\mathbf{x}) - D(\mathbf{x})\mathbf{z}, \tag{9.39a}$$
$$\mathbf{0} = \mathbf{h}(\mathbf{x}). \tag{9.39b}$$

Here $D(\mathbf{x})$ is any bounded matrix function such that HD, where $H = \mathbf{h_x}$, is boundedly invertible for all t. The systems (9.38) and (9.39) have the same solutions for $\mathbf{x}(t)$. The exact solution of (9.39) gives $\mathbf{z}(t) \equiv \mathbf{0}$, but this is no longer true in general for a numerical discretization of this system. Note that the DAE (9.39) is not the same as the original DAE (9.30) in the case when the latter is the source of the system (9.38). The choice of the matrix function D in (9.39) defines the direction of the projection onto the constraint manifold. A common choice is $D = H^T$, which yields an orthogonal projection.[44]

[44]Note that in the case of mechanical systems (9.37) we would like to avoid the lower left block of H if at all possible; see Exercise 9.10.

Indeed, there are applications where simply integrating the ODE is a perfectly valid and useful approach. The numerical solution does not precisely satisfy the constraints then, but it is close to satisfying (9.38b) within the integration tolerance. But in other applications the invariant cannot simply be ignored. This is the case when there are special reasons for insisting that the error in (9.38b) be much smaller than the error in (9.38a), or when the problem is more stable on the manifold than off it.

The latter reason applies in the case of a DAE index reduction. To see this, imagine a nonsingular transformation of variables

$$\mathbf{q} \rightarrow \begin{pmatrix} \boldsymbol{\psi} \\ \boldsymbol{\phi} \end{pmatrix} = \begin{pmatrix} \mathbf{g}(\mathbf{q}) \\ \tilde{\mathbf{g}}(\mathbf{q}) \end{pmatrix}$$

such that $\tilde{\mathbf{g}}_{\mathbf{q}}$ is orthogonal to G^T. Now, the differentiations of the constraints, i.e., (9.32), yield

$$\boldsymbol{\psi}'' = \mathbf{0},$$

and this equation has a double eigenvalue 0. This indicates a mild instability, because if $\psi(0) = \epsilon_1$, $\psi'(0) = \epsilon_2$, and $\psi'' = 0$, then $\psi(t) = \epsilon_1 + \epsilon_2 t$; i.e., perturbations grow linearly in time. The instability, known as a *drift off* the constraint manifold, is a *result of the differentiations* (i.e., it is not present in the original DAE, hence not in the equivalent ODE restricted to the manifold).

Stabilization

Rather than converting the ODE to a DAE, which carries the penalty of having to solve the resulting DAE, we can consider *stabilizing* or *attenuating* the ODE (9.38a) with respect to the invariant set $\mathcal{M} = \{\mathbf{x} : \mathbf{h}(\mathbf{x}) = \mathbf{0}\}$. The ODE

$$\mathbf{x}' = \hat{\mathbf{f}}(\mathbf{x}) - \gamma F(\mathbf{x})\mathbf{h}(\mathbf{x}) \tag{9.40}$$

obviously has the same solutions as (9.38a) on $\mathcal{M}$ (i.e., when $\mathbf{h}(\mathbf{x}) = \mathbf{0}$). It also has the desired stability behavior if HF is positive definite and the positive parameter γ is large enough. In fact, we can easily apply a Lyapunov-type argument (see Exercises 2.3–2.4) to obtain

$$\begin{aligned} \frac{1}{2}\frac{d}{dt}\mathbf{h}^T\mathbf{h} &= \mathbf{h}^T\mathbf{h}' = \mathbf{h}^T H(\hat{\mathbf{f}} - \gamma F\mathbf{h}) \\ &\leq (\gamma_0 - \gamma\lambda_0)\mathbf{h}^T\mathbf{h}, \end{aligned}$$

where γ_0 is a constant such that, using the Euclidean vector norm,

$$|H\hat{\mathbf{f}}(\mathbf{x})| \leq \gamma_0 |\mathbf{h}(\mathbf{x})| \tag{9.41}$$

for all $\mathbf{x}$ near $\mathcal{M}$, and λ_0 is the smallest eigenvalue of the positive definite matrix function HF.

Thus, asymptotic stability of the constraint manifold results for any $\gamma > \gamma_0/\lambda_0$. What this means is that any trajectory of (9.40) starting from some initial value near $\mathcal{M}$ will tend towards satisfying the constraints, i.e., towards the manifold. Moreover, this attenuation is *monotonic*:

$$|\mathbf{h}(\mathbf{x}(t+\alpha))| \le |\mathbf{h}(\mathbf{x}(t))| \tag{9.42}$$

for any $t,\ \alpha \ge 0$.

To get a grip on the values of γ_0 and λ_0, note that often $\gamma_0 = 0$ in (9.41), in which case the invariant is called an *integral invariant* (because for any $\mathbf{x}(t)$ near $\mathcal{M}$ satisfying (9.38a) it transpires that $\frac{d}{dt}\mathbf{h} = \mathbf{0}$, hence $\mathbf{h}(\mathbf{x}(t))$ is constant). For the mechanical system (9.30) it can be shown that $\gamma_0 = 1$ (Exercise 9.8). Also, if we choose

$$F(\mathbf{x}) = D(HD)^{-1},$$

where $D(\mathbf{x})$ is as before in (9.39), then $HF = I$, hence $\lambda_0 = 1$.

If the system is not stiff, then (9.40) can be integrated by an explicit method from the Runge–Kutta or Adams families, which is often faster than the implicit methods of Section 10.1.

Example 9.10

We consider again the simple pendulum in Cartesian coordinates and apply the MATLAB standard initial value problem (IVP) solver to the ODE of Example 9.9. Starting from $\mathbf{q}(0) = (1,0)^T$, $\mathbf{v}(0) = (0,-5)^T$, the solver is accurate enough and the problem simple enough that the unit circle is obtained in the $\mathbf{q}$-phase space to at least four significant digits. Then we repeat the calculations from the starting points $\mathbf{q}(0) = (1, \pm .5)^T$ and the same $\mathbf{v}(0)$. The resulting curves are depicted in Figure 9.3a.

Next we modify the ODE according to (9.40), with

$$D^T = H = \begin{pmatrix} 2q_1 & 2q_2 & 0 & 0 \\ v_1 & v_2 & q_1 & q_2 \end{pmatrix}$$

and $\gamma = 10$, and repeat these integrations. The results are depicted in Figure 9.3b. Of course, for the starting values which do not satisfy $|\mathbf{q}(0)|_2 = 1$, the exact solution of the stabilized ODE is different from the original, but the figure clearly indicates how the unit circle becomes attractive for the latter ODE system, even when the initial values are significantly perturbed. ♦

One of the earliest stabilization methods proposed in the literature was due to J. Baumgarte [15]. In this method, the acceleration-level constraints are replaced by a linear combination of the constraints on the acceleration, velocity, and position levels:

$$\mathbf{0} = \mathbf{g}'' + \gamma_1 \mathbf{g}' + \gamma_2 \mathbf{g}. \tag{9.43}$$

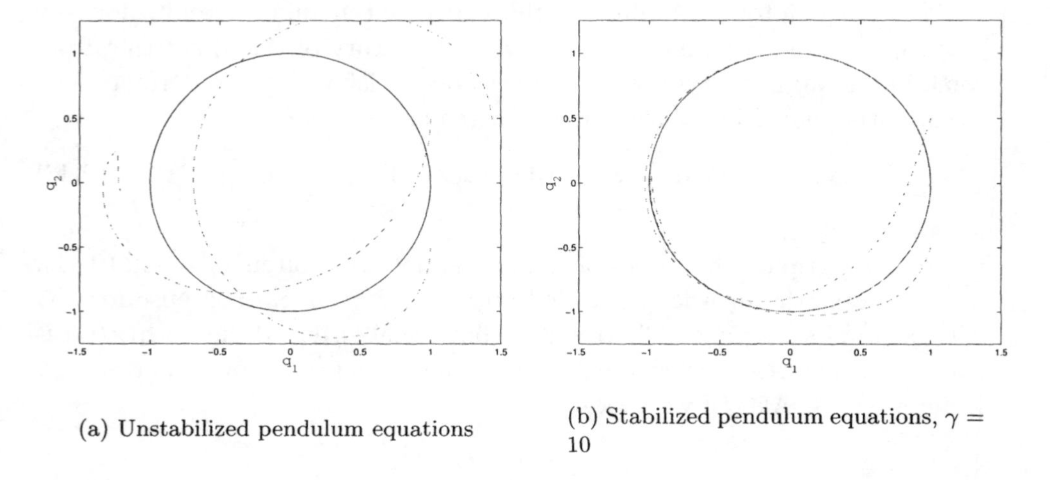

(a) Unstabilized pendulum equations

(b) Stabilized pendulum equations, $\gamma = 10$

Figure 9.3: *Perturbed (dashed lines) and unperturbed (solid line) solutions for Example* 9.9.

The parameters γ_1 and γ_2 are chosen such that the polynomial equation

$$x^2 + \gamma_1 x + \gamma_2 = 0$$

has two negative roots; thus the ODE (9.43) for $\mathbf{g}$ is stable. This stabilizes the invariant set $\mathcal{M}$. The system (9.30a), (9.30b), (9.43) is a semi-explicit index-1 DAE. It can be made into an ODE upon elimination of $\boldsymbol{\lambda}$ and be subsequently solved numerically by standard methods. But the choice of the parameters has proved tricky in practice. Exercise 9.11 and Section 10.2 elaborate more on this.

9.2.3 State Space Formulation

The differentiations of the constraints of the given high-index DAE (9.30) yield an ODE (9.34) with an inflated dimension, as we have seen. Even though the number of degrees of freedom of the system is $m - 2l$, we have in (9.34) m ODEs, and in (9.35) an additional $2l$ algebraic equations. Rather than stabilizing the invariant, another approach is to use these algebraic equations to define a reduced set of unknowns, obtaining an ODE system of the minimal size $m - 2l$. The main difficulty with this idea arises in the presence of highly nonlinear terms.

Suppose that R is a rectangular, constant matrix such that, together with the constraint Jacobian G, we obtain a nonsingular matrix with a bounded inverse:

$$\left\| \begin{pmatrix} R \\ G \end{pmatrix}^{-1} \right\| \leq K\,. \tag{9.44}$$

Defining the change of variables

$$\mathbf{u} = R\mathbf{q}, \quad \mathbf{w} = G\mathbf{q}, \tag{9.45}$$

we get

$$\mathbf{q} = \begin{pmatrix} R \\ G \end{pmatrix}^{-1} \begin{pmatrix} \mathbf{u} \\ \mathbf{w} \end{pmatrix}.$$

We can now use the constraints defining the invariant set, i.e., $\mathbf{g}(\mathbf{q}) = \mathbf{0}$ and $G\mathbf{v} = \mathbf{0}$, to express $\mathbf{w}$ as a function of $\mathbf{u}$, and hence $\mathbf{q}$ in terms of $\mathbf{u}$. For $\mathbf{u}$ we then obtain, upon multiplying the equations of motion by R, an underlying ODE of size (when converted to a first-order system) $m - 2l$,

$$\mathbf{u}'' = RM^{-1}(\mathbf{f} - G^T\boldsymbol{\lambda}), \tag{9.46}$$

where $\boldsymbol{\lambda}$ is given by (9.33).

There are two popular choices for R. The first is such that the unknowns $\mathbf{u}$ form a subset of the original $\mathbf{q}$, i.e., the columns of R are either unit vectors or $\mathbf{0}$. This has the advantage of simplicity. Note, however, that we cannot expect in general that one such choice of R will be good for all t, in the sense that (9.44) will remain valid with a moderate constant K. This *coordinate partitioning* has to be monitored and modified as necessary. The other choice is to make R orthogonal to $M^{-1}G^T$. This eliminates $\boldsymbol{\lambda}$ in (9.46), but introduces additional complications into the calculation due to the varying R.

The attraction of this approach is the small ODE system that is obtained and the elimination of any drift off the constraint manifold. On the negative side, this approach involves a somewhat messier algebra and is less transparent. The transformation nonsingularity (9.44) must be monitored and any wrinkle in the constraint manifold might have to be fully reflected here, even if it could be otherwise ignored.

9.3 Modeling with DAEs

The closing decades of the 20th century have seen many scientists recognize that their mathematical models are in fact instances of DAEs. Such a recognition has often carried with it the benefit of affording a new, sometimes revealing, computational look at the old problem.

Note, however, that whereas a sensible formulation of a mathematical model as an initial value ODE is typically followed simply by its numerical solution using some appropriate code, DAE formulations may require more user attention and intervention, combining the processes of problem formulation and numerical solution. Since high-index DAEs are all unstable, we know already before advancing to Chapter 10 that attempting to discretize them *directly* may adversely affect the resulting numerical scheme. The reformulations of the problem discussed in the previous section are done with numerical implementations in mind. In the extreme, a DAE would be converted to an ODE, but bear in mind that this may be cumbersome to carry out and costly to work with.

Consider a DAE system and its various index reductions and reformulations studied in Section 9.2. The exact solution satisfies all such equations, but numerical discretizations generally result in nonzero residuals. When a semi-explicit DAE such as (9.6) is discretized and solved numerically, it is automatically assumed that the ODE (9.6a) will be solved approximately while the algebraic constraints will be satisfied (almost) exactly. The residual in (9.6b) is essentially set to $\mathbf{0}$, while that of (9.6a) is only kept small (at the level of the truncation error). The relative importance of these residuals changes when index reduction is applied prior to discretization.

The situation is similar for an ODE with an invariant (9.38). Once a particular formulation is discretized, greater importance is placed on the constraints than on the ODE, in the sense described above.

Satisfying the constraints (including the hidden ones) exactly is in some instances precisely what one wants, and in most other cases it provides a helpful (e.g., stabilizing), or at least harmless, emphasis. State space methods (Section 9.2.3) tacitly assume that this constraint satisfaction is indeed desired, and they provide no alternative for when this is not the case.

Yet, there are also instances where such an emphasis is at odds with the natural flow of the ODE. In such cases one may be better off not insisting on satisfying constraints too accurately. Such examples arise when we apply the method of lines (Example 1.3) for a partial differential equation (PDE), allowing the spatial mesh points to be functions of time and attempting to move them as the integration in time proceeds as part of the solution process in order to meet some error equidistribution criteria which are formulated as algebraic equations (this is called a *moving mesh* method). The emphasis may then be wrongly placed, because obtaining an accurate solution to the PDE is more important than satisfying a precise mesh distribution criterion. One is better off using the DAE to devise other, clever moving mesh schemes, instead of solving it directly. Rather than dwelling on this further, we give another such example.

Example 9.11

Recall from Section 2.5 that the Hamiltonian[45] $e(\mathbf{q}, \mathbf{v})$ is constant in a Hamiltonian system given by

$$\begin{aligned} \mathbf{q}' &= \nabla_{\mathbf{v}} e, \\ \mathbf{v}' &= -\nabla_{\mathbf{q}} e, \end{aligned}$$

[45]In this chapter and the next we use e rather than H to denote the Hamiltonian, to avoid a notational clash with $H = \mathbf{h_x}$.

where $e(\mathbf{q}, \mathbf{v})$ does not depend explicitly on time t. So, this ODE system has the invariant

$$e(\mathbf{q}(t), \mathbf{v}(t)) - e(\mathbf{q}(0), \mathbf{v}(0)) = 0 \quad \forall t.$$

The system is in the form (9.38). To enforce the preservation of the invariant (conservation of energy), we can write it as a Hessenberg index-2 DAE (9.39) with $D = H^T$. This gives

$$\begin{aligned} \mathbf{q}' &= \nabla_{\mathbf{v}} e - (\nabla_{\mathbf{q}} e) z, \\ \mathbf{v}' &= -\nabla_{\mathbf{q}} e - (\nabla_{\mathbf{v}} e) z, \\ e(\mathbf{q}, \mathbf{v}) &= e(\mathbf{q}(0), \mathbf{v}(0)) . \end{aligned}$$

Note that the DAE has one degree of freedom less than the original ODE. It is sometimes very helpful to stabilize the solution with respect to this invariant; see, e.g., Example 10.8.

But when the Hamiltonian system is highly oscillatory, e.g., in the case of Example 9.8 with $0 < \varepsilon \ll 1$, the projected DAE is poorly balanced. (Roughly, large changes in z are required to produce a noticeable effect in the ODE for $\mathbf{v}$, but they then strongly affect the ODE for $\mathbf{q}$.) The observed numerical effect is that the best direct numerical discretizations of the DAE (which are necessarily implicit) require that the step size h satisfy (at best) $h = O(\sqrt{\varepsilon})$, or else the Newton iteration does not converge. With this step-size restriction, the explicit leapfrog discretization (Exercise 4.11) of the ODE is preferred.

A complete discussion of this example is beyond the scope of this presentation. Let us simply state that there are also other reasons why imposing energy conservation during a large-step integration of highly oscillatory problems is not necessarily a good idea. ♦

Of course, we do not mean to discourage the reader from using DAE models and solvers, what with having spent a quarter of our book on them! Rather, we wish to encourage careful thought on the problem formulation, whether it is based on an ODE or a DAE model.

9.4 Notes and References

A more detailed development of the DAE theory contained in Section 9.1 can be found in the books by Brenan, Campbell, and Petzold [19], Hairer and Wanner [52], and Griepentrog and März [46]. See also the survey paper [66]. However, unlike in the previous theory chapters, the material in this one is not a strict subset of any of the references.

There is an extensive theory for linear DAEs with constant coefficients which we have chosen not to develop here. For an introduction and further references, see [19]. Be careful not to confuse constant-coefficient DAEs with more general, linear DAEs.

It is interesting that, in contrast to the situation with ODEs, theorems on existence and uniqueness of solutions of nonlinear DAEs did not appear until relatively recently. Most of these results are due to Rabier and Rheinboldt [75, 76]. The theory is based on a differential geometric approach; see also [66] and references therein.

There have been many definitions of index in the literature, most of which have been shown to be equivalent or at least closely related, for the classes of problems to which they apply. The concept which we have defined here is a refinement of the *differential index*. In [49] and [52], a related concept called the *perturbation index* was introduced, which is directly motivated by the loss of smoothness in solutions to higher-index DAEs, as discussed in Section 9.1.2. However, we chose to restrict the perturbation analysis to linear(ized) DAEs (see Exercise 9.3).

Underlying the index definition, and more generally our DAE discussion, is the assumption that any matrix function which is eventually nonsingular after certain manipulations has this property independently of t. For example, in the semi-explicit form (9.13) we have considered either the case when $\mathbf{g_z}$ is nonsingular for all t or the case when it is singular for all t. This fundamental assumption breaks down for *singular DAEs*, where this matrix becomes singular at some isolated points t. (For example, in (9.47a), consider the case where $a(t)$ varies and changes sign at some points.) The situation can become much more complex, and a variety of phenomena may occur, for nonlinear, singular DAEs. The solution may remain continuous or it may not [74, 6]. See also Exercises 10.5 and 10.16.

Some of the material covered in Section 9.2 is curiously missing from the usual DAE books. We refer for more to [2, 31, 3, 15, 93, 79]. Generalized coordinate partitioning methods were introduced in [95], and tangent plane parameterization methods were implemented in [72].

9.5 Exercises

9.1. A square matrix is said to be in (block, upper-) *Hessenberg form* if it has the sparsity structure depicted in Figure 9.4. Can you guess why "DAEs in Hessenberg form" have been endowed with this name?

9.2. Consider the two-point BVP

$$\varepsilon u'' = au' + b(t)u + q(t), \tag{9.47a}$$

$$u(0) = b_1, \quad u(1) = b_2, \tag{9.47b}$$

where $a \neq 0$ is a constant and b, q are continuous functions, all $O(1)$ in magnitude.

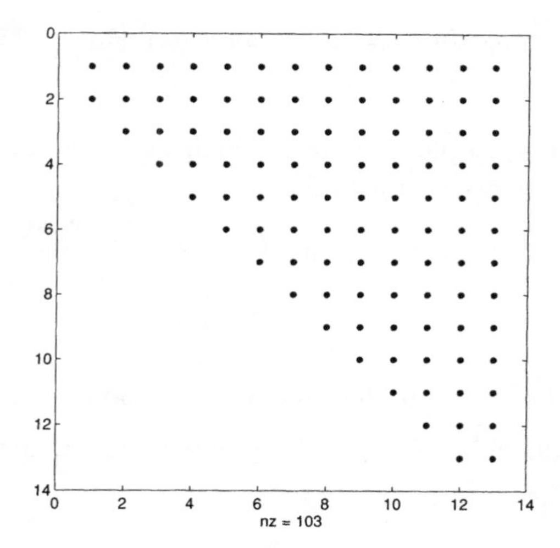

Figure 9.4: *A matrix in Hessenberg form.*

(a) Write the ODE in first-order form for the variables $y_1 = u$ and $y_2 = \varepsilon u' - au$.

(b) Letting $\varepsilon \to 0$, show that the limit system is an index-1 DAE.

(c) Show that only one of the boundary conditions in (9.47) is needed to determine the reduced solution (i.e., the solution of the DAE). Which one?

9.3. Consider the DAE [49]

$$\begin{aligned} y_1' &= y_3 y_2' - y_2 y_3', \\ 0 &= y_2, \\ 0 &= y_3 \end{aligned}$$

with $y_1(0) = 1$.

(a) Show that this DAE has index 1.

(b) Show that if we add to the right-hand side the perturbation

$$\boldsymbol{\delta}(t) = (0, \epsilon \sin \omega t, \epsilon \cos \omega t)^T,$$

which is bounded in norm by a small ϵ, the perturbed solution $\hat{\mathbf{y}}(t)$ satisfies $\hat{y}_1' = \epsilon^2 \omega$, which is unbounded as $\omega \to \infty$. The stability bound is seen to depend on $\boldsymbol{\delta}'$, as is typical for index-2 rather than index-1 problems.

(c) Show that if we add a similar perturbation to the linearization around the solution $\mathbf{y}(t)$ for $\mathbf{z} = (z_1, z_2, z_3)^T$:

$$\begin{aligned} z_1' &= z_3 y_2' + y_3 z_2' - z_2 y_3' - y_2 z_3', \\ 0 &= z_2, \\ 0 &= z_3, \end{aligned}$$

then the perturbed $\hat{\mathbf{z}}$ is bounded in terms of $\|\boldsymbol{\delta}\|$, like an index-1 DAE solution should be.

9.4. Construct an example of a DAE which for some initial conditions has index 1 and for others index 3.

9.5. Consider the IVP for the implicit ODE

$$(y')^2 = y^2, \quad y(0) = 1.$$

(a) Show that this problem has two solutions.

(b) Write down a corresponding Hessenberg index-1 DAE with two solutions.

9.6. The following equations describe a chemical reaction [69, 19]:

$$\begin{aligned} C' &= K_1(C_0 - C) - R, \\ T' &= K_1(T_0 - T) + K_2 R - K_3(T - T_C), \\ 0 &= R - K_3 e^{-K_4/T} C, \end{aligned}$$

where the unknowns are the concentration $C(t)$, the temperature $T(t)$, and the reaction rate per unit volume $R(t)$. The constants K_i and the functions C_0 and T_0 are given.

(a) Assuming that the temperature of the cooling medium $T_C(t)$ is also given, what is the index of this DAE? Is it in Hessenberg form?

(b) Assuming that $T_C(t)$ is an additional unknown, to be determined such that an additional equation specifying the desired product concentration

$$C = u$$

for a given $u(t)$ be satisfied, what is the index of this DAE? Is it in Hessenberg form?

9.7. Given a general linear DAE (9.26) with $E(t)$ decomposed as in (9.27), apply the decoupling transformation into semi-explicit form, give a condition for the DAE to have index 1, and formulate a precise stability condition.

9.8. (a) Writing the mechanical system (9.34)–(9.35) in the notation (9.38), find $H\hat{\mathbf{f}}$ and a bound on γ_0 in (9.41).

(b) Show that the velocity constraints (9.31) alone define an invariant manifold for (9.34). What are $\mathbf{h}$, H, and $H\hat{\mathbf{f}}$ then?

(c) Show that the position constraints (9.30c) alone *do not* define an invariant manifold for (9.34).

9.9. Let $r = \sqrt{x_1^2 + x_2^2}$ and consider the ODE [31]

$$\begin{aligned} x_1' &= x_2 + x_1(r^2 - 1)^{1/3} r^{-2}, \\ x_2' &= -x_1 + x_2(r^2 - 1)^{1/3} r^{-2}. \end{aligned}$$

(a) Show that

$$h(\mathbf{x}) = r^2 - 1 = 0$$

defines an invariant set for this ODE.

(b) Show that there is no finite $\gamma_0 > 0$ for which (9.41) holds.

9.10. Consider the mechanical system with holonomic constraints written as an ODE with invariant (9.34)–(9.35).

(a) Write down the equivalent Hessenberg index-2 DAE (9.39) with

$$D = \begin{pmatrix} G^T & 0 \\ 0 & G^T \end{pmatrix}.$$

(b) This D simplifies H^T of (9.37) in an obvious manner. Verify that HD is nonsingular.

(c) Show that by redefining $\boldsymbol{\lambda}$ the system you obtained can be written as

$$\begin{aligned} \mathbf{q}' &= \mathbf{v} - G^T \boldsymbol{\mu}, \\ M\mathbf{v}' &= \mathbf{f} - G^T \boldsymbol{\lambda}, \\ \mathbf{0} &= \mathbf{g}(\mathbf{q}), \\ \mathbf{0} &= G\mathbf{v}. \end{aligned}$$

This system is called the *stabilized index*-2 formulation [44].

9.11. (a) Write down the system resulting from Baumgarte's [15] stabilization (9.43) applied to the index-3 mechanical system (9.30).

(b) Consider the index-2 mechanical system given by (9.30a), (9.30b), (9.31). This is representative of nonholonomic constraints, where velocity-level constraints are not integrable into a form like (9.30c). Write down an appropriate Baumgarte stabilization

$$\mathbf{h}' + \gamma \mathbf{h} = \mathbf{0}$$

for the index-2 mechanical system and show that it is equivalent to stabilization of the invariant (9.40) with

$$F = \begin{pmatrix} 0 \\ M^{-1}G^T(GM^{-1}G^T)^{-1} \end{pmatrix}.$$

(c) However, Baumgarte's technique (9.43) for the index-3 problem is not equivalent to the stabilization (9.40). Show that the monotonicity property (9.42) does not hold here.

Chapter 10

Numerical Methods for Differential-Algebraic Equations

Numerical approaches for the solution of differential-algebraic equations (DAEs) can be divided roughly into two classes: (i) direct discretizations of the given system and (ii) methods which involve a reformulation (e.g., index reduction), combined with a discretization.

The desire for as direct a discretization as possible arises because a reformulation may be costly, it may require more input from the user, and it may involve more user intervention. The reason for the popularity of reformulation approaches is that, as it turns out, direct discretizations are essentially limited in their utility to index-1 and semi-explicit index-2 DAE systems.

Fortunately, most DAEs encountered in practical applications either are index-1 or, if higher index, can be expressed as a simple combination of Hessenberg systems. The worst-case difficulties described in Section 10.1.1 below do not occur for these classes of problems. On the other hand, the most robust direct applications of numerical ordinary differential equation (ODE) methods do not always work as well as one might hope, even for these restricted classes of problems. We will outline some of the difficulties, as well as the success stories, in Section 10.1.

We will consider two classes of problems:

- fully implicit index-1 DAEs in the general form

$$\mathbf{0} = \mathbf{F}(t, \mathbf{y}, \mathbf{y}'); \tag{10.1}$$

- index-2 DAEs in pure, or Hessenberg, form

$$\mathbf{x}' = \mathbf{f}(t, \mathbf{x}, \mathbf{z}), \tag{10.2a}$$

$$\mathbf{0} = \mathbf{g}(t, \mathbf{x}). \tag{10.2b}$$

Recall that the class of semi-explicit index-2 DAEs

$$\mathbf{x}' = \mathbf{f}(t, \mathbf{x}, \mathbf{z}), \tag{10.3a}$$

$$\mathbf{0} = \mathbf{g}(t, \mathbf{x}, \mathbf{z}) \tag{10.3b}$$

is equivalent to the class of fully implicit index-1 DAEs via the transformations (9.7) and (9.8), although the actual conversion of DAEs from one form to another may come with a price of an increased system size. For the DAE (10.3), $\mathbf{z}$ are algebraic variables which may be index-1 or index-2, whereas for the Hessenberg form the variables in $\mathbf{z}$ are all index-2 (which is why we say that the DAE is *pure* index-2).

Although there are in some cases convergence results available for numerical methods for Hessenberg DAEs of higher index, there are practical difficulties in the implementation which make it difficult to construct robust codes for such DAEs. For a DAE of index greater than 2 it is usually best to use one of the index-reduction techniques of the previous chapter to rewrite the problem in a lower-index form. The combination of this with a suitable discretization is discussed in Section 10.2.

10.1 Direct Discretization Methods

To motivate the methods in this section, consider the *regularization* of the DAE (10.3), where (10.3b) is replaced by the ODE

$$\varepsilon \mathbf{z}' = \mathbf{g}(t, \mathbf{x}, \mathbf{z}), \tag{10.4}$$

which depends on a small parameter $0 \le \varepsilon \ll 1$. Despite the promising name, we do not intend to actually carry out this regularization, unless special circumstances such as for a singular DAE (e.g., Exercise 10.16) require it, because the obtained very stiff ODE (10.3a), (10.4) is typically more cumbersome to solve than the DAE (recall, e.g., Example 9.8).[46] But this allows us to consider suitable ODE methods. Observe that:

- since the regularized ODE is very stiff, it is natural to consider methods for stiff ODEs for the direct discretization of the limit DAE;
- ODE discretizations which have stiff decay are particularly attractive: to recall (Section 3.5), any effect of an artificial initial layer which the regularization introduces can be skipped, fast ODE modes are approximated well at mesh points, and so the passage to the limit of $\varepsilon \to 0$ in (10.4) is smooth and yields a sensible discretization for the DAE.

[46] For this reason we may also assume that the regularized ODE problem is stable under given initial or boundary conditions. Were the regularized problem to be actually solved, the term $\varepsilon \mathbf{z}'$ might have to be replaced by a more general $\varepsilon B \mathbf{z}'$, where, e.g., $B = -\mathbf{g}_{\mathbf{z}}$.

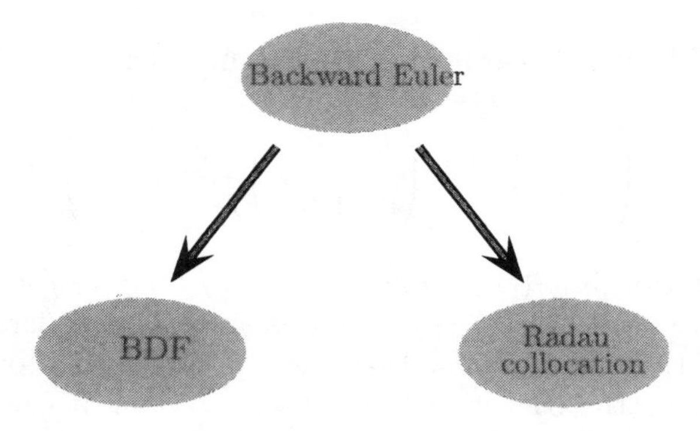

Figure 10.1: *Methods for the direct discretization of DAEs in general form.*

The rest of this section is therefore devoted to the direct application of ODE methods to low-index DAEs. All the winning methods have stiff decay, but this property alone is not sufficient. For initial value DAEs which are cumbersome to transform, and especially for DAEs whose underlying ODE is stiff, the backward differentiation formulae (BDF) and Radau collocation methods discussed in this section are the overall methods of choice. We thus begin with the simplest method of this kind, the backward Euler method, and then consider its extension to higher order via BDF or Radau methods; see Figure 10.1.

10.1.1 A Simple Method: Backward Euler

Consider the general DAE

$$\mathbf{0} = \mathbf{F}(t, \mathbf{y}, \mathbf{y}') .$$

The idea of a direct discretization is simple: approximate $\mathbf{y}$ and $\mathbf{y}'$ by a discretization formula like multistep or Runge–Kutta. Applying the backward Euler method to this DAE, we obtain

$$\mathbf{0} = \mathbf{F}\left(t_n, \mathbf{y}_n, \frac{\mathbf{y}_n - \mathbf{y}_{n-1}}{h_n}\right) . \tag{10.5}$$

This gives, in general, a system of m nonlinear equations for $\mathbf{y}_n$ at each time step n.

Unfortunately, this simple method does not always work. In the worst case, there are simple higher-index DAE systems with well-defined solutions for which the backward Euler method, and in fact all other multistep and Runge–Kutta methods, are unstable or not even applicable.

Example 10.1
Consider the following linear index-2 DAE, which depends on a parameter η,

$$\begin{pmatrix} 0 & 0 \\ 1 & \eta t \end{pmatrix} \mathbf{y}' + \begin{pmatrix} 1 & \eta t \\ 0 & 1+\eta \end{pmatrix} \mathbf{y} = \begin{pmatrix} q(t) \\ 0 \end{pmatrix}. \tag{10.6}$$

The exact solution is $y_1(t) = q(t) + \eta t q'(t)$, $y_2(t) = -q'(t)$, which is well defined for all values of η. The problem is stable for moderate values of η. Yet, if $\eta = -1$, we show below that there is no solution of the equations defining $\mathbf{y}_n$ using the backward Euler discretization. It can be shown (see Exercise 10.1) that the backward Euler method is unstable when $\eta < -0.5$.

Let us analyze this problem. To transform to semi-explicit form, define $u = y_1 + \eta t y_2$, $v = y_2$, hence

$$\mathbf{y} = \begin{pmatrix} 1 & -\eta t \\ 0 & 1 \end{pmatrix} \begin{pmatrix} u \\ v \end{pmatrix}.$$

We readily obtain

$$u' + v = 0, \quad u = q(t),$$

for which the backward Euler method gives

$$u_n = q(t_n), \quad v_n = -\frac{q(t_n) - u_{n-1}}{h}$$

(note that a *forward* Euler method makes no sense here). Thus, provided that we start with a consistent initial value for u, i.e., $u_0 = q(0)$, we have

$$v_n = -q'(t_n) + O(h),$$

which is all that one can expect from a first-order method for ODEs.

This is in marked contrast to what happens when applying backward Euler directly to (10.6),

$$\begin{pmatrix} 0 & 0 \\ 1 & \eta t_n \end{pmatrix} \frac{\mathbf{y}_n - \mathbf{y}_{n-1}}{h} + \begin{pmatrix} 1 & \eta t_n \\ 0 & 1+\eta \end{pmatrix} \mathbf{y}_n = \begin{pmatrix} q(t_n) \\ 0 \end{pmatrix}.$$

Defining

$$\begin{pmatrix} u_n \\ v_n \end{pmatrix} = \begin{pmatrix} 1 & \eta t_n \\ 0 & 1 \end{pmatrix} \mathbf{y}_n,$$

we get from this latter discretization

$$u_n = q(t_n), \quad (1+\eta) v_n = (q(t_n) - q(t_{n-1}))/h.$$

We see that, while u_n is reproduced exactly, v_n is undefined when $\eta = -1$ and has $O(1)$ error when $\eta \neq 0$.

The transformation to semi-explicit form *decouples* the solution components $\mathbf{y}$ into differential and algebraic variables. The backward Euler discretization works well for the decoupled problem. But in general, a direct discretization of nondecoupled DAEs *of index higher than* 1 is not recommended. ♦

For the remainder of Section 10.1 we thus consider only index-1 or semi-explicit index-2 DAEs.

For the simplest class of nonlinear DAEs, namely, semi-explicit index-1,

$$\mathbf{x}' = \mathbf{f}(t, \mathbf{x}, \mathbf{z}), \tag{10.7a}$$

$$\mathbf{0} = \mathbf{g}(t, \mathbf{x}, \mathbf{z}), \tag{10.7b}$$

where $\mathbf{g}_\mathbf{z}$ is nonsingular, it is easy to see that the backward Euler method retains all of its properties (i.e., order, stability, and convergence) from the ODE case. First, we recall that by the implicit function theorem, there exists a function $\tilde{\mathbf{g}}$ such that

$$\mathbf{z} = \tilde{\mathbf{g}}(t, \mathbf{x}).$$

(Let us assume, for simplicity, that there is only one such $\tilde{\mathbf{g}}$, so what is depicted in Exercise 9.5 does not happen.) Thus the DAE (10.7) is equivalent to the ODE

$$\mathbf{x}' = \mathbf{f}(t, \mathbf{x}, \tilde{\mathbf{g}}(t, \mathbf{x})). \tag{10.8}$$

Now, consider the backward Euler method applied to (10.7),

$$\frac{\mathbf{x}_n - \mathbf{x}_{n-1}}{h_n} = \mathbf{f}(t_n, \mathbf{x}_n, \mathbf{z}_n), \tag{10.9a}$$

$$\mathbf{0} = \mathbf{g}(t_n, \mathbf{x}_n, \mathbf{z}_n). \tag{10.9b}$$

Solving for $\mathbf{z}_n$ in (10.9b) and substituting into (10.9a) yields

$$\frac{\mathbf{x}_n - \mathbf{x}_{n-1}}{h_n} = \mathbf{f}(t_n, \mathbf{x}_n, \tilde{\mathbf{g}}(t_n, \mathbf{x}_n)), \tag{10.10}$$

which is just the backward Euler discretization of the underlying ODE (10.8). Hence we can conclude from the analysis for the nonstiff case in Section 3.2 that the backward Euler method is first-order accurate, stable, and convergent for semi-explicit index-1 DAEs.

For fully implicit index-1 DAEs, the convergence analysis is a bit more complicated. It is possible to show that for an index-1 DAE, there exists time- (and solution-) dependent transformation matrices in a neighborhood of the solution, which locally decouple the linearized system into differential and algebraic parts. Convergence and first-order accuracy of the method on the differential part can be shown via the techniques of Section 3.2. The backward Euler method is exact for the algebraic part. The complications arise mainly due to the time-dependence of the decoupling transformations, which enters into the stability analysis. (Recall that for fully implicit *higher-index* DAEs, time-dependent coupling between the differential and algebraic parts of the system can ruin the method's stability, as demonstrated in Example 10.1. Fortunately, for index-1 systems, it only complicates the convergences analysis; however, it may affect some stability properties of the method.) See Section 10.3.1 for pointers to further details.

The convergence result for backward Euler applied to fully implicit index-1 DAEs extends to semi-explicit index-2 DAEs in an almost trivial way. Making use of the transformation (9.8), it is easy to see that solving the index-1 system (9.8) by the backward Euler method gives exactly the same solution for $\mathbf{x}$ as solving the original semi-explicit index-2 system (10.3) by the same method. A separate argument must be made concerning the accuracy of the algebraic variables $\mathbf{z}$. For starting values which are accurate to $O(h)$, it turns out that the solution for $\mathbf{z}$ is accurate to $O(h)$, after two steps have been taken.

For nonlinear problems of the form (10.5) a Newton iteration for $\mathbf{y}_n$, starting from an approximation $\mathbf{y}_n^0$ based on information from previous steps, yields for the $(\nu+1)$st iterate,

$$\mathbf{y}_n^{\nu+1} = \mathbf{y}_n^{\nu} - \left(\frac{1}{h_n}\frac{\partial \mathbf{F}}{\partial \mathbf{y}'} + \frac{\partial \mathbf{F}}{\partial \mathbf{y}}\right)^{-1} \mathbf{F}\left(t_n, \mathbf{y}_n^{\nu}, \frac{\mathbf{y}_n^{\nu} - \mathbf{y}_{n-1}}{h_n}\right). \tag{10.11}$$

Note that, in contrast to the ODE case, the iteration matrix is not simply dominated by an $h_n^{-1}I$ term. We discuss the implication of this in Section 10.1.4.

10.1.2 BDF and General Multistep Methods

The constant step-size BDF method applied to a general nonlinear DAE of the form (10.1) is given by

$$\mathbf{F}\left(t_n, \mathbf{y}_n, \frac{1}{\beta_0 h}\sum_{j=0}^{k} \alpha_j \mathbf{y}_{n-j}\right) = \mathbf{0}, \tag{10.12}$$

where β_0 and α_j, $j = 0, 1, \ldots, k$, are the coefficients of the BDF method.

Most of the available software based on BDF methods addresses the fully implicit index-1 problem. Fortunately, many problems from applications naturally arise in this form. There exist convergence results underlying the methods used in these codes which are a straightforward extension of the results for backward Euler. In particular, the k-step BDF method of fixed step size h for $k < 7$ converges to $O(h^k)$ if all initial values are correct to $O(h^k)$ and if the Newton iteration on each step is solved to accuracy $O(h^{k+1})$. This convergence result has also been extended to variable step-size BDF methods, provided that they are implemented in such a way that the method is stable for standard ODEs. See the discussion in Section 5.5. As with backward Euler, this convergence result extends to semi-explicit index-2 DAEs via the transformation (9.8). A separate argument must be made concerning the accuracy of the algebraic variable $\mathbf{z}$. For starting values which are accurate to $O(h^k)$, it turns out that the solution for $\mathbf{z}$ is accurate to $O(h^k)$ after $k+1$ steps have been taken.

There has been much work on developing convergence results for general multistep methods. For general index-1 DAEs and for Hessenberg index-2 DAEs, the coefficients of the multistep methods must satisfy a set of order conditions which is in addition to the order conditions for ODEs, to attain order greater than 2. It turns out that these additional order conditions are satisfied by BDF methods.

You may wonder if all this additional complication is really necessary: why not simply write (10.1) as (10.3), then consider (10.3b) as the limit of (10.4)? Then apply the known theory for BDF from the ODE case!?

The answer is that there is no such a priori known convergence theory in the ODE case. The basic convergence, accuracy, and stability theory of Chapters 3, 4, and 5 applies to the case $h \to 0$, whereas here we must always consider $\varepsilon \ll h$. Indeed, since any DAE of the form (10.3) can be "treated" this way, regardless of index, we cannot expect much in general in view of the negative results in Chapter 9 for higher-index DAEs. For an ODE system (10.3a)–(10.4) whose limit is an index-2 DAE (10.3), convergence results as stated above do apply. But these results are not easier to obtain for the ODE: on the contrary, the very stiff ODE case is generally more difficult.

Example 10.2

To check the convergence and accuracy of BDF methods, consider the simple linear example,

$$\begin{aligned} x_1' &= \left(\alpha - \frac{1}{2-t}\right)x_1 + (2-t)\alpha z + \frac{3-t}{2-t}, \\ x_2' &= \frac{1-\alpha}{t-2}x_1 - x_2 + (\alpha-1)z + 2e^t, \\ 0 &= (t+2)x_1 + (t^2-4)x_2 - (t^2+t-2)e^t, \end{aligned}$$

where α is a parameter. This DAE is in pure index-2 form (10.2). For the initial conditions $x_1(0) = x_2(0) = 1$ we have the exact solution

$$x_1 = x_2 = e^t, \quad z = -\frac{e^t}{2-t}.$$

Recall that we can define $y' = z$ with some initial condition (say, $y(0) = 0$) to obtain a fully implicit index-1 DAE for $\mathbf{x} = (x_1, x_2)^T$ and y. The BDF discretization remains the same. We select $\alpha = 10$ and integrate this DAE from $t = 0$ to $t = 1$ using the first three BDF methods. In Figure 10.2 we display maximum errors in x_1 and in z for different values of h ranging

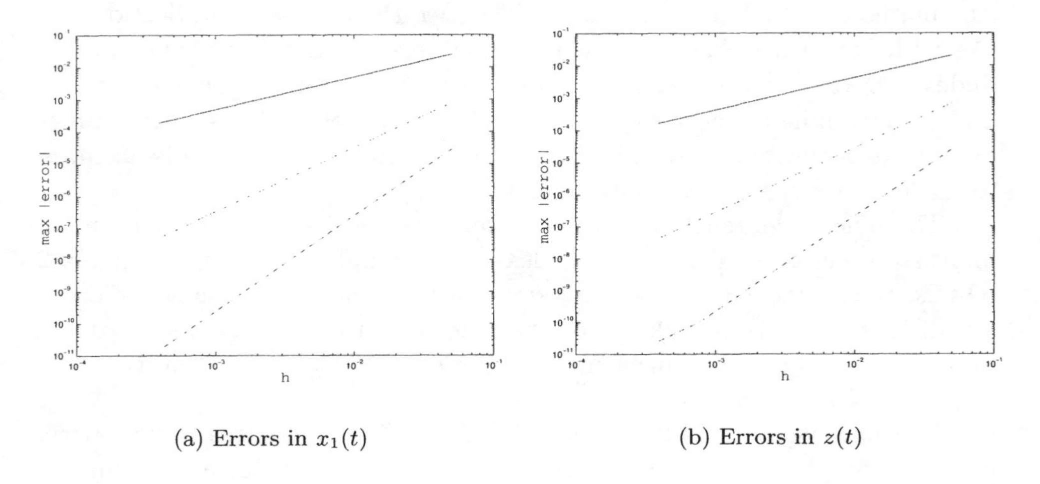

(a) Errors in $x_1(t)$

(b) Errors in $z(t)$

Figure 10.2: *Maximum errors for the first three BDF methods for Example* 10.2.

from $\frac{1}{20}$ to $\frac{1}{2560}$. We use a log-log scale, so the slopes of the curves indicate the orders of the methods. The results clearly indicate that the convergence order of the k-step BDF method is indeed k and that in absolute value the errors are pleasantly small. ♦

10.1.3 Radau Collocation and Implicit Runge–Kutta Methods

Runge–Kutta Methods

The s-stage implicit Runge–Kutta method applied to the general nonlinear DAE of the form (10.1) is defined by

$$\mathbf{0} = \mathbf{F}(t_i, \mathbf{Y}_i, \mathbf{K}_i), \tag{10.13a}$$

$$t_i = t_{n-1} + c_i h, \quad i = 1, 2, \ldots, s, \tag{10.13b}$$

$$\mathbf{Y}_i = \mathbf{y}_{n-1} + h \sum_{j=1}^{s} a_{ij} \mathbf{K}_j, \tag{10.13c}$$

and

$$\mathbf{y}_n = \mathbf{y}_{n-1} + h \sum_{i=1}^{s} b_i \mathbf{K}_i\,. \tag{10.14}$$

We assume here that the coefficient matrix $A = (a_{ij})$ is nonsingular.

For the semi-explicit problem (10.3) the formula (10.13) for the internal stages reads

$$\begin{aligned}
\mathbf{K}_i &= \mathbf{f}(t_i, \mathbf{X}_i, \mathbf{Z}_i), \\
\mathbf{X}_i &= \mathbf{x}_{n-1} + h \sum_{j=1}^{s} a_{ij} \mathbf{K}_j, \\
\mathbf{0} &= \mathbf{g}(t_i, \mathbf{X}_i, \mathbf{Z}_i).
\end{aligned}$$

For the algebraic variables $\mathbf{z}$ in this case it is often better to avoid the quadrature step implied by (10.14), because there is no corresponding integration in the DAE. This gives an advantage to stiffly accurate methods which satisfy $b_j = a_{sj}, j = 1, \ldots, s$, because for these methods the constraints are automatically satisfied at the final stage. Indeed, for such methods we have

$$\mathbf{y}_n = \mathbf{Y}_s$$

in (10.13), and (10.14) is not used. For (10.3) we then simply set $\mathbf{x}_n = \mathbf{X}_s$.

As was the case for general multistep methods, there are additional order conditions which the method coefficients must satisfy for the method to attain order greater than 2, for general index-1 and Hessenberg index-2 DAEs. For Runge–Kutta methods, there is an additional set of order conditions even for semi-explicit index-1 DAEs. We are often faced with an *order reduction*, the causes of which are closely related to the causes of order reduction for Runge–Kutta methods applied to stiff ODEs (recall Sections 4.7.3 and 8.6). This is not surprising, given the close relationship between very stiff ODEs and DAEs.

Note: It is possible to skip the remainder of this subsection, if you are interested mainly in using the methods and do not require an understanding of the causes of order reduction.

Order Reduction

To understand this order reduction, consider first the simple scalar ODE

$$\varepsilon z' = -z + q(t) \tag{10.15a}$$

and its limit DAE

$$0 = -z + q(t), \tag{10.15b}$$

to which we apply an s-stage Runge–Kutta method with a nonsingular coefficient matrix A. Using notation similar to Chapter 4, the internal stage solution values are

$$Z_i = z_{n-1} + h/\varepsilon \sum_{j=1}^{s} a_{i,j}(q(t_j) - Z_j), \quad i = 1, \ldots, s.$$

So, with $\mathbf{Z} = (Z_1, \ldots, Z_s)^T$, $\mathbf{Q} = (q(t_1), \ldots, q(t_s))^T$, we have

$$\mathbf{Z} = (\varepsilon h^{-1} I + A)^{-1}(\varepsilon h^{-1}\mathbf{1} z_{n-1} + A\mathbf{Q})$$

or

$$\mathbf{Z} = \varepsilon h^{-1} A^{-1}\mathbf{1} z_{n-1} + (I - \varepsilon h^{-1} A^{-1})\mathbf{Q} + O(\varepsilon^2 h^{-2}).$$

Letting $\varepsilon \to 0$ we get the exact DAE solution at the internal stages

$$Z_i = q(t_i), \qquad i = 1, \ldots, s. \tag{10.16}$$

At the end of the current step,

$$z_n = z_{n-1} - \frac{h}{\varepsilon}\mathbf{b}^T(\mathbf{Z} - \mathbf{Q}) = z_{n-1} - \mathbf{b}^T A^{-1}(\mathbf{1} z_{n-1} - \mathbf{Q}) + O(\varepsilon h^{-1}),$$

and for the DAE (10.15b) this gives

$$z_n = (1 - \mathbf{b}^T A^{-1}\mathbf{1}) z_{n-1} + \mathbf{b}^T A^{-1}\mathbf{Q}. \tag{10.17}$$

The recursion (10.17) for z_n converges if $|R(-\infty)| = |1 - \mathbf{b}^T A^{-1}\mathbf{1}| \le 1$, but the order of approximation of the ODE, which involves quadrature precision, may be reduced. For instance, for an s-stage collocation method, the approximate solution on the subinterval $[t_{n-1}, t_n]$ is a polynomial which interpolates $q(t)$ at the collocation points t_i of (10.13b). The local error $z_n - q(t_n)$ (assuming $z_{n-1} = q(t_{n-1})$ for a moment) is therefore strictly an interpolation error, which is $O(h^{s+1})$.

The situation is much better if the method has stiff decay, which happens when $\mathbf{b}^T$ coincides with the last row of A. In this case $c_s = 1$ necessarily, and

$$z_n = Z_s = q(t_n)$$

is exact. This can also be obtained from (10.17) upon noting that $\mathbf{b}^T A = (0, \ldots, 0, 1)$. Thus, while Gauss collocation yields a reduced local error order $O(h^{s+1})$, down from the usual order $O(h^{2s+1})$, Radau collocation yields the exact solution for (10.15b) at mesh points t_n.

Next, consider the system

$$x' = -x + q_1(t), \tag{10.18a}$$
$$\varepsilon z' = -z + x + q_2(t) \tag{10.18b}$$

and the corresponding index-1 DAE obtained with $\varepsilon = 0$. Applying the same Runge–Kutta discretization to this system and extending the notation in an obvious manner, e.g.,

$$\begin{aligned} X_i &= x_{n-1} + h\sum_{j=1}^{s} a_{i,j}(-X_j + q_1(t_j)), \\ Z_i &= z_{n-1} + \frac{h}{\varepsilon}\sum_{j=1}^{s} a_{i,j}(-Z_j + X_j + q_2(t_j)), \end{aligned}$$

we obtain for (10.18b) as $\varepsilon \to 0$,

$$
\begin{aligned}
Z_i &= X_i + q_2(t_i), \qquad i = 1, \dots, s, \\
z_n &= (1 - \mathbf{b}^T A^{-1}\mathbf{1}) z_{n-1} + \mathbf{b}^T A^{-1}(\mathbf{Q}_2 + \mathbf{X}),
\end{aligned}
$$

with an obvious extension of vector notation. Thus, the *stage accuracy* of the method, i.e., the local truncation error at each stage, enters the error in z_n unless the method has stiff decay.

A Runge–Kutta method is said to have *stage order* r if r is the minimum order of the local truncation error over all internal stages. For an s-stage collocation method, the stage order is s. For an s-stage diagonally implicit Runge–Kutta (DIRK) method, the stage order is 1. We see that for (10.18) the local error in z_n has the reduced order $r + 1$, unless the method has stiff decay. For the latter there is no reduction in order.

This result can be extended to general semi-explicit index-1 DAEs (10.3). But it does not extend to fully implicit index-1 DAEs or to higher-index DAEs. In particular, DIRK methods experience a severe order reduction for semi-explicit index-2 DAEs and hence also for fully implicit index-1 problems. This is true even for DIRK methods which have stiff decay.

The rooted tree theory of Butcher has been extended to yield a complete set of necessary and sufficient order conditions for classes of DAEs such as semi-explicit index-1, fully implicit index-1, and Hessenberg index-2 and index-3. We will not pursue this further here.

Collocation Methods

By their construction, Runge–Kutta methods which are collocation methods are not subject to such severe order limitations as DIRK methods in the DAE case. These methods were introduced in Section 4.7. For the semi-explicit DAE (10.3) we approximate $\mathbf{x}$ by a continuous piecewise polynomial $\mathbf{x}_\pi(t)$ of degree $< s + 1$ on each subinterval $[t_{n-1}, t_n]$, while $\mathbf{z}$ is approximated by a piecewise polynomial which may be discontinuous at mesh points t_n and has degree $< s$ on each subinterval (see Exercise 10.7). The convergence properties are summarized below.

Consider an s-stage collocation method of (ODE) order p, with all $c_i \neq 0$, approximating the fully implicit index-1 DAE (10.1) which has sufficiently smooth coefficients in a neighborhood of an isolated solution. Let $\gamma = 1 - \mathbf{b}^T A^{-1}\mathbf{1}$ and assume $|\gamma| \leq 1$. This method converges and the order satisfies the following conditions.

- The error in $\mathbf{y}_n$ is at least $O(h^s)$.
- If $|\gamma| < 1$ then the error in $\mathbf{y}_n$ is $O(h^{s+1})$.
- If $\gamma = -1$ and a mild mesh restriction applies then the error in $\mathbf{y}_n$ is $O(h^{s+1})$.
- If $c_s = 1$ then the error in $\mathbf{y}_n$ is $O(h^p)$.

For the semi-explicit index-2 DAE (10.3), the error results for the differential variable $\mathbf{x}$ are the same as for the index-1 system reported above. For the algebraic variable $\mathbf{z}$, the error satisfies the following conditions.

- The error in $\mathbf{z}_n$ is at least $O(h^{s-1})$.
- If $|\gamma| < 1$ then the error in $\mathbf{z}_n$ is $O(h^s)$.
- If $\gamma = -1$ and a mild mesh restriction applies then the error in $\mathbf{z}_n$ is $O(h^s)$.
- If $c_s = 1$ then the error in $\mathbf{z}_n$ is $O(h^s)$.

In particular, collocation at Radau points retains the full order $p = 2s-1$ for the differential solution components, and so this family of methods is recommended as the method of choice for general-purpose use among one-step methods for initial value DAEs and for very stiff ODEs.

Example 10.3
Figure 10.3 is a schematic depiction of a simple circuit containing linear resistors, a capacitor, voltage sources (operating voltage U_b and initial signal U_e), and two nonbipolar transistors. For the resistors and the capacitor the current relates directly to the voltage drop along the device (recall Example 9.3). For the transistors the relationship is nonlinear and is characterized by the voltage $U = U_B - U_E$ between the base and the emitter. (The third pole of the transistor is the collector C.) We use

$$\begin{aligned} I_E &= f(U) = \beta[e^{U/U_F} - 1], \\ I_C &= -\alpha I_E, \\ I_B &= (\alpha - 1)I_E, \end{aligned}$$

where $U_F = 0.026,\ \alpha = 0.99,\ \beta = 1.\text{e} - 6.$

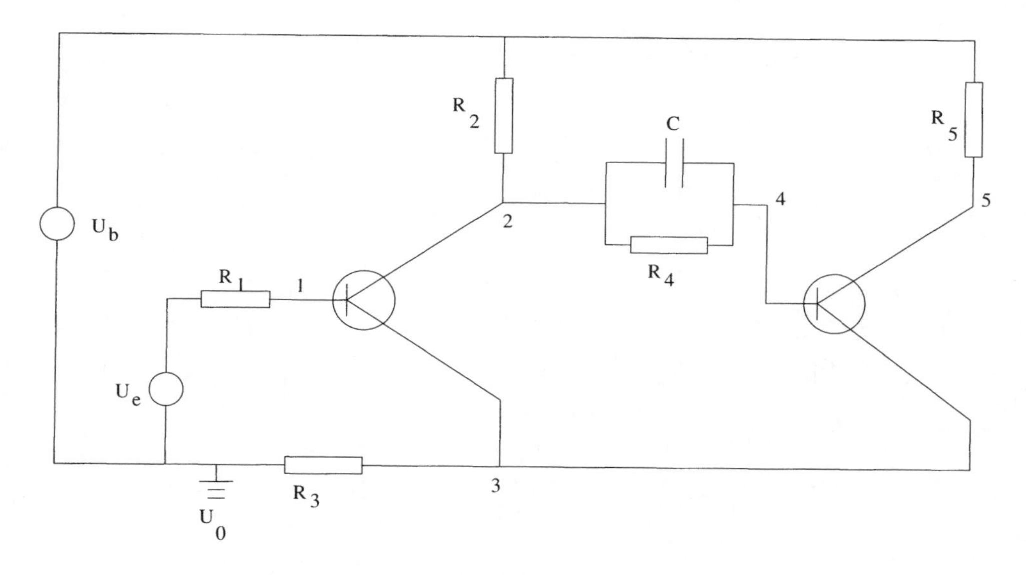

Figure 10.3: *A simple electric circuit.*

Applying Kirchoff's current law at the five nodes in sequence, we get

$$\begin{aligned}
0 &= (U_1 - U_e)/R_1 - (\alpha - 1)f(U_1 - U_3), \\
C(U_2' - U_4') &= (U_B - U_2)/R_2 + (U_4 - U_2)/R_4 - \alpha f(U_1 - U_3), \\
0 &= (U_3 - U_0)/R_3 - f(U_1 - U_3) - f(U_4 - U_3), \\
C(U_4' - U_2') &= (\alpha - 1)f(U_4 - U_3) - (U_4 - U_2)/R_4, \\
0 &= (U_4 - U_b)/R_5 + \alpha f(U_4 - U_3).
\end{aligned}$$

We use the values $U_0 = 0$ (ground voltage), $U_b = 5$, $U_e = 5\sin(2000\pi t)$, $R_1 = 200$, $R_2 = 1600$, $R_3 = 100$, $R_4 = 3200$, $R_5 = 1600$, $C = 40.\mathrm{e}-6$. (The potentials are in Volts, the resistances are in Ohms, t is in seconds.)

This is a simple index-1 DAE which, however, has scaling and sensitivity difficulties due to the exponential in the definition of f. We can obviously make it semi-explicit for the differential variable $U_2 - U_4$, but we leave the system in the fully implicit form and apply the collocation code RADAU5. Recall that this code is based on collocation at three Radau points. It applies to ODEs and DAEs of the form

$$M\mathbf{y}' = \tilde{\mathbf{f}}(t, \mathbf{y})$$

(see Section 10.3.2 and Exercise 10.15), and here we have such a form with the constant matrix

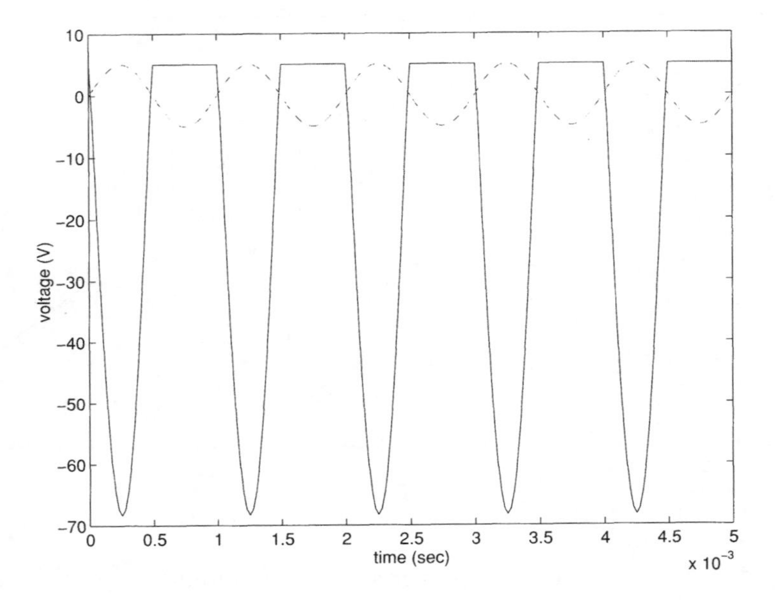

Figure 10.4: *Results for a simple electric circuit:* $U_2(t)$ *(solid line) and the input* $U_e(t)$ *(dashed line).*

$$M = \begin{pmatrix} 0 & 0 & 0 & 0 & 0 \\ 0 & C & 0 & -C & 0 \\ 0 & 0 & 0 & 0 & 0 \\ 0 & C & 0 & -C & 0 \\ 0 & 0 & 0 & 0 & 0 \end{pmatrix}.$$

For consistent initial conditions, only $U_2(0) - U_4(0)$ are free. The rest are determined by the four algebraic equations. (How four? Three are apparent; the fourth is obtained upon adding up the two equations containing derivatives, which cancels out the derivative term.) A consistent initial vector is given by

$$\mathbf{y}(0) = (0, U_b, 0, 0, U_b)^T.$$

A plot of the resulting U_2 as a function of time, as well as the input signal U_e, is given in Figure 10.4. It is seen that $U_2(t)$ has periods where it becomes large and negative. The solution is not very smooth. The code used 655 steps, of which 348 were accepted, for this simulation. The right-hand side function was evaluated almost 7000 times, but the Jacobian, only 342 times. ♦

10.1.4 Practical Difficulties

Even though there are order and convergence results for the backward Euler method (as well as for BDF and collocation at Radau points) applied to fully implicit index-1 and semi-explicit index-2 DAEs, some practical difficulties persist. Fortunately, they are not insurmountable.

Obtaining a Consistent Set of Initial Conditions

A major difference in practice between the numerical solution of ODEs and DAEs is that the solution of a DAE system must be started with a *consistent set of initial conditions.* Recall from Section 9.1 that this means that the constraints, and possibly some hidden constraints, must be satisfied at the initial point.

There are two basic types of initialization problems: when there is not enough information for a general-purpose code, and when there is too much information, or not the correct type of information, for the DAE to have a solution. To understand the first of these better, consider the simplest instance of a semi-explicit, index-1 DAE (10.7). Suppose that

$$\mathbf{x}(0) = \mathbf{x}_0$$

is provided. This is precisely the information needed to specify a solution trajectory for this problem. For an ODE, i.e., (10.7) without $\mathbf{z}$ and $\mathbf{g}$, we can use the differential equation to obtain $\mathbf{x}'(0)$ as well (denote this by $\mathbf{x}'_0$). This information is used by a general-purpose code to obtain an accurate initial guess for the Newton iteration and/or a reliable error estimate for the first time step. A general-purpose DAE solver may require[47] the value of $\mathbf{z}_0$. This is for three reasons: to completely specify the solution at $t = 0$, to provide an initial guess for the variant of Newton's iteration used to find $\mathbf{z}_1$, and to compute $\mathbf{x}'_0$ from

$$\mathbf{x}'_0 = \mathbf{f}(0, \mathbf{x}_0, \mathbf{z}_0).$$

The solution process for $\mathbf{z}_0$ consists in this case of solving the nonlinear equations

$$\mathbf{0} = \mathbf{g}(0, \mathbf{x}_0, \mathbf{z}_0),$$

given $\mathbf{x}_0$. Unlike in later steps, where we have $\mathbf{z}_{n-1}$ to guess $\mathbf{z}_n$ initially, here we must face a "cold start." This can be done with an off-the-shelf nonlinear equation solver. Also, some of the newer BDF software offers this as an initialization option. The implementation requires little in the way of additional information from the user, and can exploit structure in the iteration matrix, making use of the same linear algebra methods which are used in subsequent time-stepping.

Note that the above does not address the question of finding *all* initial values $\mathbf{z}_0$, in case there is more than one isolated solution for these nonlinear algebraic equations. An extension of this procedure is given in Exercise 10.6. Another consistent initialization problem, that of finding initial values of the

[47]Note that $\mathbf{z}_0$ is not needed for the exact solution. Moreover, this value is never used in a simple calculation like for Example 10.2.

solution variables such that the system starts in a steady state, is present already for ODEs and discussed in Exercise 5.8.

Consistent initialization of more general index-1 DAEs involves more difficulties, because the differential and the algebraic variables are not separated. Thus, information that should be determined internally by the system may be specified externally (i.e., in addition to the DAE system).

Example 10.4
For the semi-explicit index-1 DAE

$$\begin{aligned} u' &= -(u+v)/2 + q_1(t), \\ 0 &= (u-v)/2 - q_2(t), \end{aligned}$$

it is clear that a prescribed u_0 determines $v_0 = u_0 - 2q_2(0)$, and then $u_0' = -(u_0 + v_0)/2 + q_1(0)$.

But now, let $u = y_1 + y_2$, $v = y_1 - y_2$. This yields the DAE

$$\begin{aligned} y_1' + y_2' + y_1 &= q_1(t), \\ y_2 &= q_2(t). \end{aligned}$$

To get an isolated solution to the DAE, we need to specify $y_1(0) + y_2(0)$. But we cannot specify $y_1(0)$ and $y_2(0)$ arbitrarily, because $y_2(0) = q_2(0)$ is already determined. Specifying $y_1(0)$, it is not possible to solve directly for the remaining initial values as we did in the semi-explicit case. Instead, we can only find $y_1'(0) + y_2'(0) = q_1(0) - y_1(0)$. To find $y_1'(0)$ and $y_2'(0)$ individually, we need also the information from the derivative of the constraint, namely, $y_2'(t_0) = q_2'(t_0)$. ♦

The situation gets more complex, of course, for higher-index problems. Recall that consistent initial conditions for higher-index systems must satisfy the hidden constraints which are derivatives of the original constraints.

Example 10.5
Consider once more the simple pendulum in Cartesian coordinates. The equations (9.21) for this index-3 Hessenberg DAE are given in Example 9.7.

Note at first that $\mathbf{q}(0)$ cannot be specified arbitrarily: given, e.g., $q_1(0)$, the value of $q_2(0) = \pm\sqrt{1 - q_1(0)^2}$ is determined up to a sign. Then, from the hidden constraint (9.22) the specification of one of the components of $\mathbf{v}(0)$ also determines the other. In other words, the user's specification of $\mathbf{q}(0)$ and $\mathbf{v}(0)$ must satisfy the constraints (9.21e) and (9.22).

This then determines $\mathbf{q}'(0)$ by (9.21a)–(9.21b) and $\lambda(0)$ according to (9.23). Finally, $\mathbf{v}'(0)$ is determined by (9.21c)–(9.21d), although this may be considered less necessary. ♦

To make this task easier for non-Hessenberg DAEs (especially in large applications such as electric circuits; see Example 9.3), methods and software are available which use graph-theoretic algorithms to determine the minimal set of equations to be differentiated in order to solve for the consistent initial values. Initialization for general index-1 systems and for higher-index systems is often handled on a case-by-case basis.

Ill-Conditioning of Iteration Matrix

Another difficulty, which shows up already in the solution of index-1 DAEs but is more serious for index-2 systems, concerns the linear system to be solved at each Newton iteration. For explicit ODEs, as $h_n \to 0$ the iteration matrix tends to the identity.[48] For index-1 and Hessenberg DAEs, the condition number of the iteration matrix is $O(h_n^{-p})$, where p is the index. To illustrate, consider the backward Euler method applied to the semi-explicit index-1 DAE (10.7). The iteration matrix is

$$\begin{pmatrix} h_n^{-1}I - \mathbf{f_x} & -\mathbf{f_z} \\ -\mathbf{g_x} & -\mathbf{g_z} \end{pmatrix}.$$

It is easy to see that the condition number of this matrix is $O(h_n^{-1})$. For small h_n, this can lead to failure of the Newton iteration. However, scaling can improve the situation. In this case, multiplying the constraints by h_n^{-1} yields an iteration matrix whose condition number no longer depends on h_n^{-1} in this way. For Hessenberg index-2 systems the conditioning problem can be partially fixed by scaling of both the algebraic variables and the constraints (see Exercise 9.2).

Error Estimation for Index-2 DAEs

Recall that in modern BDF codes, the errors at each step are estimated via a weighted norm of a divided difference of the solution variables. For ODEs, this norm is taken over all the solution variables. This type of error estimate still works for fully implicit index-1 DAEs, but it is not appropriate for index-2 problems, as illustrated by the following example.

Example 10.6
Consider the simple index-2 DAE

$$\begin{aligned} y_1 &= q(t), \\ y_2 &= y_1', \end{aligned}$$

[48]Of course, for very stiff ODEs the term $h_n^{-1}I$ which appears in the iteration matrix does not help much, because there are larger terms which dominate. The situation for a very stiff ODE is similar to that of the limit DAE.

solved by the backward Euler method to give

$$\begin{aligned} y_{1,n} &= q(t_n), \\ y_{2,n} &= \frac{y_{1,n} - y_{1,n-1}}{h_n} = \frac{q(t_n) - q(t_{n-1})}{h_n}. \end{aligned}$$

The truncation error is estimated via the second divided difference of the numerical solution which, for the algebraic variable y_2, yields

$$\begin{aligned} \text{EST} &= h_n(h_n + h_{n-1})[y_{2,n}, y_{2,n-1}, y_{2,n-2}] \\ &= h_n(h_n + h_{n-1}) \left(\frac{\frac{y_{2,n}-y_{2,n-1}}{h_n} - \frac{y_{2,n-1}-y_{2,n-2}}{h_{n-1}}}{h_n + h_{n-1}} \right) \\ &= h_n \left(\frac{\frac{q(t_n)-q(t_{n-1})}{h_n} - \frac{q(t_{n-1})-q(t_{n-2})}{h_{n-1}}}{h_n} \right. \\ &\qquad \left. - \frac{\frac{q(t_{n-1})-q(t_{n-2})}{h_{n-1}} - \frac{q(t_{n-2})-q(t_{n-3})}{h_{n-2}}}{h_{n-1}} \right). \end{aligned} \tag{10.19}$$

For an ODE, or even for the differential variable y_1 of this example, EST $\to 0$ as $h_n \to 0$ (all previous step sizes are fixed). However, (10.19) yields for the error estimate of the algebraic variable

$$\begin{aligned} \lim_{h_n \to 0} \text{EST} &= \lim_{h_n \to 0} \frac{q(t_n) - q(t_{n-1})}{h_n} - \frac{q(t_{n-1}) - q(t_{n-2})}{h_{n-1}} \\ &= q'(t_{n-1}) - \frac{q(t_{n-1}) - q(t_{n-2})}{h_{n-1}}, \end{aligned}$$

which in general is nonzero. Thus, the error estimate for this variable cannot be decreased to zero by reducing the step size. This can lead to repeated error test failures. The approximation to y_2 is actually much more accurate than the error estimate suggests. ♦

The problem can be fixed by eliminating the algebraic variables (and in particular the index-2 variables) from the error test. In fact, it has been shown that this strategy is safe in the sense that it does not sacrifice the accuracy of the lower-index (or differential) variables, which control the time-evolution of the system. We note that the algebraic variables should not be removed from the Newton convergence test.

Given the difficulties encountered for direct DAE discretization methods and our recommendation not to apply such methods for DAEs beyond semi-explicit index-2, we must also emphasize again that, on the other hand, such direct discretization methods are important in practice. Index reduction may be necessary at times, but it is often not a desirable medicine! One class of applications where this is important is in large circuit simulation, as discussed in Example 9.3. We saw a small instance in Example 10.3. Other such examples often arise in chemical engineering and in a variety of applications involving the method of lines. For large problems, which arise routinely in practice, a conversion to explicit ODE form can be a disaster if as a result the sparsity structure of the matrices involved is lost.

10.1.5 Specialized Runge–Kutta Methods for Hessenberg Index-2 DAEs

The methods discussed in this section apply to Hessenberg index-2 problems (10.2) and not to the more general form of (10.3). The structure of the pure index-2 system is exploited to achieve gains which are not possible for the perturbed ODE (10.4).

Projected Runge–Kutta Methods

As we have seen in Chapter 8, one-sided formulas like Radau collocation without upwinding are not well suited for the solution of general boundary value problems (BVPs). Since a stable BVP can have solution modes which decrease rapidly in both directions, a symmetric method is preferred, or else such modes must be explicitly decoupled. The Gauss collocation methods have been particularly successful for the solution of ODE BVPs. However, these methods do not have stiff decay, and when implemented in a straightforward manner as described in Section 10.1.3, they suffer a severe order reduction for Hessenberg index-2 DAEs. In general, the midpoint method is accurate only to $O(1)$ for the index-2 variable $\mathbf{z}$ in (10.2). There are additional difficulties for these methods applied to Hessenberg index-2 DAEs, including potential instability and the lack of a nice, local error expansion. Fortunately, all of these problems can be eliminated by altering the method to include a projection onto the constraint manifold at the end of each step. Thus, not only $\mathbf{z}_\pi(t)$ but also $\mathbf{x}_\pi(t)$, the piecewise polynomial approximating $\mathbf{x}(t)$, may become discontinuous at points t_n (see Exercises 10.8 and 10.9).

Let $\mathbf{x}_n$, $\mathbf{z}_n$ be the result of one step, starting from $\mathbf{x}_{n-1}, \mathbf{z}_{n-1}$, of an implicit Runge–Kutta method (10.13) applied to the Hessenberg index-2 DAE (10.2). Rather than accepting $\mathbf{x}_n$ as the starting value for the next step, the *projected Runge–Kutta* method modifies $\mathbf{x}_n$ at the end of each step so as to satisfy

$$\begin{aligned} \hat{\mathbf{x}}_n &= \mathbf{x}_n + \mathbf{f}_{\mathbf{z}}(t_n, \mathbf{x}_n, \mathbf{z}_n)\boldsymbol{\lambda}_n, && (10.20a) \\ \mathbf{0} &= \mathbf{g}(t_n, \hat{\mathbf{x}}_n). && (10.20b) \end{aligned}$$

(The extra variables $\boldsymbol{\lambda}_n$ are needed for the projection only. They are not saved.) Then set $\mathbf{x}_n \leftarrow \hat{\mathbf{x}}_n$ and advance to the next step.

Note that for a method with stiff decay, (10.20b) is already satisfied by $\mathbf{x}_n$, so there is no need to project. For collocation, the projection gives the methods essentially the same advantages that Radau collocation has without the extra projection. In particular, *projected collocation methods* achieve superconvergence order for $\mathbf{x}$ at the mesh points. The solution for $\mathbf{z}$ can be determined from the solution for $\mathbf{x}$, and to the same order of accuracy, via a postprocessing step.

Projected collocation at Gauss points has order $2s$ and is useful for boundary value DAEs.

Half-Explicit Runge–Kutta Methods

For many applications, a fully implicit discretization method is not warranted. For example, many mechanical systems are essentially nonstiff and can, with the exception of the constraints, be handled via explicit methods. One way to accommodate this is via *half-explicit Runge–Kutta methods.* The methods obtained share many attributes with the methods to be described in the next section.

The half-explicit Runge–Kutta method is defined, for a semi-explicit DAE (10.3), by

$$\begin{aligned}
\mathbf{X}_i &= \mathbf{x}_{n-1} + h\sum_{j=1}^{i-1} a_{ij}\mathbf{f}(t_j, \mathbf{X}_j, \mathbf{Z}_j), \\
\mathbf{0} &= \mathbf{g}(t_i, \mathbf{X}_i, \mathbf{Z}_i), \quad i = 1, \dots, s, \\
\mathbf{x}_n &= \mathbf{x}_{n-1} + h\sum_{i=1}^{s} b_i\mathbf{f}(t_i, \mathbf{X}_i, \mathbf{Z}_i), \\
\mathbf{0} &= \mathbf{g}(t_n, \mathbf{x}_n, \mathbf{z}_n).
\end{aligned} \tag{10.21}$$

Thus, at each stage i, $\mathbf{X}_i$ is evaluated explicitly and a smaller nonlinear system is solved for $\mathbf{Z}_i$.

For semi-explicit index-1 DAEs, the order of accuracy is the same as for ODEs. In fact, the method is not very different from the corresponding explicit Runge–Kutta method applied to the ODE $\mathbf{x}' = \mathbf{f}(t, \mathbf{x}, \mathbf{z}(\mathbf{x}))$. For semi-explicit index-2 systems in Hessenberg form, there is generally order reduction, but higher-order methods of this type have been developed.

10.2 Methods for ODEs on Manifolds

The numerical solution of differential systems where the solution lies on a manifold defined explicitly by algebraic equations is a topic of interest in its own right. It also provides a useful approach for solving DAEs.

As in Section 9.2.2, consider the nonlinear differential system

$$\mathbf{x}' = \hat{\mathbf{f}}(\mathbf{x}) \tag{10.22a}$$

and assume for simplicity that for each initial value vector $\mathbf{x}(0) = \mathbf{x}_0$, there is a unique $\mathbf{x}(t)$ satisfying (10.22a). Suppose in addition that there is an *invariant set* $\mathcal{M}$ defined by the algebraic equations

$$\mathbf{0} = \mathbf{h}(\mathbf{x}) \tag{10.22b}$$

such that if $\mathbf{h}(\mathbf{x}_0) = \mathbf{0}$ then $\mathbf{h}(\mathbf{x}(t)) = \mathbf{0}$ for all t. There are various approaches possible for the numerical solution of (10.22).

1. Solve the stabilized ODE (9.40) numerically, using one of the discretization methods described in earlier chapters. The question of choosing the stabilization parameter γ arises. As it turns out, the best choice of γ typically depends on the step size, though; see Exercise 10.11 and Example 10.7.

2. Rather than discretizing (9.40), it turns out to be cheaper and more effective to stabilize the discrete dynamical system, i.e., to apply the stabilization at the end of each step. Thus, an ODE method is applied at each step to (10.22a). This step is followed by a *poststabilization* or a *coordinate projection* step to bring the numerical solution closer to satisfying (10.22b), not unlike the projected Runge–Kutta methods of the previous section.

3. The "automatic" approach attempts to find a discretization for (10.22a) which automatically satisfies the equations (10.22b) as well. This is possible when the constraints are at most quadratic; see Exercises 4.15–4.16.

Of these approaches, we now concentrate on poststabilization and coordinate projection.

10.2.1 Stabilization of the Discrete Dynamical System

If the ODE is not stiff then it is desirable to use an explicit discretization method, but to apply stabilization at the end of the step. This is reminiscent of half-explicit Runge–Kutta methods (10.21). Suppose we use a one-step method of order p with a step size h for the given ODE (without a stabilization term). Thus, if at time t_{n-1} the approximate solution is $\mathbf{x}_{n-1}$, application of the method gives

$$\tilde{\mathbf{x}}_n = \boldsymbol{\phi}_h^f(\mathbf{x}_{n-1})$$

as the approximate solution at t_n (e.g., forward Euler: $\boldsymbol{\phi}_h^f(\mathbf{x}_{n-1}) = \mathbf{x}_{n-1} + h\hat{\mathbf{f}}(\mathbf{x}_{n-1})$).

The *poststabilization* approach modifies $\tilde{\mathbf{x}}_n$ at the end of the time step to produce $\mathbf{x}_n$, which better approximates the invariant's equations:

$$\tilde{\mathbf{x}}_n = \boldsymbol{\phi}_h^f(\mathbf{x}_{n-1}), \tag{10.23a}$$

$$\mathbf{x}_n = \tilde{\mathbf{x}}_n - F(\tilde{\mathbf{x}}_n)\mathbf{h}(\tilde{\mathbf{x}}_n). \tag{10.23b}$$

The stabilization matrix function F was mentioned already in (9.40), and its selection is further discussed in Section 10.2.2.

Example 10.7
For the scalar ODE with invariant

$$\begin{aligned} x' &= \psi'(t), \\ 0 &= x - \psi(t) \end{aligned}$$

with $x(0) = \psi(0)$, where ψ is a given, sufficiently differentiable function, the exact solution is $x = \psi(t)$.

The poststabilization procedure based, e.g., on forward Euler,

$$\begin{aligned} \tilde{x}_n &= x_{n-1} + h\psi'(t_{n-1}), \\ x_n &= \tilde{x}_n - (\tilde{x}_n - \psi(t_n)), \end{aligned}$$

produces the exact solution for this simple example.

Consider, on the other hand, the stabilization (9.40). Here it gives the stabilized differential equation

$$x' = \psi'(t) - \gamma(x - \psi(t)) .$$

For $\gamma = 0$ the invariant is stable but not asymptotically stable, while for $\gamma > 0$ $\mathcal{M}$ is asymptotically stable, with the monotonicity property (9.42) holding.

But this asymptotic stability does not necessarily guarantee a vanishing drift: consider forward Euler with step size h applied to the stabilized ODE

$$x_n = x_{n-1} + h[\psi'(t_{n-1}) - \gamma(x_{n-1} - \psi(t_{n-1}))] .$$

The best choice for γ is the one which yields no error accumulation. This is obtained for $\gamma = 1/h$, giving

$$x_n = \psi(t_{n-1}) + h\psi'(t_{n-1}) .$$

(Note that this γ depends on the discretization step size.) So, the drift

$$z_n - \psi(t_n) = -\frac{h^2}{2}\psi''(t_{n-1}) + O(h^3) ,$$

although second order in h, may not decrease and may even grow arbitrarily with h fixed, if ψ'' grows. Such is the case, for instance, for $\psi(t) = \sin t^2$ as t grows. ♦

For the poststabilization to be effective, we must design F such that

$$\|I - HF\| \leq \rho < 1, \tag{10.24}$$

where $H = \mathbf{h_x}$. It has been shown, assuming (i) sufficient smoothness near the manifold $\mathcal{M}$ and (ii) that either $\rho = O(h)$ or (9.42) holds, that for an ODE method of (nonstiff) order p:

- the global error satisfies

$$\mathbf{x}_n - \mathbf{x}(t_n) = O(h^p) \tag{10.25}$$

(i.e., the stabilization does not change the method's global order, in general);

- there is a constant K that depends only on the local solution properties such that

$$|\mathbf{h}(\mathbf{x}_n)| \le K(\rho h^{p+1} + h^{2(p+1)})\,; \tag{10.26}$$

- if $HF = I$ then

$$|\mathbf{h}(\mathbf{x}_n)| = O(h^{2(p+1)})\,. \tag{10.27}$$

Example 10.8

Recall the modified Kepler problem of Exercise 4.19 (with the notation e in place of H for the Hamiltonian there). For the unmodified problem, $\alpha = 0$ and the solution has period 2π. Thus, the error in the solution can be simply measured at integer multiples of 2π. In Table 10.1 we record results using poststabilization with $F = H^T(HH^T)^{-1}$ (denoted "pstab") and also using an explicit second-order Runge–Kutta method with and without poststabilization (denoted "pstab-eRK" and "eRK," respectively), and the projected midpoint method of Section 10.1.5 and Exercise 10.9 ("proj-midpt") applied to the projected invariant formulation (9.39). All runs are with uniform time steps h and $\beta = 0.6$. Note that the projected midpoint method has better stability properties and preserves the invariant, but the symmetry of the original ODE is lost.

We observe the second-order accuracy of all methods considered and the invariant's accuracy order $2(p+1) = 6$ of the poststabilization methods. The stabilization methods improve the constant of the global error, compared to their unstabilized counterparts, but not the order. The cheapest method here, for the given range of time integration and relative to the quality of the results, is the poststabilized explicit method. The projected midpoint method is more expensive than the rest and is not worth its price, despite being most accurate for a given step size.

Note that the midpoint method loses all significant digits for $h = .01\pi$ before reaching $t = 50\pi$. The pointwise error does not explode, however, but remains $O(1)$. Also, the error in the Hamiltonian remains the same, depending only on the step size h, not on the interval length. Calculations with the poststabilized midpoint method up to $t = 2000\pi$ yield similar conclusions regarding the invariant's error for it as well (but not for the poststabilized explicit Runge–Kutta method, where a smaller step size is found necessary). ♦

Method	h	$\|q_2(2\pi)\|$	$\|q_2(4\pi)\|$	$\|q_2(20\pi)\|$	$\|q_2(50\pi)\|$	$\\|\mathbf{h}\\|_\infty$
midpt	$.01\pi$	.16	.30	.72	.10	.42e-2
eRK	$.01\pi$	.12	.18	.67	.52	.36e-1
pstab-midpt	$.01\pi$	.54e-2	.11e-1	.54e-1	.13	.81e-7
pstab-eRK	$.01\pi$	.40e-2	.81e-2	.40e-1	.10	.15e-6
proj-midpt	$.01\pi$	14e-2	.28e-2	.14e-1	.34e-1	0
midpt	$.001\pi$	.16e-2	.32e-2	.16e-1	.40e-1	.42e-4
eRK	$.001\pi$	.15e-2	.29e-2	.12e-1	.20e-1	.41e-4
pstab-midpt	$.001\pi$	.54e-4	.11e-3	.54e-3	.14e-2	.83e-13
pstab-eRK	$.001\pi$	.40e-4	.81e-4	.40e-3	.10-2	.86e-13
proj-midpt	$.001\pi$	.14e-4	.29e-4	.14e-3	.36e-3	0

Table 10.1: *Errors for Kepler's problem using various second-order methods.*

A closely related stabilization method is the *coordinate projection method.* Here, following the same unstabilized ODE integration step as before

$$\tilde{\mathbf{x}}_n = \phi_h^f(\mathbf{x}_{n-1}),$$

we determine $\mathbf{x}_n$ as the minimizer of $|\mathbf{x}_n - \tilde{\mathbf{x}}_n|_2$ such that

$$\mathbf{0} = \mathbf{h}(\mathbf{x}_n)\,.$$

There is a constrained least squares minimization problem to be solved for $\mathbf{x}_n$ at each step n. As it turns out, the poststabilization method (10.23) with $F = H^T(HH^T)^{-1}$ (for which obviously $HF = I$) coincides with one Newton step for this local minimization problem.[49] An analogue of the relationship between these two stabilization methods would be using a PECE version of a predictor-corrector as in Section 5.4.2, compared to iterating the corrector to convergence using functional iteration. In particular, the two methods almost coincide when the step size h is very small.

For this reason, there has been a tendency in the trade to view the two methods of poststabilization and coordinate projection as minor variants of each other. There is an advantage in efficiency for the poststabilization method, though. Note that the choice of F is more flexible for the poststabilization method, that (10.27) implies that the first Newton iteration of the coordinate projection method is already accurate to $O(h^{2(p+1)})$, and that no additional iteration at the current time step is needed for maintaining this accuracy level of the invariant in later time steps.

[49] An energy norm $|\mathbf{x}_n - \tilde{\mathbf{x}}_n|_A^2 = (\mathbf{x}_n - \tilde{\mathbf{x}}_n)^T A(\mathbf{x}_n - \tilde{\mathbf{x}}_n)$ for a positive definite matrix A can replace the 2-norm in this minimization, with a corresponding modification in F; see Exercise 10.13. The error bound (10.27) still holds for the outcome of one Newton step.

Exercise 4.17 provides another example of poststabilization (which coincides with coordinate projection) in action.

10.2.2 Choosing the Stabilization Matrix F

The smaller $\|I - HF\|$ is, the more effective the poststabilization step. The choice $F = H^T(HH^T)^{-1}$, which was used in Example 10.8 above, or more generally the choice corresponding to one Newton step of coordinate projection $F = D(HD)^{-1}$, achieves the minimum $HF = I$.

However, choices of F satisfying $HF = I$ may be expensive to apply. In particular, for the Euler–Lagrange equations (9.30), it is desirable to avoid the complicated and expensive matrix $\frac{\partial(G\mathbf{v})}{\partial \mathbf{q}}$. Such considerations are application-dependent. To demonstrate possibilities, let us continue with the important class of Euler–Lagrange equations and set

$$B = M^{-1}G^T.$$

Note that inverting (or rather, decomposing) GB is necessary already to obtain the ODE with invariant (10.22).

If we choose for the index-3 problem

$$F = \begin{pmatrix} B(GB)^{-1} & 0 \\ 0 & B(GB)^{-1} \end{pmatrix} \tag{10.28}$$

(or the sometimes better choice

$$F = \begin{pmatrix} G^T(GG^T)^{-1} & 0 \\ 0 & G^T(GG^T)^{-1} \end{pmatrix}$$

which, however, requires an additional cost), then

$$HF = \begin{pmatrix} I & 0 \\ L & I \end{pmatrix}, \qquad L = \frac{\partial(G\mathbf{v})}{\partial \mathbf{q}} B(GB)^{-1},$$

so $HF \neq I$.

Note, however, that

$$(I - HF)^2 = 0.$$

The effect of $HF = I$ can therefore be achieved by applying poststabilization with the cheap F of (10.28) *twice*. The decomposition (or "inversion") needed for evaluating F is performed once and this is frozen for further application at the same time step (possibly a few time steps).

The application to multibody systems with holonomic constraints is then as given in Algorithm 10.1. In the case of nonholonomic constraints, the DAE is index-2 and only one application of F per step is needed.

Algorithm 10.1. Poststabilization for Multibody Systems

1. *Starting with* $(\mathbf{q}_{n-1}, \mathbf{v}_{n-1})$ *at* $t = t_{n-1}$*, use a favorite ODE integration method* $\boldsymbol{\phi}_h^f$ *(e.g., Runge–Kutta or multistep) to advance the system*

$$\begin{aligned} \mathbf{q}' &= \mathbf{v}, \\ M(q)\mathbf{v}' &= \mathbf{f}(\mathbf{q}, \mathbf{v}) - G^T(q)\boldsymbol{\lambda}, \\ \mathbf{0} &= G(q)\mathbf{v}' + \frac{\partial(G\mathbf{v})}{\partial \mathbf{q}}\mathbf{v} \end{aligned}$$

by one step. Denote the resulting values at t_n *by* $(\tilde{\mathbf{q}}_n, \tilde{\mathbf{v}}_n)$.

2. *Poststabilize using* F *of* (10.28)*:*

$$\begin{pmatrix} \hat{\mathbf{q}}_n \\ \hat{\mathbf{v}}_n \end{pmatrix} = \begin{pmatrix} \tilde{\mathbf{q}}_n \\ \tilde{\mathbf{v}}_n \end{pmatrix} - F(\tilde{\mathbf{q}}_n, \tilde{\mathbf{v}}_n)\mathbf{h}(\tilde{\mathbf{q}}_n, \tilde{\mathbf{v}}_n).$$

3. *Set*

$$\begin{pmatrix} \mathbf{q}_n \\ \mathbf{v}_n \end{pmatrix} = \begin{pmatrix} \hat{\mathbf{q}}_n \\ \hat{\mathbf{v}}_n \end{pmatrix} - F(\tilde{\mathbf{q}}_n, \tilde{\mathbf{v}}_n)\mathbf{h}(\hat{\mathbf{q}}_n, \hat{\mathbf{v}}_n).$$

Example 10.9
Consider a two-link planar robotic system with a prescribed path for its end effector (the "robot's hand"). Thus, one end of a rigid rod is fixed at the origin, and the other is connected to another rigid rod with rotations allowed in the $(x - y)$-plane. Let θ_1 be the angle that the first rod makes with the horizontal axis, and let θ_2 be the angle that the second rod makes with respect to the first rod (see Figure 10.5). The masses of the rods are denoted by m_i and their lengths are denoted by l_i. The coordinates of the link between the rods are given by

$$x_1 = l_1 c_1, \qquad y_1 = l_1 s_1,$$

and those of the "free" end are

$$x_2 = x_1 + l_2 c_{12}, \qquad y_2 = y_1 + l_2 s_{12},$$

where $c_i = \cos\theta_i$, $s_i = \sin\theta_i$, $c_{12} = \cos(\theta_1 + \theta_2)$, $s_{12} = \sin(\theta_1 + \theta_2)$.

Referring to the notation of the Euler–Lagrange equations (9.30), we let

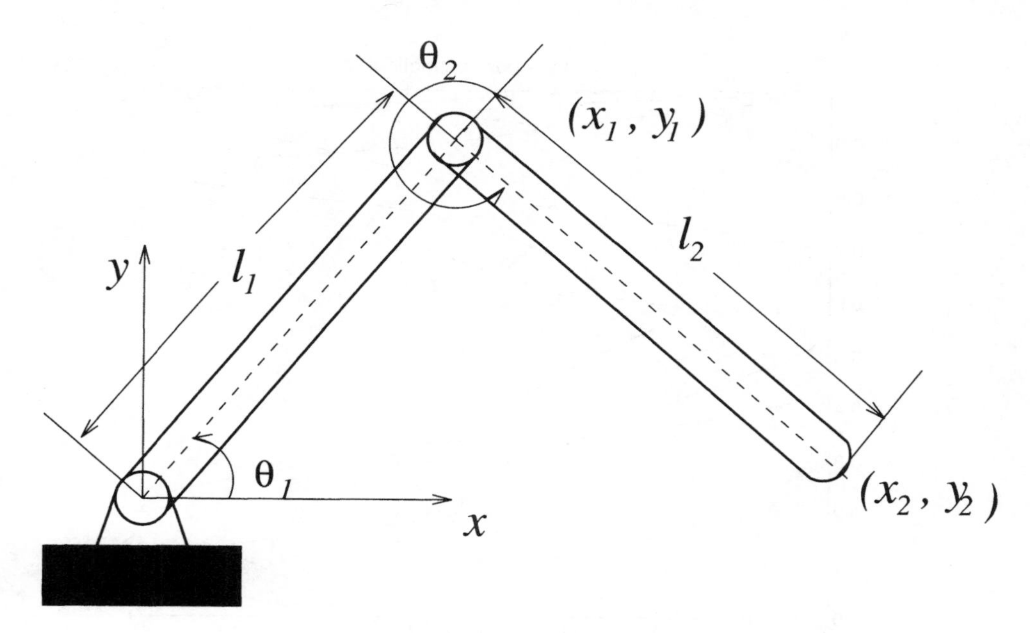

Figure 10.5: *Two-link planar robotic system.*

$\mathbf{q} = (\theta_1, \theta_2)^T$ and obtain

$$M = \begin{pmatrix} m_1 l_1^2/3 + m_2(l_1^2 + l_2^2/3 + l_1 l_2 c_2) & m_2(l_2^2/3 + l_1 l_2 c_2/2) \\ m_2(l_2^2/3 + l_1 l_2 c_2/2) & m_2 l_2^2/3 \end{pmatrix},$$

$$\mathbf{f} = \begin{pmatrix} -m_1 g l_1 c_1/2 - m_2 g(l_1 c_1 + l_2 c_{12}/2) \\ -m_2 g l_2 c_{12}/2 \end{pmatrix} + \begin{pmatrix} m_2 l_1 l_2 s_2/2(2\theta_1'\theta_2' + (\theta_2')^2) \\ -m_2 l_1 l_2 s_2 (\theta_1')^2/2 \end{pmatrix}.$$

In the following simulation we use the data

$$m_1 = m_2 = 36kg, \quad l_1 = l_2 = 1m, \quad g = 9.81m/s^2,$$
$$\theta_1(0) = 70^o, \quad \theta_2(0) = -140^o, \quad \theta_1'(0) = \theta_2'(0) = 0.$$

So far we do not have constraints $\mathbf{g}$. Indeed, for a double pendulum the equations of motion form an implicit ODE (or an index-0 DAE), because the topology of this simple mechanical system has no closed loops and we are using relative (minimal) coordinates to describe the system. But now we prescribe some path constraint on the position of (x_2, y_2), and this yields, in turn, a constraint force $G^T\lambda$ as well. We choose the constraint

$$y_2(t) = \sin^2(t/2)$$

(for y_2 expressed in terms of $\mathbf{q}$ as described above). The obtained constrained path for (x_2, y_2) is depicted in Figure 10.6. In this case the constraint forces become large at a few distinct times.

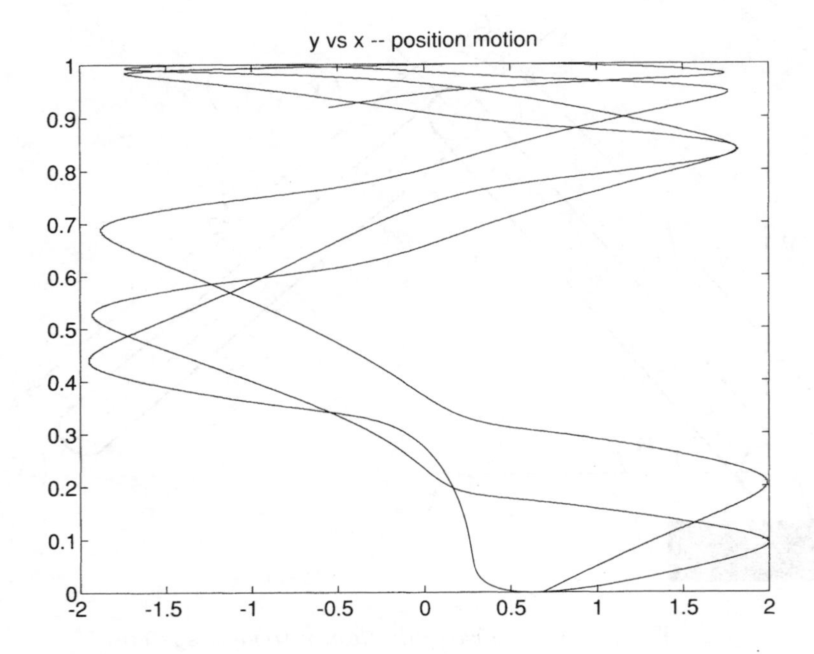

Figure 10.6: *Constraint path for* (x_2, y_2).

In Table 10.2 we record the measured drifts, i.e., the error in the path constraint ("drift-position") and in its derivative ("drift-velocity"), based on runs up to $b = 10s$ using an explicit Runge–Kutta scheme of order 2 with a constant step size h. We record results using Baumgarte's technique (9.43), denoting it "Baum(γ_1, γ_2)," and also using various choices of F for poststabilization: "S-pos" stands for stabilizing only with respect to the position constraints $\mathbf{g}(t, \mathbf{q}) = \mathbf{0}$; "S-vel" stands for stabilizing only with respect to the velocity constraints $\mathbf{g}' = \mathbf{0}$; "S-both" stands for using F of (10.28) once; "S-both^2" is the choice recommended in Algorithm 10.1; and finally "S-full" uses $F = H^T(HH^T)^{-1}$.

Note that without stabilization the computation blows up for $h = 0.01$. The Baumgarte stabilization is not as effective as the S-stabilizations, especially for the case $h = .01$. Other parameters (γ_1, γ_2) tried do not yield significantly better results. The choice of Algorithm 10.1 shows drift-convergence order $6 = 2(p + 1)$ and, given that it is much cheaper than S-full and not much more expensive than the other choices for F, we conclude that S-both^2 gives the most bang for the buck here. ♦

10.3 Software, Notes, and References

10.3.1 Notes

Of course, scientists encountered the need to numerically solve mathematical models involving differential equations with constraints and implicit differen-

h	Stabilization	Drift-velocity	Drift-position
.01	Baum(0, 0)	*	*
.01	Baum(12, 70)	.51	.25e-1
.01	S-full	.72e-5	.63e-6
.01	S-both	.31e-1	.68e-5
.01	S-both^2	.20e-3	.68e-6
.01	S-vel	.60e-14	.78e-2
.01	S-pos	.73	.28e-2
.001	Baum(0, 0)	.66e-4	.56e-4
.001	Baum(12, 70)	.60e-3	.38e-4
.001	S-full	.39e-10	.53e-11
.001	S-both	.41e-4	.46e-11
.001	S-both^2	.20e-9	.78e-15
.001	S-vel	.43e-14	.58e-4
.001	S-pos	.44e-2	.88e-10

Table 10.2: *Maximum drifts for the robot arm; * denotes an error overflow.*

tial equations many decades, if not centuries, ago. But the recognition that DAE classes are worth being considered as such, in order to methodically derive good numerical methods and software, is relatively recent.

The original idea for discretizing DAEs directly with suitable ODE methods was described in the landmark 1971 paper of Gear [42]. He used BDF methods and applied them to problems of the type discussed in Example 9.3. This was followed in the 1980s by a deluge of efforts to design and analyze numerical methods and to write general-purpose software for DAEs. The direct discretization methods described in Section 10.1 are covered in more detail in Hairer and Wanner [52] and Brenan, Campbell, and Petzold [19].

We have chosen not to discuss convergence results for numerical methods applied directly to index-3 DAEs. However, there are convergence results for some numerical methods (both BDF and Runge–Kutta) applied to Hessenberg DAEs of index greater than 2; see [45, 52].

We have noted that direct discretization methods are not applicable to general, higher-index DAEs. Campbell [28, 29] has developed least-squares-type methods for such problems which may be viewed as automatic index reduction. The methods require differentiation of the original DAE, which is accomplished by an automatic differentiation software package such as ADIFOR [16]. Using these and similar ideas, initialization schemes for general DAE systems have been constructed.

The first results on order reduction for general multistep methods applied to higher-index DAEs were given by R. März; see [46, 52] for a summary and for further references.

More details and proofs for Sections 10.2.1 and 10.2.2 can be found in Chin [31] and [2]. See also [3], which is the source for Examples 10.8 and 10.9.

Example 10.3 was taken from [58]. Other examples from various applications in the literature are formulated as exercises below.

10.3.2 Software

Excellent and widely available software exists for the solution of initial value problems (IVPs) and BVPs in DAEs. Here we briefly outline some of the available codes. With the exception of software for mechanical systems, they all apply to stiff ODEs as well (and if you read the first few pages of this chapter carefully, then you should be able to understand why this is natural).

IVPs

- The code DASSL by Petzold uses the fixed-leading-coefficient form of the BDF formulae to solve general index-1 DAEs; see [19] for details. Versions for large-scale problems (called DASPK) and for sensitivity analysis are also available.
- The code RADAU5 by Hairer and Wanner [52] is based on the three-stage Radau collocation method. It solves DAEs of the form

$$M\mathbf{y}' = \tilde{\mathbf{f}}(t, \mathbf{y}), \tag{10.29}$$

  where M is a constant, square matrix which may be singular; see Exercise 10.15. The code is applicable to problems of index-1, -2 or -3, but the user must specify which variables are higher-index (this implies in particular a special structure).
- There are many codes, both commercial and publicly available, which are designed specifically for simulating **constrained mechanical systems**. They use many of the methods mentioned here, including Baumgarte stabilization, poststabilization and coordinate projection, and various coordinate partitioning methods. The code MEXX by Lubich et al. [64] is based on a half-explicit extrapolation method which we have not covered and implements fast linear algebra techniques for tree-structured mechanical systems.

BVPs

- The code COLDAE by Ascher and Spiteri [12] uses projected Gauss collocation to extend COLNEW [13] for boundary value, semi-explicit index-2 DAEs in the form (10.3). An additional singular value decomposition decouples algebraic variables of different indexes if needed.

10.4 Exercises

10.1. Show that the implicit Euler method is unstable for the DAE (10.6) if $\eta < -0.5$.

10.2. Consider the backward Euler method applied to the Hessenberg index-2 DAE (10.2).

(a) Show that the condition number of the iteration matrix is $O(h^{-2})$.

(b) How should the equations and variables be scaled to reduce the condition number to $O(1)$?

(c) What are the implications of scaling the variables on the accuracy one can expect in these variables from the linear system solver?

10.3. Set $\varepsilon = 0$ in Example 9.8 and solve the resulting DAE numerically. You may use any (justifiable) means you like, including index reduction and use of an appropriate software package. Plot the solution and compare with Figure 9.2. Discuss.

10.4. Consider two linked bars of length l_i and mass m_i, $i = 1, 2$. One end of one bar is fixed at the origin, allowing only rotational motion in the plane (as in Figure 10.5). The other end of the other bar is constrained to slide along the x-axis.

The equations of motion form a nonlinear index-3 DAE of the form (9.30). Using redundant, absolute coordinates, let u_i, v_i, ϕ_i be the coordinates of the center of mass of the ith bar. Then define

$$\mathbf{q} = (u_1, v_1, \phi_1, u_2, v_2, \phi_2)^T,$$

$$M = \mathrm{diag}\{m_1, m_1, m_1 l_1^2/3, m_2, m_2, m_2 l_2^2/3\},$$

$$\mathbf{f} = (0, -9.81, 0, 0, -9.81, 0)^T,$$

$$\mathbf{g} = \begin{pmatrix} u_1 - l_1/2\cos\phi_1 \\ v_1 - l_1/2\sin\phi_1 \\ u_2 - 2u_1 - l_2/2\cos\phi_2 \\ v_2 - 2v_1 - l_2/2\sin\phi_2 \\ l_1\sin\phi_1 + l_2\sin\phi_2 \end{pmatrix},$$

$$G = \mathbf{g_q} = \begin{pmatrix} 1 & 0 & l_1/2\sin\phi_1 & 0 & 0 & 0 \\ 0 & 1 & -l_1/2\cos\phi_1 & 0 & 0 & 0 \\ -2 & 0 & 0 & 1 & 0 & l_2/2\sin\phi_2 \\ 0 & -2 & 0 & 0 & 1 & -l_2/2\cos\phi_2 \\ 0 & 0 & l_1\cos\phi_1 & 0 & 0 & l_2\cos\phi_2 \end{pmatrix}.$$

(a) Following the lines of Example 10.9, derive a more compact formulation of the slider-crank mechanism in relative coordinates, leading to only two ODEs and one constraint. What are the advantages and disadvantages of each formulation?

(b) Set $m_1 = m_2 = 1$, $l_1 = 1$, $\phi_1(0) = \frac{7\pi}{4}$, and $\phi_1'(0) = 0$. Compute and plot the solution for $b = 70$ and each of the two cases (i) $l_2 = 1.1$ and (ii) $l_2 = 0.9$. Your simulation method should use the formulation in absolute coordinates given above, and combine index reduction and some stabilization with an ODE solver or a lower-index DAE solver.

Explain the qualitatively different behavior observed for the different values of l_2.

10.5. This exercise continues the previous one. Set the various parameters at the same values as above, except $l_2 = l_1 = 1$. Then the last row of G vanishes, i.e., a singularity occurs, each time the periodic solution crosses a point where the two bars are upright, i.e., $(\phi_1, \phi_2) = (\frac{\pi}{2}, \frac{3\pi}{2})$.

(a) Use the same method you have used in the previous exercise to integrate this problem, despite the singularity. Explain your observed results. [What you obtain may depend on the numerical method you use and the error tolerance you prescribe, so you are on your own: make sure the program is debugged before attempting to explain the results.]

(b) Explain why a stabilization method which stabilizes only with respect to the velocity constraints $G\mathbf{q}' = \mathbf{0}$ would do significantly worse here than a method which stabilizes also with respect to the position constraints $\mathbf{g} = \mathbf{0}$. [Hint: You should have solved Exercise 10.4(b) before attempting this one.]

10.6. Consider a semi-explicit index-1 DAE of the form

$$
\begin{aligned}
\mathbf{f}(t, \mathbf{x}, \mathbf{z}, \mathbf{x}') &= \mathbf{0}, \\
\mathbf{g}(t, \mathbf{x}, \mathbf{z}) &= \mathbf{0},
\end{aligned}
$$

where the matrices $\mathbf{f}_{\mathbf{x}'}$ and $\mathbf{g}_{\mathbf{z}}$ are square and nonsingular.

(a) Show that to specify a solution trajectory the initial value information needed is $\mathbf{x}(0) = \mathbf{x}_0$.

(b) The initialization problem is to find $\mathbf{x}_0' = \mathbf{x}'(0)$ and $\mathbf{z}_0 = \mathbf{z}(0)$. Describe a solution algorithm.

10.7. Consider the index-1 DAE (10.3) and the two implicit midpoint methods

$$
\begin{aligned}
\frac{\mathbf{x}_n - \mathbf{x}_{n-1}}{h} &= \mathbf{f}\left(t_{n-1/2}, \frac{\mathbf{x}_n + \mathbf{x}_{n-1}}{2}, \frac{\mathbf{z}_n + \mathbf{z}_{n-1}}{2}\right), \qquad (10.30) \\
\mathbf{0} &= \mathbf{g}\left(t_{n-1/2}, \frac{\mathbf{x}_n + \mathbf{x}_{n-1}}{2}, \frac{\mathbf{z}_n + \mathbf{z}_{n-1}}{2}\right)
\end{aligned}
$$

and

$$\begin{aligned}
\frac{\mathbf{x}_n - \mathbf{x}_{n-1}}{h} &= \mathbf{f}\left(t_{n-1/2}, \frac{\mathbf{x}_n + \mathbf{x}_{n-1}}{2}, \mathbf{z}_{n-1/2}\right), \\
\mathbf{0} &= \mathbf{g}\left(t_{n-1/2}, \frac{\mathbf{x}_n + \mathbf{x}_{n-1}}{2}, \mathbf{z}_{n-1/2}\right).
\end{aligned} \tag{10.31}$$

In the second method $\mathbf{z}(t)$ is approximated by a constant $\mathbf{z}_{n-1/2}$ on each subinterval $[t_{n-1}, t_n]$, so the resulting approximate solution $\mathbf{z}_\pi(t)$ is discontinuous at mesh points t_n.

(a) Find an example to show that (10.31) has better stability properties than (10.30). [This may be challenging.]

(b) Design an a posteriori process, i.e., a process that starts after the solution to (10.31) has been calculated, to improve the approximate values of $\mathbf{z}$ to be second-order accurate at mesh points. Test this on your example.

10.8. Consider the Hessenberg index-2 DAE (10.2) and the midpoint method (10.31) applied to it.

(a) Show that the global error is second order on a uniform mesh (i.e., using a constant step size) but only first order on an arbitrary mesh.

(b) What is the condition on the mesh to achieve second-order accuracy?

10.9. (a) Describe the projected midpoint method based on (10.31) and show that the obtained solution $\mathbf{x}_n$ is second-order accurate.

(b) Consider the following modification of (10.31):

$$\begin{aligned}
\frac{\mathbf{x}_n - \mathbf{x}_{n-1}}{h} &= \mathbf{f}\left(t_{n-1/2}, \frac{\mathbf{x}_n + \mathbf{x}_{n-1}}{2}, \mathbf{z}_{n-1/2}\right), \\
\mathbf{0} &= \mathbf{g}(t_n, \mathbf{x}_n),
\end{aligned} \tag{10.32}$$

where $\mathbf{g}(0, \mathbf{x}_0) = \mathbf{0}$ is assumed. Investigate the properties of this method and compare it to the projected midpoint method.

[This exercise is significantly more difficult than the previous two.]

10.10. (a) Apply the midpoint method, the projected midpoint method, and the method given by (10.32) to the problem of Example 10.2 with the same data. Describe your observations.

(b) Attempt to explain your observations.

[This may prove difficult; if in distress, see [9].]

10.11. Consider the Hessenberg index-3 DAE

$$\begin{aligned} y' &= x, \\ x' &= -z + \phi(t), \\ 0 &= y - \psi(t), \end{aligned}$$

where $\phi(t), \psi(t)$ are known, smooth functions.

(a) Formulate this DAE as an ODE with invariant.

(b) Discretize the stabilized ODE (9.40) with $F = H^T(HH^T)^{-1}$ using forward Euler. What is the best choice for γ?

(c) Formulate the Baumgarte stabilization (9.43) for this simple problem and discretize it using forward Euler. Try to figure out a best choice for the parameters γ_1 and γ_2 for this case.

[This latter task should prove somewhat more difficult [3].]

10.12. Mechanical systems with holonomic constraints yield index-3 DAEs, as we have seen. Mechanical systems with nonholonomic constraints involve constraints "on the velocity level," such as $G\mathbf{v} = \mathbf{0}$, but which cannot be integrated into constraints involving generalized positions $\mathbf{q}$ alone. So, mechanical systems with nonholonomic constraints yield index-2 DAEs.

Now, every budding mechanical engineer doing robotics knows that systems with nonholonomic constraints are more complex and difficult than systems with holonomic constraints, whereas every budding numerical analyst knows that index-3 DAEs are harder than index-2 DAEs.

Who is right, the engineers or the numerical analysts? Explain.

10.13. The coordinate projection method for an ODE with invariant (10.22), using an energy norm based on a symmetric positive definite matrix A, is defined as follows.

- At time step n we use an ODE method for (10.22a) to advance from $\mathbf{x}_{n-1}$ to $\tilde{\mathbf{x}}_n$.
- Then $\mathbf{x}_n$ is determined as the solution of the constrained least squares problem

$$\begin{aligned} \min_{\mathbf{x}_n} \quad & \frac{1}{2}(\mathbf{x}_n - \tilde{\mathbf{x}}_n)^T A(\mathbf{x}_n - \tilde{\mathbf{x}}_n) \\ \text{subject to} \quad & \mathbf{h}(\mathbf{x}_n) = \mathbf{0}. \end{aligned}$$

Consider one Newton step linearizing $\mathbf{h}$ to solve this nonlinear system, starting from $\mathbf{x}_n^0 = \tilde{\mathbf{x}}_n$.

(a) Show that this step coincides with a poststabilization step (10.23) with

$$F = A^{-1}H^T(HA^{-1}H^T)^{-1}.$$

(b) Assuming that the ODE discretization method has order p, show that one Newton step brings the solution to within $O(h^{2(p+1)})$ of satisfying the constraints (10.22b).

(c) For the Euler–Lagrange equations, explain why it may be advantageous to choose $A = M$, where M is the mass matrix.

10.14. Consider the Hamiltonian system describing the motion of a "stiff reversed pendulum" [10],

$$\begin{aligned}\mathbf{q}' &= \mathbf{p},\\ \mathbf{p}' &= -(r(\mathbf{q}) - r_0)\boldsymbol{\nabla}_{\mathbf{q}} r - \varepsilon^{-2}(\phi(\mathbf{q}) - \phi_0)\boldsymbol{\nabla}_{\mathbf{q}}\phi,\end{aligned}$$

where $\mathbf{q} = (q_1, q_2)^T$, $\mathbf{p} = (p_1, p_2)^T$, $r = |\mathbf{q}|_2$, $\boldsymbol{\nabla}_{\mathbf{q}} r = r^{-1}\mathbf{q}$, $\phi = \arccos(q_1/r)$, $\boldsymbol{\nabla}_{\mathbf{q}}\phi = r^{-2}(-q_2, q_1)^T$ (all functions of t of course). Set $r_0 = 1$, $\phi_0 = \pi/4$, $\mathbf{q}(0) = \frac{1}{\sqrt{2}}(1,1)^T$, $\mathbf{p}(0) = (1,0)^T$. Also let $e_S(t) = \frac{1}{2}[(\boldsymbol{\nabla} r^T \mathbf{p})^2 + (r - r_0)^2]$ and $\Delta e_S = \max_{t\in[0,5]} |e_S(0) - e_S(t)|$.

(a) Use an initial value ODE solver to integrate this highly oscillatory IVP for $\varepsilon = 10^{-1}$, 10^{-2}, and 10^{-3}. (Also try $\varepsilon = 10^{-4}$ if this is not getting too expensive.) Record the values of Δe_S and conjecture the value of this quantity in the limit $\varepsilon \to 0$.

(b) Consider the DAE obtained as in Example 9.8,

$$\begin{aligned}\mathbf{q}' &= \mathbf{p},\\ \mathbf{p}' &= -(r(\mathbf{q}) - r_0)\boldsymbol{\nabla}_{\mathbf{q}} r - (\boldsymbol{\nabla}_{\mathbf{q}}\phi)\lambda,\\ 0 &= \phi(\mathbf{q}) - \phi_0\,.\end{aligned}$$

Solve this DAE numerically subject to the same initial conditions as above and calculate Δe_S. Compare to the conjectured limit from (a). Conclude that this DAE is *not* the correct limit DAE of the highly oscillatory ODE problem [80]!

10.15. Many ODE problems arise in practice in the form (10.29), where M is a constant, possibly singular matrix. Let

$$M = U \begin{pmatrix} \Sigma & 0 \\ 0 & 0 \end{pmatrix} V^T$$

denote the singular value decomposition of M, where U and V are orthogonal matrices and $\Sigma = \mathrm{diag}(\sigma_1, \dots, \sigma_{m-l})$, with $\sigma_1 \geq \sigma_2 \geq \dots \geq \sigma_{m-l} > 0$.

(a) Show that (10.29) can be written in the semi-explicit form (10.3), where

$$\begin{pmatrix} \mathbf{x} \\ \mathbf{z} \end{pmatrix} = V^T \mathbf{y}$$

and $\mathbf{f}$ and $\mathbf{g}$ are defined in terms of $\tilde{\mathbf{f}}$, U, V, and Σ.

(b) Show that, since the transformation V is constant and well conditioned, any Runge–Kutta or multistep discretization applied to (10.29) corresponds to an equivalent method applied to (10.3).

10.16. [Parts (c) and (d) of this exercise are difficult and time-consuming.]

The following equations describe a simple steady state, one-dimensional, unipolar hydrodynamic model for semiconductors in the isentropic case [7]:

$$\phi' = \rho E - \alpha J, \tag{10.33a}$$
$$E' = \rho - 1, \tag{10.33b}$$
$$\phi = J^2/\rho + \rho, \tag{10.33c}$$
$$\rho(0) = \rho(b) = \bar{\rho}. \tag{10.33d}$$

The constants J, α, b, and $\bar{\rho} > 1$ are given. Although you don't need to know the physical interpretation to solve the exercise, we note that $\rho(t)$ is the electron density and $E(t)$ is the (negative) electric field. The independent variable t is a space variable. This model corresponds to a current-driven $\rho^+ \rho \rho^+$ device, and we concentrate on one ρ-region.

The flow is *subsonic* where $\rho > J$ and *supersonic* where $\rho < J$. Setting $\bar{\rho} > J$, the question is if *transonic* flows are possible; i.e., is there a region $(a_\rho, b_\rho) \subset (0, b)$ where $\rho < J$?

(a) Show that for the subsonic or supersonic cases, (10.33) is an index-1, boundary value DAE.

(b) Propose a computational approach for this boundary value DAE in the strictly subsonic or the supersonic case. Solve for the values $J = 1/2, \alpha = 0, \bar{\rho} = 3, b = 10.3$. [You may either write your own program or use appropriate software.]

(c) For the transonic case, $\rho(t) - J$ crosses from positive to negative and back to positive (at least once), and the simple DAE model breaks down. The solution may be discontinuous at such crossing points. The (Rankine–Hugoniot) condition for a shock of this sort to occur at $t = t_0$ is that the jump in ϕ vanish across such a point. Using phase plane analysis, show that such transonic solutions exist for suitable parameter values!

[This part of the exercise is suitable only for those who really like the challenge of analysis.]

(d) For the numerical solution of the transonic case it seems best to abandon the DAE and regularize the problem: replace (10.33a) by

$$\phi' = \rho E - \alpha J - \varepsilon \rho''$$

(a nontrivial, physically based choice which allows ϕ to be simply evaluated, no longer one of the primary unknowns) and append the boundary condition

$$\frac{\partial}{\partial \rho}\phi(\bar{\rho}, J)\rho'(0) = \bar{\rho}E(0) - \alpha J.$$

Solve the obtained boundary value ODE for $J = 2, \alpha = 0, \bar{\rho} = 3, b = 10.3$, and $\varepsilon = 10^{-3}$. Plot the solution, experiment further with different values of b, and discuss.

[Be warned that this may be challenging: expect interior sharp layers where the DAE solution jumps. We suggest using a good software package and employing a continuation method (Section 8.4), starting with $\varepsilon = 1$ and gradually reducing it.]

Bibliography

[1] V.I. Arnold. *Mathematical Methods of Classical Mechanics.* Springer-Verlag, New York, 1978.

[2] U. Ascher. Stabilization of invariants of discretized differential systems. *Numer. Algorithms*, 14:1–24, 1997.

[3] U. Ascher, H. Chin, L. Petzold, and S. Reich. Stabilization of constrained mechanical systems with DAEs and invariant manifolds. *J. Mech. Struct. Machines*, 23:135–158, 1995.

[4] U. Ascher, H. Chin, and S. Reich. Stabilization of DAEs and invariant manifolds. *Numer. Math.*, 67:131–149, 1994.

[5] U. Ascher, J. Christiansen, and R. Russell. Collocation software for boundary value ODE's. *ACM Trans. Math Software*, 7:209–222, 1981.

[6] U. Ascher and P. Lin. Sequential regularization methods for nonlinear higher index DAEs. *SIAM J. Sci. Comput.*, 18:160–181, 1997.

[7] U. Ascher, P. Markowich, P. Pietra, and C. Schmeiser. A phase plane analysis of transonic solutions for the hydrodynamic semiconductor model. *Math. Models Methods Appl. Sci.*, 1:347–376, 1991.

[8] U. Ascher, R. Mattheij, and R. Russell. *Numerical Solution of Boundary Value Problems for Ordinary Differential Equations.* SIAM, Philadelphia, second edition, 1995.

[9] U. Ascher and L. Petzold. Projected implicit Runge–Kutta methods for differential-algebraic equations. *SIAM J. Numer. Anal.*, 28:1097–1120, 1991.

[10] U. Ascher and S. Reich. *The midpoint scheme and variants for Hamiltonian systems: advantages and pitfalls.* SIAM J. Sci. Comput., to appear, 1997.

[11] U. Ascher, S. Ruuth, and B. Wetton. Implicit-explicit methods for time-dependent partial differential equations. *SIAM J. Numer. Anal.*, 32:797–823, 1995.

[12] U. Ascher and R. Spiteri. Collocation software for boundary value differential-algebraic equations. *SIAM J. Sci. Comput.*, 15:938–952, 1994.

[13] G. Bader and U. Ascher. A new basis implementation for a mixed order boundary value ODE solver. *SIAM J. Sci. Comput.*, 8:483–500, 1987.

[14] R. Barrett, M. Beary, T. Chan, J. Demmel, J. Donald, J. Dongarra, V. Eijkhout, R. Pozo, C. Romaine, and H. Van der Vorst. *Templates for the Solution of Linear Systems.* SIAM, Philadelphia, 1996.

[15] J. Baumgarte. Stabilization of constraints and integrals of motion in dynamical systems. *Comput. Methods Appl. Mech.*, 1:1–16, 1972.

[16] C. Bischof, A. Carle, G. Corliss, A. Griewank, and P. Hovland. ADIFOR - generating derivative codes from FORTRAN programs. *Scientific Programming*, 1:11–29, 1992.

[17] G. Bock. Recent advances in parameter identification techniques for ODE. In P. Deuflhard and E. Hairer, editors, *Numerical treatment of inverse problems.* Birkhauser, Boston, 1983.

[18] R.W. Brankin, I. Gladwell, and L.F. Shampine. RKSUITE: *A Suite of Runge–Kutta Codes for the Initial Value Problem for ODEs.* Report 92-s1, Dept. of Mathematics, SMU, Dallas, TX, 1992.

[19] K. Brenan, S. Campbell, and L. Petzold. *Numerical Solution of Initial-Value Problems in Differential-Algebraic Equations.* SIAM, Philadelphia, second edition, 1996.

[20] F. Brezzi and M. Fortin. *Mixed and Hybrid Finite Element Methods.* Springer-Verlag, New York, 1991.

[21] P.N. Brown, G.D. Byrne, and A.C. Hindmarsh. VODE, a variable-coefficient ODE solver. *SIAM J. Sci. Stat. Comput.*, 10:1038–1051, 1989.

[22] A. Bryson and Y.C. Ho. *Applied Optimal Control.* Ginn and Co., Waltham, MA, 1969.

[23] K. Burrage. *Parallel and Sequential Methods for Ordinary Differential Equations.* Oxford University Press, Oxford, UK, 1995.

[24] K. Burrage and J.C. Butcher. Stability criteria for implicit Runge–Kutta methods. *SIAM J. Numer. Anal.*, 16:46–57, 1979.

[25] K. Burrage, J.C. Butcher, and F. Chipman. An implementation of singly-implicit Runge–Kutta methods. *BIT*, 20:452–465, 1980.

[26] J.C. Butcher. *The Numerical Analysis of Ordinary Differential Equations.* Wiley, New York, 1987.

[27] M.P. Calvo, A. Iserles, and A. Zanna. *Numerical Solution of Isospectral Flows.* Technical report, DAMTP, Cambridge, UK, 1995.

[28] S.L. Campbell. Least squares completions of nonlinear differential-algebraic equations. *Numer. Math.*, 65:77–94, 1993.

[29] S.L. Campbell. Numerical methods for unstructured higher-index DAEs. *Ann. Numer. Math.*, 1:265–278, 1994.

[30] J.R. Cash and M.H. Wright. *User's Guide for* TWPBVP*: A Code for Solving Two-Point Boundary Value Problems.* Technical report, on line in NETLIB, 1996.

[31] H. Chin. *Stabilization Methods for Simulations of Constrained Multibody Dynamics.* Ph.D. thesis, Institute of Applied Mathematics, Univ. of British Columbia, 1995.

[32] W.A. Coppel. *Dichotomies in Stability Theory.* Lecture Notes in Math. 629, Springer-Verlag, New York, 1978.

[33] G. Dahlquist. A special stability problem for linear multistep methods. *BIT*, 3:27–43, 1963.

[34] G. Dahlquist. Error analysis for a class of methods for stiff nonlinear initial value problems. In *Numerical Analysis, Dundee*, Springer-Verlag, New York, 1975, pp. 50–74.

[35] C. de Boor. Good approximation by splines with variable knots. ii. In *Lecture Notes in Math.* 353, Springer-Verlag, New York, 1973.

[36] C. de Boor and B. Swartz. Collocation at Gaussian points. *SIAM J. Numer. Anal.*, 10:582–606, 1973.

[37] L. Dieci, R.D. Russell, and E.S. Van Vleck. Unitary integrators and applications to continuous orthonormalization techniques. *SIAM J. Numer. Anal.*, 31:261–281, 1994.

[38] E. Doedel and J. Kernevez. Software for continuation problems in ordinary differential equations. *SIAM J. Numer. Anal.*, 25:91–111, 1988.

[39] J.R. Dormand and P.J. Prince. A family of embedded Runge–Kutta formulae. *J. Comput. Appl. Math.*, 6:19–26, 1980.

[40] E. Fehlberg. Low order classical Runge–Kutta formulas with step size control and their application to some heat transfer problems. *Computing*, 6:61–71, 1970.

[41] L. Fox. *The Numerical Solution of Two-Point Boundary Value Problems in Ordinary Differential Equations.* Oxford University Press, Oxford, UK, 1957.

[42] C.W. Gear. The simultaneous numerical solution of differential-algebraic equations. *IEEE Trans. Circuit Theory*, CT-18:89–95, 1971.

[43] C.W. Gear. *Numerical Initial Value Problems in Ordinary Differential Equations.* Prentice-Hall, Englewood Cliffs, NJ, 1973.

[44] C.W. Gear, G. Gupta, and B. Leimkuhler. Automatic integration of the Euler-Lagrange equations with constraints. *J. Comput. Appl. Math.*, 12/13:77–90, 1985.

[45] C.W. Gear and J.B. Keiper. The analysis of generalized BDF methods applied to Hessenberg form DAEs. *SIAM J. Numer. Anal.*, 28:833–858, 1991.

[46] E. Griepentrog and R. März. *Differential-Algebraic Equations and Their Numerical Treatment.* Teubner, Leipzig, 1986.

[47] J. Guckenheimer and P. Holmes. *Nonlinear Oscillations, Dynamical Systems, and Bifurcations of Vector Fields.* Springer-Verlag, New York, 1983.

[48] W. Hackbusch. *Iterative Solution of Large Sparse Systems of Equations.* Springer-Verlag, New York, 1994.

[49] E. Hairer, Ch. Lubich, and M. Roche. *The Numerical Solution of Differential-Algebraic Systems by Runge–Kutta Methods, Volume* 1409. Springer-Verlag, New York, 1989.

[50] E. Hairer, S.P. Norsett, and G. Wanner. *Solving Ordinary Differential Equations I: Nonstiff Problems.* Springer-Verlag, New York, second edition, 1993.

[51] E. Hairer and D. Stoffer. Reversible long term integration with variable step sizes. *SIAM J. Sci. Comput.*, 18:257–269, 1997.

[52] E. Hairer and G. Wanner. *Solving Ordinary Differential Equations II: Stiff and Differential-Algebraic Problems.* Springer-Verlag, New York, 1991.

[53] N.A. Haskell. The dispersion of surface waves in multilayered media. *Bull. Seismol. Soc. Amer.*, 43:17–34, 1953.

[54] P. Henrici. *Discrete Variable Methods in Ordinary Differential Equations.* Wiley, New York, 1962.

[55] M. Hirsch, C. Pugh, and M. Shub. *Invariant manifolds, Volume* 583. Springer-Verlag, New York, 1976.

[56] T.E. Hull, W.H. Enright, and K.R. Jackson. *User's Guide for* DVERK—*A Subroutine for Solving Non-Stiff ODEs.* Report 100, Dept. of Computer Science, Univ. of Toronto, 1975.

[57] A. Jameson. Computational transonics. *Comm. Pure Appl. Math.*, XLI:507–549, 1988.

[58] W. Kampowsky, P. Rentrop, and W. Schmidt. Classification and numerical simulation of electric circuits. *Surv. Math. Ind.*, 2:23–65, 1992.

[59] H.B. Keller. *Numerical Solution of Two Point Boundary Value Problems.* SIAM, Philadelphia, 1976.

[60] B.L.N. Kennett. *Seismic Wave Propagation in Stratified Media.* Cambridge University Press, Cambridge, UK, 1983.

[61] W. Kutta. Beitrag zur näherungsweisen integration totaler differentialgleichungen. *Z. Math. Phys.*, 46:435–453, 1901.

[62] J.D. Lambert. *Numerical Methods for Ordinary Differential Systems.* Wiley, New York, 1991.

[63] M. Lentini and V. Pereyra. An adaptive finite difference solver for nonlinear two-point boundary value problems with mild boundary layers. *SIAM J. Numer. Anal.*, 14:91–111, 1977.

[64] Ch. Lubich, U. Nowak, U. Pohle, and Ch. Engstler. MEXX—*Numerical Software for the Integration of Constrained Mechanical Multibody Systems.* Preprint sc 92-12, ZIB Berlin, 1992.

[65] J.B. Marion and S.T. Thornton. *Classical Dynamics of Particles and Systems.* Harcourt Brace Jovanovich, Orlando, FL, third edition, 1988.

[66] R. März. Numerical methods for differential-algebraic equations. *Acta Numerica*, 1:141–198, 1992.

[67] R.M.M. Mattheij and J. Molnaar. *Ordinary Differential Equations in Theory and Practice.* Wiley, Chichester, UK, 1996.

[68] R.M.M. Mattheij and G.W.M. Staarink. *Implementing Multiple Shooting for Nonlinear BVPs.* Rana 87-14, EUT, Eindhoven, the Netherlands, 1987.

[69] C.C. Pantelides. The consistent initialization of differential-algebraic systems. *SIAM J. Sci. Comput.*, 9:213–231, 1988.

[70] V. Pereyra and G. Sewell. Mesh selection for discrete solution of boundary value problems in ordinary differential equations. *Numer. Math.*, 23:261–268, 1975.

[71] L.R. Petzold, L.O. Jay, and J. Yen. Numerical solution of highly oscillatory ordinary differential equations. *Acta Numerica*, 6:437–484, 1997.

[72] F. Potra and W. Rheinboldt. On the numerical solution of the Euler-Lagrange equations. *Mech. Structures Mach.*, 19:1–18, 1991.

[73] A. Prothero and A. Robinson. On the stability and accuracy of one-step methods for solving stiff systems of ordinary differential equations. *Math. Comp.*, 28:145–162, 1974.

[74] P. Rabier and W. Rheinboldt. On the computation of impasse points of quasilinear differential algebraic equations. *Math. Comp.*, 62:133–154, 1994.

[75] P.J. Rabier and W.C. Rheinboldt. A general existence and uniqueness theorem for implicit differential algebraic equations. *Differential Integral Equations*, 4:563–582, 1991.

[76] P.J. Rabier and W.C. Rheinboldt. A geometric treatment of implicit differential-algebraic equations. *J. Differential Equations*, 109:110–146, 1994.

[77] M. Rao. *Ordinary Differential Equations Theory and Applications.* Edward Arnold, London, 1980.

[78] S. Reddy and N. Trefethen. Stability of the method of lines. *Numer. Math.*, 62:235–267, 1992.

[79] W.C. Rheinboldt. Differential-algebraic systems as differential equations on manifolds. *Math. Comp.*, 43:473–482, 1984.

[80] H. Rubin and P. Ungar. Motion under a strong constraining force. *Comm. Pure Appl. Math.*, 10:65–87, 1957.

[81] C. Runge. Ueber die numerische auflösung von differentialgleichungen. *Math. Ann.*, 46:167–178, 1895.

[82] J.M. Sanz-Serna and M.P. Calvo. *Numerical Hamiltonian Problems.* Chapman and Hall, London, 1994.

[83] T. Schlick, M. Mandziuk, R.D. Skeel, and K. Srinivas. *Nonlinear Resonance Artifacts in Molecular Dynamics Simulations.* Manuscript, 1997.

[84] M.R. Scott and H.A. Watts. Computational solution of linear two-point boundary value problems. *SIAM J. Numer. Anal.*, 14:40–70, 1977.

[85] L.F. Shampine. *Numerical Solution of Ordinary Differential Equations.* Chapman and Hall, London, 1994.

[86] L.F. Shampine and M.K. Gordon. *Computer Solution of Ordinary Differential Equations.* W. H. Freeman, San Francisco, 1975.

[87] L.F. Shampine and H.A. Watts. The art of writing a Runge–Kutta code, part i. In J.R. Rice, editor, *Mathematical Software III*, Academic Press, New York, 1977, pp. 257–275.

[88] I. Stakgold. *Green's Functions and Boundary Value Problems.* Wiley, New York, 1979.

[89] H. Stetter. *Analysis of Discretization Methods for Ordinary Differential Equations.* Springer-Verlag, New York, 1973.

[90] G. Strang and G. Fix. *An Analysis of the Finite Element Method.* Prentice-Hall, Englewood Cliffs, NJ, 1973.

[91] J.C. Strikwerda. *Finite Difference Schemes and Partial Differential Equations.* Wadsworth & Brooks/Cole, Pacific Grove, CA, 1989.

[92] S.H. Strogatz. *Nonlinear Dynamics and Chaos.* Addison-Wesley, Reading, MA, 1994.

[93] A.M. Stuart and A.R. Humphries. *Dynamical Systems and Numerical Analysis.* Cambridge University Press, Cambridge, UK, 1996.

[94] J.H. Verner. Explicit Runge–Kutta methods with estimates of the local truncation error. *SIAM J. Numer. Anal.*, 15:772–790, 1978.

[95] R.A. Wehage and E.J. Haug. Generalized coordinate partitioning for dimension reduction in analysis of constrained dynamic systems. *J. Mech. Design*, 104:247–255, 1982.

[96] R. Weiss. The convergence of shooting methods. *BIT*, 13:470–475, 1973.

[97] S.J. Wright. Stable parallel algorithms for two-point boundary value problems. *SIAM J. Sci. Comput.*, 13:742–764, 1992.

Index